1992

TELECOMMUNICATIONS

W9-DGV-700

An Introduction to Electronic Media

FOURTH EDITION

TELECOMMUNICATIONS
An Introduction to Electronic Media

LYNNE S. GROSS

Wm. C. Brown Publishers

Book Team

Editor *Stan Stoga*
Developmental Editor *Jane F. Lambert*
Production Editor *Connie Balius-Haakinson*
Designer *Eric Engelby*
Art Editor *Margaret Rose Buhr*
Photo Editor *Robin Storm*
Visuals Processor *Amy L. Saffran*

 Wm. C. Brown Publishers

President *G. Franklin Lewis*
Vice President, Publisher *Thomas E. Doran*
Vice President, Operations and Production *Beverly Kolz*
National Sales Manager *Virginia S. Moffat*
Group Sales Manager *Eric Ziegler*
Executive Editor *Edgar J. Laube*
Director of Marketing *Kathy Law Laube*
Marketing Manager *Carla Aspelmeier*
Managing Editor, Production *Colleen A. Yonda*
Manager of Visuals and Design *Faye M. Schilling*
Production Editorial Manager *Julie A. Kennedy*
Production Editorial Manager *Ann Fuerste*
Publishing Services Manager *Karen J. Slaght*

WCB Group

President and Chief Executive Officer *Mark C. Falb*
Chairman of the Board *Wm. C. Brown*

Main Cover Photograph: Eric Engelby

Cover insets (left to right): © Scott Smith/Sports File Photography,
© Yvonne Hemsey/Gamma Liaison, © Viacom, © Sygma

Back Cover Photograph: Big Bird © 1991 Jim Henson Productions, Inc.
Used by permission Children's Television Workshop

Line art by Diphrent Strokes, Inc.: Plates 1–10, Figures 5.14, 6.3, 6.14,
6.17, 8.1, 8.2, 8.3, 8.4, 8.5, 8.6, 9.3, 10.6, 12.1, 13.2, 14.11, 14.12, 14.13,
14.21, 15.6, 16.1, and 16.3.

Library of Congress Catalog Card Number: 91–70345

ISBN 0–697–08647–X

Printed in the United States of America by Wm. C. Brown Publishers,
2460 Kerper Boulevard, Dubuque, IA 52001

10 9 8 7 6 5 4 3 2 1

To my husband, Paul

CONTENTS

PURPOSE

Telecommunications is one of the most potent forces in the world today. It influences society as a whole, and it influences every one of us as an individual. As each year passes, telecommunications grows in scope. The early pioneers of radio would never recognize today's vast array of electronic media—broadcast television, cable TV, direct broadcast satellite, and videocassettes, just to name a few. Neither would they recognize the structure that has evolved in such areas as regulation, advertising, and audience measurement. They would marvel that their early concepts of equipment have led to such developments as audiotape recorders, cameras, videotape recorders, digital effects generators, editors, computer graphics, and satellites. If they could see the quantity and variety of programming available today, they might not recognize that it all began with amateurs listening for radio signals on their "primitive" crystal sets.

All indications are that telecommunications will continue to change at a rapid pace. As it does, it will further affect society. All people, whether they be individuals working in the telecommunications field or individual members of society, have a right to become involved with media and have an obligation to understand why people need to interact with the media. Some knowledge of the background and structure of the industry is an essential basis for this understanding.

A major goal of this book is to provide just that kind of knowledge so that intelligent decisions about the role of telecommunications can be made both by those who are practitioners in the field and those who are members of the general society.

Organization of the Book

This is the fourth edition of this book, the first appearing in 1983, the second in 1986, and the third in 1989. A number of major changes have occurred in telecommunications since 1989, so corresponding changes needed to be made in this book.

Chapter 1 in Part 1 has been substantially rewritten and now includes several sections on ethics, in addition to the sections that deal with social significance. Both social impact and ethics are themes that are carried throughout the book and should be considered in conjunction with all the chapters. Because the international scene is becoming so important within the telecommunications field, this subject has been moved from the back of the book to Chapter 2. In this way, students should have a base of comparison against which to view American broadcasting.

PREFACE

Part 2 deals with the various electronic media forms—Commercial Radio (Chapter 3), Commercial Television (Chapter 4), Public Broadcasting (Chapter 5), Cable Television and Competing Media (Chapter 6), and Personal and Organizational Telecommunications (Chapter 7). All of these chapters have been significantly updated and have incorporated material concerning the newer electronic media. For example, low-power TV and teletext are in Chapter 4 as forms of commercial television; MMDS, SMATV, and DBS are in Chapter 6; and Chapter 7 contains material on VCRs and video discs. In fact, Chapter 7 is reorganized to acknowledge the fast-growing and significant uses being made of telecommunications within corporations and educational institutions, and by individuals armed with their camcorders.

Because financial and business considerations are such an important part of the electronic media these days, this material has been moved forward to Part 3. A brand-new chapter on business practices starts off this section, which includes advice to students concerning employment. Updated chapters on advertising and audience measurement follow.

Part 4 deals with programming. Program types have remained essentially the same as they were three years ago, but programming practices have changed greatly.

In Part 5, Chapter 14 on production equipment has been almost totally rewritten to take into account new developments in equipment. Chapter 15, dealing with distribution, now contains a separate section dealing with high definition television.

The final part of the book, Part 6, covers regulation. The two chapters (Chapter 16—Regulatory Bodies and Chapter 17—Laws and Regulations) have been somewhat reorganized, and new cases and regulatory philosophies have been included.

All the chapters should lead the reader to assess the strengths and weaknesses of the particular subject being discussed. Information relating to future directions telecommunications may possibly take is also woven throughout the appropriate chapters.

Special Features
Each part of the book begins with an overall statement that relates the chapters to one another. Each chapter begins with a pertinent quote and a short introduction.

Each chapter conclusion summarizes major points but does so in an organizational manner slightly different from that given within the chapter. For example, if the chapter is ordered chronologically, the conclusion may be organized in a topical manner. This should help the reader form a gestalt of the material presented.

Further aids in understanding the material are the thought questions at the end of each chapter. These questions do not have "correct" answers, but rather are intended to lead the reader to form his or her own judgments. Discussions centering around these questions will indicate that varying opinions surround telecommunications issues.

Chapters are broken down into major divisions, and marginal notes appear within each division. Each marginal note highlights the main subject being discussed in the adjacent paragraph or paragraphs. Taken together, these notes serve as review points for the reader. Throughout the text, important words are boldfaced. These, too, should aid learning. Some of the boldfaced words are defined in the glossary, while others are names of important people or organizations.

The chapters may be read in any sequence; however, some of the terms that are defined early in the book may be unfamiliar to people who read later chapters first. The glossary can help overcome this problem. It includes important technical terms that the reader may want to review from time to time, as well as terms that are not necessary to an understanding of the text but that may be of interest to the reader.

Chapter notes, which appear at the end of the book, are extensive and provide many sources for further study of particular subjects.

The photographs and charts that appear throughout the book supplement the textual information.

SUPPLEMENTARY MATERIALS

Student Study Guide
A student study guide has been prepared to help students learn the material thoroughly. The chapters in the study guide correspond to the chapters in the text itself. Each chapter in the study guide is divided into six sections—overview, key terms and concepts, study questions, sample test questions, research sources, and answers to sample test questions. This material can be examined before, during, or after the reading of the text material itself.

Instructor's Manual
The instructor's manual available with *Telecommunications* offers two sample course outlines that can be adapted to semesters or quarters, as well as learning objectives, suggested lecture topics/activities, films and tapes, overheads, test questions, and a bibliography for each text chapter.

TestPak

WCB TestPak is a computerized system that enables you to make up customized exams quickly and easily. Test questions can be found in the Test Item File, which is printed in your instructor's manual or as a separate packet. For each exam you may select up to 250 questions from the file and either print the test yourself or have wcb print it.

Classroom Tape Material

In order to aid in the teaching of the class, wcb has collected a number of videotapes to be used in conjunction with this class. For more information about these tapes, contact your sales representative.

Overhead Transparencies

Also, in order to aid the teaching of the class, a number of overhead transparencies have been prepared. Many of these are charts that contain figures likely to be outdated quickly. These will be updated every year, so instructors can always have the latest information to present to the students. Qualified adopters can obtain these overheads by contacting Wm. C. Brown sales representatives.

ACKNOWLEDGMENTS

This book represents the combined efforts of many people, including the following reviewers who offered excellent suggestions.

Prudence Faxon, California State Polytechnic University
Linda Fuller, Worcester State College
Peter R. Gershon, Indiana State University
Linda Rhodes, California State University-Sacramento

In addition, I would like to thank the Book Team at Wm. C. Brown for their patience and suggestions. And I would like to thank my husband and three sons for their advice and encouragement while I was working on the text.

Lynne Schafer Gross

Introduction

Telecommunications is a powerful force in society. Radio and television permeate our lives, yet many generations of people existed without electronic media. The pervasive influence has occurred in a short space of time, but its intensity compensates for its youth.

In foreign countries many of the forms of electronic media have existed for an even shorter period of time. But they have grown to a point where worldwide media interaction promises to be a major force and major issue of the coming decades.

PART 1

Social and Ethical Implications of the Electronic Media

Introduction

The influence that telecommunications exerts upon our society is obviously extensive. The mere ability to communicate instantaneously affects the process of communication. Beyond this, the permeation of opinions, emotions, and even fads can often be attributed to various elements of the media.

The pervasiveness of radio and television, whether applauded or condemned, cannot be denied. The influence extends from the individual through the social structure, the economy, technology, and politics. All of this places a responsibility on individuals within the telecommunications field to act in an ethical manner that does not betray the confidence placed in them.

Television is less a means of communication (the imparting or interchange of thoughts, opinions, and information by speech, writing, or signs) than it is a form of communion (act of sharing or holding in common; participation, association; fellowship).

Richard Schickel
The Urban Review

A Rationale for Study

Everyone has an opinion about radio or television fare, and everyone can exhibit a certain amount of expertise about a force that is seen and heard on a daily basis.

Then why study this field? Some of the answers to this question are obvious. Anyone who is aiming toward a career in this area will profit from an intimate knowledge of the history and inner workings of the industry. Radio and television are highly competitive fields, and those armed with knowledge have a greater chance for career survival than those who are naive about the inner workings and interrelationships of networks, stations, cable TV facilities, advertisers, unions, program suppliers, telephone companies, the government, and a host of other organizations that affect the actions and programming of the industry.

On a broader scope, individuals owe it to themselves to understand the messages, tools, and communication facilities that belong to our society because they are so crucial in shaping our lives. Rare is the individual who has not been emotionally touched or repulsed by a scene on TV. Rare, too, is the individual who has never formed, reinforced, or changed an opinion on the basis of a presentation seen or heard on one of the electronic media. A knowledge of the communications industry and its related areas can lead to a greater understanding of how this force can influence and affect both individual lives and the structure of society as a whole.

In addition, telecommunications is a fast-paced, fascinating industry worthy of study in its own right. It is associated with glamour and excitement (and power and greed), both on-screen and off. Although, in reality, the day-to-day workings of the industry can be as mundane as any other field, the fact that it is a popular art that includes the rich and famous makes it of special interest. The ramifications of the power that the electronic media exert over society is most deserving of study.

The Broad Context

Studying this field used to be fairly simple. There were two media—radio and television, and together they were called **broadcasting.** As time progressed, broadcasting was divided into two categories—commercial and public (originally called educational). These two coexisted fairly harmoniously because public broadcasting was small and not really a threat to its commercial kin. In fact, it often relieved commercial broadcasting of its more onerous public service requirements, because the commercial broadcasters could point out that public broadcasting served that interest.

Then in the mid-1970s a number of other media came to the fore to challenge radio and TV, creating an alphabet soup that included CATV, VCR, DBS, MMDS, SMATV, and LPTV. The word *broadcasting* no longer seemed to apply because that word implied a wide dissemination of information through the airwaves. Many of these other media were sending information through wires, and cable TV was even going around touting its **narrowcasting** because its programs were intended for specific audience members. For a while these new forms were referred to as new media or new technologies, so people studied broadcasting and the new media. Also, during the 1970s many companies began using television, particularly for training. This was referred to as *industrial TV,* but it was not studied to any great degree.

In the 1980s when the new media weren't so new anymore, they began being referred to as developing technologies, but some of them didn't develop very well. In fact, a number of them just plain died. Generally, the term **electronic media** was used to describe broadcasting and the newer competitive forces, but sometimes the word **telecommunications** was used to label the entire group, including industrial TV, which, by now, had changed its name to *corporate TV.*

Now in the 1990s, the field of study may be broadening even more. Video gear has been developed to the point where just about anyone can afford it and can use it to create quality pictures and sound. This democratization of video seems to be leading to a new field called *personal video.* In addition, the telephone and computer have teamed up and may be on the verge of creating a new round in the information evolution. In a way, this brings broadcasting full circle. Radio can be seen to have its antecedents in the telephone because, at one point, the telephone was seen as a mass medium and the radio as an individual, private medium. In 1877, a song called "The Wondrous Telephone" contained the following lyrics: "You stay at home and listen to the lecture in the hall, Or hear the strains of music from a fashionable ball!"[1] The original idea for the telephone was that it would deliver words and music to large groups of people. When radio was first developed, many people tried to invent ways to make the signals private so that two people could have their own confidential conversation.

Of course, over the years the two media switched roles—telephones being the private medium and radio becoming the mass medium. The two also went their separate ways academically and socially. Rarely were they studied in the same curriculum, and rarely did people trained for broadcasting obtain jobs in the telephone industry. The social, economic, and political issues affecting each were quite dissimilar.

Then along came the computer and a device called a modem. When this modem was attached between a computer and a telephone, data generated by the computer could be sent over phone wires to another computer. Suddenly information sent through the telephone was appearing on what looked like (or were) television screens. Some of the information being transmitted over this computer-telephone system was not private, but was intended for anyone in the population who wanted it or was willing to pay for it, including those in corporations. It included news, stock market quotes, sports, and other information traditionally provided by radio and TV, as well as newspapers and magazines.

The word *telecommunications* was somewhat taken over by the telephone industry to encompass both the old telephone services and all the new data transmission and other fancy services the computer enabled the telephone to undertake.

Now the telephone, computer, radio and TV broadcasting, cablecasting, personal video, corporate video (now sometimes called *organizational video*), and the alphabet soup of newer technologies seem to be merging into some form of information supplier as yet undetermined. The most common word used to encompass all of this is *telecommunications,* but the word or even the concept could change drastically in the near future.

The words *telecommunications, electronic media,* and *broadcasting* are used somewhat interchangeably in this book because, at the time of writing, the general use of all three words is somewhat ambiguous. The book will concentrate, however, on commercial radio, commercial television, public radio, public television, cable television and its competitors, and organizational and personal video. It will also deal with the telephone and computer, but mainly as they relate to and are becoming merged with the media taught in the traditional radio and TV curriculum.

Statistics of Pervasiveness and Change

This force, by whatever name—telecommunications, electronic media, or broadcasting—is very pervasive in our society. On the sending end, there are about 10,700 radio stations and approximately 1,400 TV stations in the country. In addition, there are almost 11,000 cable TV systems. (See plate 1 for more detail.) On the receiving end, 99 percent of households have radios, 98 percent have TVs, 93 percent have telephones, 59 percent subscribe to cable TV, and 73 percent own videocassette recorders.[2] (See plate 2 for more detail.)

More important, people don't just own radios, TVs, and VCRs; they use them. The average household has TV on about seven hours a day, with the average person watching about four hours a day. The average person also tunes in radio about three hours a day.[3] (See plates 3 through 7 for more detail.)

Because of the fast-paced nature of the electronic media business, many characteristics of the industry are subject to rapid change. This can be seen by what happened to the TV industry during the decade of the 1980s. (See plate 8.)

The beginning of the 1980s saw a television field dominated by three networks. By the end of the decade, these networks had competition from many sides, primarily cable TV, independent TV stations, and VCRs. The average share of network prime-time viewership went from 85 percent in 1980 to 67 percent in 1989. Not surprisingly, the average number of channels available per household rose from 8 to 28, cable TV penetration increased 37 percent, and almost three-quarters of American households equipped themselves with a VCR during the 1980s.[4]

What all this meant was that TV was going from a mass appeal medium, where most people in the country watched the same programming at the same relative time, to a more fractionalized medium that was appealing to smaller groups of consumers. No longer did the three network programming chiefs call the shots as to what the people would watch and when they would watch it. People could tape programs off the air and watch them whenever they wanted. Or, instead of watching NBC, CBS, or ABC, they could watch one of the many alternatives.

Thanks to the remote control, which was in more than 50 percent of homes by the end of the 1980s,[5] people could switch channels from the comfort of their easy chairs. Before remote controls, networks could rely on the fact that much of the audience that tuned in at the beginning of the evening would stay with the same station because they did not feel like getting up and changing the channel. Now, with the remote control some people watch two or more programs at once.[6] Even more people switch to a new channel the minute commercials are on. This channel switching phenomenon has become known as **grazing.**

remote control

Although radio and TV income rose during the decade, it did not keep pace with costs, causing the networks to cut back in many ways. The networks were far from dead by the end of the 1980s, but they certainly did not wield the power that they had in previous decades. So far in the 1990s, the networks appear to be continuing a decline. However, given the volatile nature of the electronic media business, anything could happen.

Communication Models

Various models have been designed to explain the communication function. One model that can apply to both personal communications and telecommunications is presented by Bert E. Bradley.[7] This model shows communication as an ongoing process that includes a source, a message, a channel, a receiver, barriers, and feedback.

For telecommunications, the **source** is composed of the total number of people needed to communicate a message. In radio or television, this is usually a large number that includes actors, camera operators, a producer, an audio technician, the station manager, and all the other positions necessary for programming to be successful.

source

Of course, some people are more important sources than are other people. For example, the people who decide which news stories will be broadcast, which will receive the most time, and which will be presented at the beginning or end of the newscast are more important in the communication process than are the camera operators who focus on the anchorpersons.

Some communication models refer to the people who make important decisions as **gatekeepers.** They do, in effect, control what information passes through.

The people who act as the source are concerned with sending a **message.** The message will have a purpose—some messages are intended to inform, some are designed to entertain, and some are planned to persuade. Transmission of a message is the primary reason for communication although, obviously, both the message and its process of transmission are often far from perfect.

message

Messages are transmitted through **channels.** For telecommunications, some of the most common channels are radio, broadcast television, and cable

channel

Figure 1.1
Communication model.

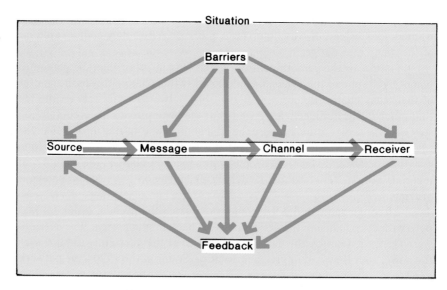

TV. Various devices and processes can influence the effectiveness of a channel. For example, instant replay, which is a particular quality of the TV medium, enables sportscasting to communicate to the fans in ways that other channels, such as radio and newspapers, cannot.

receiver

The **receiver** of the message is the audience. For some forms of telecommunications, primarily broadcast TV, the audience is a composite of individuals that is generally large, heterogeneous, and fairly anonymous. Cable systems that narrowcast are seeking a more homogeneous audience. For corporate video, the receiver might be even narrower—for example, the people who will be selling a new line of women's clothes may be targeted. In all these instances, the people representing the source send their message through the channel to the receiver.

barriers

However, many **barriers** can obstruct the communication process or reduce its effectiveness. The source can have built-in physical or psychological barriers. For example, if all the people representing the source live and work in California, they may not have a proper frame of reference to design a message that will communicate to people in Massachusetts. If the majority of people working in a newsroom are Democrats, they may not communicate in a way that is considered appropriate by Republicans. Individual biases and experiences can affect how the source sends its messages.

Messages also have barriers. A poorly written joke may not entertain. An informational piece delivered in Spanish will not communicate to someone who does not understand that language. A campaign speech intended to persuade might only inform.

Static is an example of a physical barrier that can affect the communication ability of a channel. The government, through its regulations, can also create barriers, whether advantageous or disadvantageous. If a TV station feels it must give equal time to opposing viewpoints on a controversial issue, this restriction might inhibit the station from discussing the issue at all. Consumer groups that threaten boycotts can affect the behavior of the channel. On the other hand, a message can be distorted simply because information presented on radio or TV takes on added importance. For example, sometimes criminal acts that are fairly minor get blown out of proportion because they receive extensive coverage.

The receiver, too, is affected by many barriers. Some of these are in the receiver's background or environment. For example, a person who has lived in France might perceive a program about French politics in an entirely different way than a person who has never even visited France. A person listening to a radio station in a hot room might react differently to the music than someone listening in a cold room. Interruptions or distractions, such as a phone ringing or dinner being placed on the table, can also be barriers for the receiver.

The message, channel, and receiver all serve as foundations of **feedback**   feedback
to the source. If no one laughs at a joke during rehearsal, this can be a signal to the creator that it is not funny. Managers of affiliated radio and TV stations often express their opinions to the people who provide the network programs. But the receivers are the major source of feedback for those who constitute the source. Ratings are the primary form of feedback, and they can definitely alter the source as is evidenced by the cancellation of shows with poor ratings. But fan letters and complaints also count. Fans have been instrumental in keeping such shows as "Cagney and Lacey," "Star Trek," "Starman," and "Scarecrow and Mrs. King" on the air (at least for a little while) after networks have said they were going to cancel them.[8]

There are also barriers to feedback. People can laugh out of politeness, not bother to complain, or fill out a rating form incorrectly.

Despite all the barriers and the inefficiencies of feedback, the telecommunications process does work, and most messages do proceed from the source to the receiver with some degree of clarity.

Electronic Media Research

A great deal of research about the media has been undertaken by college professors and research institutes. Rather than dealing exclusively with numbers and percentages, academic studies attempt to explore underlying phenomena. For example, they try to determine **cause and effect** (does watching violence on TV cause husbands to be more violent toward their wives than they would otherwise be?) or **uses and gratifications** (are people in their sixties more likely to use radio to learn about current news than people in their twenties?).

Sometimes these studies are conducted in a **laboratory** setting (people   methods
might be brought to a university classroom where they are given a pretest about their feelings toward American Indians, are shown a program about Indians, and then given a posttest on their feelings), or they are conducted in the **field** (for example, researchers interview family members and observe them

watching TV in their own homes in order to determine whether fathers, mothers, or children are most likely to determine what TV programs will be watched).

importance

All of these types of studies are particularly helpful because they enable media practitioners, regulators, advertisers, individuals, and social agents to understand the effects of telecommunications on society and, hence, to provide better service to the public.

difficulties

Electronic media research is often difficult to administer, mainly because radio and television are such pervasive forces in our world today. If one is going to have an experimental group that watches television, how can a group of people who never watch television be found to act as a control for the experiment? How can effects of radio and television be isolated when people are constantly under the influence of so many other social factors such as family, school, and church? How can effects of the media be determined when individuals vary so greatly among themselves in terms of reaction?

Despite the difficulties inherent in such questions, a great deal has been learned about media and society from academic research. One area researchers have explored extensively involves the effects of television on children. When "Sesame Street" first aired a great deal of research was conducted to assess its effects. These studies found, among other things, that "Sesame Street" viewers performed better at reading than did children who did not watch the programs.[9] Other studies have explored parent-child co-viewing, children's perceptions of television reality, and uses of television by children with different cultural backgrounds.[10]

children and TV

TV research also explored the effect of violence on children. Generally this research was not conclusive. Some studies found that violence did affect children adversely, while other studies found that it did not.[11]

In one study, forty-four third and fourth graders were randomly divided into two groups, both of which were shown a new trailer on the school grounds and were told that it was used for kindergarten children. The experimenter pointed out a TV camera on the wall and said that it would take pictures of all that took place inside the trailer. One group of children was then taken into a room and shown a violent western, while the control group was taken into another room and not shown any film. The experimenter then said he had to go to see the principal and asked each group of third and fourth graders to watch the kindergarten children for him. He turned on the TV monitor that showed a still empty trailer and asked both groups to watch the room and come to him in the principal's office if anything went wrong. Both groups were shown a videotape of two little children coming into the trailer, starting to play, and then getting into a fight and pushing each other until they apparently broke the camera. The group of children who had seen the violent western took significantly longer to seek adult help than did the children who had not seen the film. The conclusion of this study was that exposure to TV violence taught children to accept aggression as a way of life.[12]

In another study, kindergartners and first graders were shown a frightening movie scene. Some of the children were told they could put their hands over their eyes if they wanted to; others were told they could turn off the TV if the movie was too scary; and others were simply shown the tape. Afterwards they were asked questions to determine their emotional reaction to and interpretations of the scene. Those who had been told they could put their hands over their eyes (although very few of them did) had less fearful reactions and less threatening interpretations than the other two groups. There was no difference between those who had been told they could turn off the TV and those that had been told nothing.[13]

Another subject that has been researched frequently is television news. news For example, one study tested redundancy and recaps by showing similar news stories to four groups of people. One group saw stories for which the pictures and words reinforced each other, and recaps of the stories were given right after the stories. Another group was shown redundant pictures and words but no recaps; the third group had recaps but no redundancy; and the fourth group had neither redundancy nor recaps. The study concluded that redundancy helps recall of stories, and recaps lead to improved understanding.[14]

Other studies in the news area have dealt with the gatekeeper phenomenon and local television news, attention given to stories about Third World countries, and geographic biases in network TV news.[15] other areas

These subjects of academic research—TV and children, and news—are illustrative but by no means the only types of research done concerning telecommunications. Other common research subjects are radio age groups, effects of sex and violence on TV, broadcast economics, advertising, minorities and women, news coverage, and political influence. Research journals such as *Journal of Broadcasting and Electronic Media* and *Journal of Communication* are excellent sources for keeping up to date on the latest research in the field.

Not all academic research is of an experimental nature. Historical and biographical research is also undertaken so that a clear understanding of the past and the people who created this past can help pave the way for the future. Researchers scour primary materials, such as letters written by individuals who were instrumental in broadcasting's history and documents printed throughout the years, in order to uncover new facts and interpretations. historical and
biographical

Recently, radio and television archives have been established in various places around the country. In these archives both early and recent programs are collected so that they can be viewed by scholars and interested members of the public. In some instances this viewing is purely for information, entertainment, or nostalgia, but more and more these programs are being utilized for sociological research. For example, situation comedies can depict the changing complexion of families throughout the last several decades, and public affairs programs can document the changing views of liberals and conservatives.[16] archives

Individual Effects

Individual preoccupation with telecommunications is enormous as evidenced by the statistics on viewership. But what effect does all of this viewing and listening have on the individual? This is not an easy question to answer. For one thing, as previously noted, difficulty exists in trying to isolate the effects of radio and TV on an individual from the effects of other elements of society, such as family, school, and church. What is more, different individuals are affected in different ways. What follows is a discussion summarizing some of the effects, both pro and con, that have been espoused with varying degrees of commitment by consumers, critics, and practitioners of the media.

education

Television can be an educational force for many individuals, not only through purposeful educational programs, but also through documentaries and entertainment programs. A movie that is set in a foreign country can unobtrusively teach a viewer about the customs and physical features of that country. News programs can make everyone more knowledgeable about current events than was the case in pretelevision days. Science fiction can convey minilessons in astronomy or physics. Commercials can inform people of new products that may meet their specific needs. Some people even claim that game shows are educational because of the content of their questions.

Yet, if what is presented is educational, then much of it is negatively educational. Television, with its happy families who solve all problems in thirty minutes, does not represent reality. A person from a deprived background can be frustrated and discontent if television is considered the only window to the "real world." By fictionalizing for the sake of drama, television practitioners can "teach" that life is far more exciting and rewarding than the average individual's humdrum existence. Individuals who react to this with aggressive self-pity can be dangerous to themselves and to society.

The mere act of watching TV, regardless of program content, can be harmful to individuals. Watching TV is a passive, sedentary antiactivity that often requires little or no physical or mental activity on the part of the viewer, except perhaps occasionally pushing a button on the remote control in order to "graze." People who watch TV in this manner have been referred to as "couch potatoes." The time spent watching aimless hours of TV subtracts from the time an individual has for more meaningful activities, such as study and human interaction. In some cases watching TV becomes a major factor in procrastination or even psychic paralysis.

Particularly dangerous is watching TV as opposed to watching programs. Many individuals turn on the TV set and watch whatever is on regardless of its quality or interest to them. Rather than planning to watch programs that they particularly want to see, they flip the dial until they hit upon the **least objectionable fare**—a program that may not really appeal to them, but one that is more desirable than any other programs on at that time.

using time

Would they be better off using the time for something more constructive?

Yet television viewing makes some people's time worthwhile. The old and infirm can use TV as something to live for, something to pass the time, something to keep them in touch with the world. Others can use radio or TV to unwind from a hectic day. All individuals need time for pure relaxation—a time to escape vicariously and temporarily from their trials and tribulations. Why not use least objectionable fare to fill these times?

Radio and television provide role models, especially for the young. A youth who watches a medical show and decides to become a doctor and a child who joins a sports team after listening to an interview with a sports star both profit. Minority children and adults can develop increased self-esteem by watching minority individuals in significant roles. In a time when many children are from broken homes or homes in which they see little of their working parents, positive role models are needed. role models

But what if the role models are negative? What if a role model is a loose-living amoral person or an aggressive fugitive? Certainly, with the drama for drama's sake fare that appears, role models can be both oversimplified and dangerously stereotyped. Many programs, particularly the older reruns, portray mothers as passive, fathers as buffoons, and children as insensible patsies. Worse yet are portrayals of servile blacks and lazy Mexicans, all of which can affect the impressions of children, or adults for that matter, concerning personality traits and social positions of races, sexes, or creeds of people.

Individuals do tend to select and retain those elements of a radio or TV program that reinforce previously held beliefs, attitudes, and values. A person who feels that women are and should be passive will notice those programs or parts of programs that demonstrate women's passivity and tend to ignore program material to the contrary, even though this may be the crux of the message. In this way TV, even though it may not be trying to do so, can magnify stereotypes.

The vast number of channels available to individuals creates new issues. Some praise this because it gives more options; others say it just provides more of the same old junk. Others feel it creates media overload and actually prevents people from finding material they can learn from or enjoy because they have too many choices. Of course, ultimately all the negatives of electronic media programming can be controlled by the individual who has the power to turn off the TV or radio. media overload

Sociological Effects

Television can have a personal and intimate effect upon a person, almost as if he or she were in a vacuum. But in reality people do not live in vacuums; they live in a society where they must interact with one another through various organized methods. Television affects these sociological organizations and, of course, the individuals within them.

One of the backbones of our social structure is the family. In the days of early radio, several generations of a family would live in close proximity, forming a secure social structure for one another. However, as transportation and communication improved, family ties weakened, and the young often left home to seek their fortune in other parts of the country or world. The once "ideal" family of father-provider, mother-homemaker, son, and daughter, which family was prevalent even when television began, is now a distinct minority. Women are a large element of the work force, and many children live in one-parent homes. Throughout the transformations of the family during the past seventy years, radio and then television have been depicting family life through both comedy and drama. Has this depiction been realistic? Has it been beneficial?

Comedy, because it is constantly in pursuit of a laugh, and drama, because it aims for suspense, cannot be truly realistic without being boring. No families lead lives that are a laugh a minute or a trauma a moment, but TV programs are produced this way in order to sustain interest. On the other hand, a perusal of family radio shows of the 1930s compared with family TV shows of the 1990s shows a vast difference in family values and structures that are reflected, at least to some degree, in the changing society. Some blame TV for the high divorce rate by pointing out that the shows centering around divorced parents seem to be condoning and even glamorizing divorce. Others point to the "old-fashioned" programs that emphasize family closeness and integrity and say that these are role models for present families.

The mere presence of a television set within a family also causes controversy. Is this "box" preventing family members from talking and interacting with one another, or is it giving them subjects of common interest about which to talk? If family members were not watching television, would they be engaged in other family activities, or would they be spread out among other, perhaps destructive, activities? Does the music blaring from a radio increase the generation gap, or does it serve as a mild form of teenage independence that prevents other more dangerous forms of rebellion against parental domination?

Arguments can be made for all points of view on these questions, and certainly no one answer is correct. Families, like individuals, differ in and among themselves, but each family, regardless of its structure, does have the obligation to control television rather than let television control the family.

schools

Schools are also sociological institutions that all people encounter for at least part of their lives. The quality of education and the quality of school graduates has frequently been a heated subject that has drawn radio and television into its midst. There are those who feel that television actually stimulates an increased intellectual awakening in young people. They point to the fact that when a famous story is dramatized on television, the libraries are inundated with requests from people who want to read the book. Teachers sometimes assign television programs as homework and use them as a basis for an intellectual lesson.

Countering this, however, are critics who point out that children do not read as well as they used to. They say this is because children watch endless cartoons that make no demands upon reading or spelling abilities and that children waste time watching TV instead of doing homework.

fads

Society's fads are frequently created or at least fanned by the media. Everything from Little Orphan Annie rings to Madonna look-alikes have been the result of media exposure. Many "Mickey Mouse Club" fans grew into

"Star Trek" appreciators called Trekkies, and the sociological implications were the same—adoration and belonging.

The cultural aspects of society generally receive stepchild treatment on radio and TV. Public television and some of the cable channels show cultural material, but most of the available channels avoid cultural programs because they do not attract large audiences.

culture

Sports, on the other hand, receive wide play on radio and television, to the point where the sheer amount of sports programming becomes controversial. A spectrum of questions can be raised regarding whether sports programming inhibits or encourages social interaction. The "sports widow" phenomenon is bandied about by cartoonists and comedians, but because sports is such a widespread topic of conversation, watching sports can lead to sociable and sometimes heated conversations. Watching often encourages people to participate in sports, certainly not a sedentary activity. The vicarious "thrill of victory and agony of defeat" can provide people with the excitement they need in their lives and, perhaps in this way, stifle more socially unacceptable aggressive behavior. However, there is evidence that sports programming can also incite violence. For example, women's crisis centers report that, for victims of domestic violence, Super Bowl day is one of the worst days of the year.[17]

sports

When the sociological phenomenon of violence comes into discussion, the opinions about television's role become very heated. Over and over again citizens' groups, church groups, and individuals have equated rising violence and crime in the streets with violence on TV. The dramatic nature of television programming emphasizes conflict and, more often than not, this conflict is shown in the form of violence. The critics of such programming worry that it is creating a society in which violence is taken as commonplace, and its rising tide is accepted as a natural phenomenon. Worse yet, they feel that the programs are emotionally damaging and lead people toward aggressive acts that they might not contemplate if they did not see them on TV. Even more alarming, they feel that the violent programs are instructional and that those watching, including prison inmates, learn better ways of committing crimes by observing techniques and mannerisms.

violence

To counter these arguments, network executives point out that Homer's *Iliad* and Shakespeare's plays have a higher degree of violence than do modern teleplays and that violence is used as an interest holder and not an instrument intended for social destruction. People have watched violence for centuries and have not suffered apparent ill effects. But, counter the critics, several people watching one violent program once in awhile, as happened prior to TV, is not the same as millions of people watching violent programs every evening. The cumulative effect of the television watching, they feel, is what is harmful to society.

In the area of violence, as well as in other sociological areas, media practitioners must take stock of the fact that their product does have a significant impact upon society.

Technological Effects

The electronic media and technology exhibit an interesting spiraling relationship. Improved technology leads to improved communications that in turn leads to a desire for improved technology. In early times the discovery of radio waves led to the transmission of Morse code without the use of wires, and the ability to transmit Morse code led to the desire to transmit voice through this wireless means.

As each technological step is taken, new uses are found that push the technology even further. When special effects were first introduced into television in the form of corner inserts and split screens, the "gee whiz" approach of directors and technical directors led to a desire for more special effects. This led to equipment with provisions for soft wipes, reversed images, swirling digital images, and many other effects to challenge the creativity of the operators, who then thought of new effects with which to challenge the engineers. When the technology was developed to enable cable to carry twenty channels, many wondered what type of programming could possibly fill all those channels. Within a brief period of time twenty were easy to fill, and the technology was developed for more than thirty channels, which, in turn, were filled. People who happily watched TV for many years without a remote control now wonder how they could have possibly gotten along without it. And on and on goes the saga. Humans' curiosity and ingenuity chase each other around the technological maypole with the end result of either constant improvement or constant turmoil, depending on one's point of view.

Certainly, the constant technological development keeps people in the media business on their toes. Any production method or delivery system that tries to coast on its laurels will find itself standing in the dust as the newest of the new technologies rushes by.

media damage

Many entertainment and information forms have been severely injured, permanently or temporarily, by this constant rush of technological change. Vaudeville barely exists anymore, its demise due largely to radio and films. Radio experienced a skid when television came to the fore and had to readjust drastically in order to revitalize itself. Radio and television have led newspapers to cast but a shadow of their former power, and movie theaters are now threatened from videocassettes. The conventional broadcasting system looks over its shoulder at cable TV while cable TV glances furtively at direct broadcast satellite. Survival of the fittest often translates into survival of the newest and the shiniest.

changes

Technical developments definitely bring changes to the electronic media. The audiotape recorder altered radio, and the videotape recorder did likewise for TV. Programs could be produced at times other than when they were aired, and they could even be produced in small sequences out of order and then assembled by means of editing. As equipment became miniaturized, production possibilities were maximized. Now even individuals taping with their home camcorders can contribute material to network shows.

The technology exists to create a society in which people sit poised before the TV set for most of their waking hours, using it not only as a source of entertainment and information, but as a lifeline and pipeline to accomplish the routines of their business and personal lives. But the existence of the technology does not mean that it should function in this manner—the benefits and/or harms to society must be evaluated. The technological possibilities are enormous, but the social consequences must be taken into account.

Economic Effects

The exact interrelationship of broadcasting and the economy is nebulous. The actual number of dollars passing through radio, television, and cable TV, primarily in the form of advertising money, is very small, especially in relation to the social impact of the electronic media.

advertising

The indirect economic effect is large but difficult to measure. Advertising stimulates buying, but relating particular purchases to particular advertisements ranges somewhere between difficult and impossible. Rarely will a person see or hear an ad and run right out to buy the product. Advertising builds an awareness over time so that an eventual purchase could be the result of several radio or TV ads, a billboard, a magazine ad, the yellow pages, or perhaps the recommendation of a friend. An accurate figure cannot be derived for how much telecommunications contributes to the nation's actual purchases.

Taken a step further, most purchases would be made even if there were no radio and TV advertising. This advertising serves more to develop brand awareness than product awareness. People would buy bread even if it were not advertised, but advertising can lead consumers to buy brand A instead of brand B. The extent to which advertising develops false needs and leads people to buy products that are frivolous to their existence is highly criticized. Advertising is accused of creating a materialistic society that overspends on the needless and the useless, often at the expense of the worthwhile and needed.

Advertising also influences programming. Some feel that programs are merely a lure to draw an audience for commercials and that the real purpose of radio and television is advertising, not entertainment or information. Programming material is selected not on its merit but on the number of people it will deliver. In some instances the desired people are of a particular age, sex, and/or income level, as they are more likely to buy the products advertised.

Another aspect of economics is the amount of money brought in by the electronic media. Within our capitalistic society, most forms of telecommunications are profit oriented. Some forms, such as public broadcasting, are nonprofit in nature, but commercial radio and TV stations, cable TV systems, networks, advertising agencies, video stores, syndicators, and a host of other media entities are working to make a profit and a return for their shareholders.

profit

To some extent, the amount of profit is determined by the amount advertisers are willing to spend. Advertisers have, indeed, increased the amount they spend, but not in proportion to the number of places they now have to spend it. While network ad revenues did not quite double during the 1980s, cable ad revenues increased almost thirty times, and the number of channels available to the average consumer went up three and a half times. (See plate 8.) As channels proliferate, the number of advertising dollars must be spread thinner. This has made life particularly difficult for the networks because cable and other media are siphoning ad dollars from them. As their audience decreases, networks find they must charge less per ad. This cuts down on their total revenue and means that expenses must also be cut back. At present, most forms of the electronic media are experiencing the need for belt-tightening. In the past, commercial stations and networks and some of the entities of the cable and videocassette business were highly profitable "cash cows." But now, in part because of a slowing economy and in part because there are more telecommunications outlets than the market can bear, the industry is no longer as profitable as it once was.[18]

Closely allied to the economic health of the industry is the issue of profit versus public responsibility. As economic health worsens, the impulse of stations is to eliminate those elements that do not make money. But broadcasters, unlike furniture manufacturers or automobile dealers, are mandated to serve in the public interest, and it is often public-interest programming that is most unprofitable. To what extent should a station forego profit to air local-issue material that brings in little or no revenue either because it will not deliver an audience or because it is too controversial for advertisers? How much money should a network pour into covering news in remote countries even though the outcome of this news may affect many Americans? How much money should a cable TV system spend on access equipment and personnel so that members of the community can create their own programs, which other members of the community may or may not watch? Drawing limits on issues of this nature is never easy and necessitates constant give-and-take between the telecommunications industry and the public served.

The economic future of telecommunications is not assured. In the final analysis the public, by accepting or rejecting what is offered, decides the health of both the industry as a whole and the particular elements within that industry, such as individual radio stations and individual networks.

Political Effects

Would George Washington or Abraham Lincoln have been elected president if radio and television had been available in their times? No one knows, of course, but many theorize that the low-keyed George and gawky Abe would not have succeeded in an electronic age. Like it or not, candidates today must be TV personalities. A politician who shuns the TV camera shuns a major source of exposure to his or her constituency.

This applies not only to political campaigning but also to daily activities. News programs report the actions of senators, governors, and other public officials on a regular basis, and politicians who make buffoons of themselves on the evening news are not long for the world of elected officialdom.

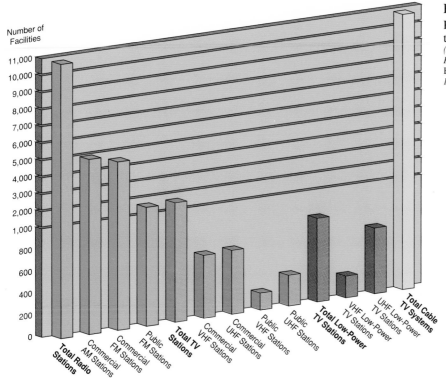

PLATE 1

Radio and television transmission facilities. *(Source: Broadcasting Publications, Inc.,* Broadcasting, *December 10, 1990, p. 115.)*

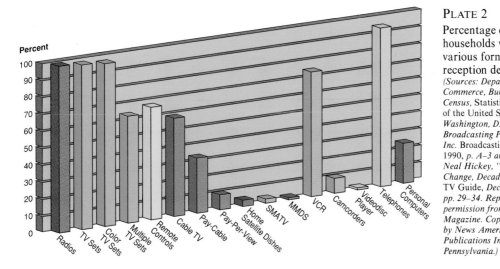

PLATE 2

Percentage of households with various forms of reception devices. *(Sources: Department of Commerce, Bureau of Census,* Statistical Abstracts of the United States, 1990, *Washington, D.C., p. 550; Broadcasting Publications, Inc.* Broadcasting Yearbook, 1990, *p. A–3 and D–3; and Neal Hickey, "Decade of Change, Decade of Choice,"* TV Guide, *December 9, 1989, pp. 29–34. Reprinted with permission from TV Guide® Magazine. Copyright © 1989 by News America Publications Inc., Radnor, Pennsylvania.)*

PLATE 3

Percentage of people using radio and TV during different time periods. *(Sources: Nielsen Media Research,* Nielsen 1990 Report on Television, *p. 5 and Southern California Broadcasting Association compilations, 1990.)*

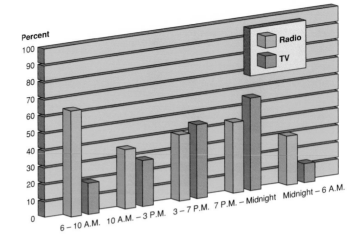

PLATE 4

Videocassette recorder activity in homes with VCRs. *(Source: Nielsen Media Research,* Nielsen 1990 Report on Television, *p. 11.)*

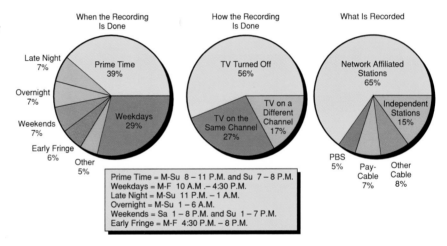

Prime Time = M-Su 8 – 11 P.M. and Su 7 – 8 P.M.
Weekdays = M-F 10 A.M. – 4:30 P.M.
Late Night = M-Su 11 P.M. – 1 A.M.
Overnight = M-Su 1 – 6 A.M.
Weekends = Sa 1 – 8 P.M. and Su 1 – 7 P.M.
Early Fringe = M-F 4:30 P.M. – 8 P.M.

PLATE 5

Hours of TV usage per week by household characteristics. *(Source: Nielsen Media Research,* Nielsen 1990 Report on Television, *p. 7.)*

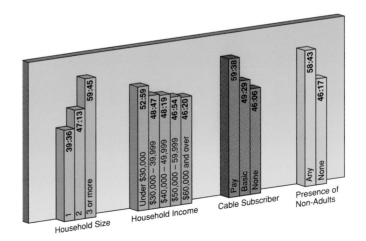

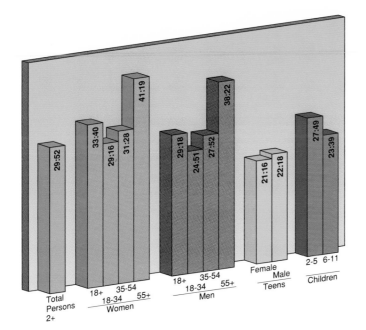

PLATE 6
Weekly viewing by gender and age. *(Source: Nielsen Media Research,* Nielsen 1990 Report on Television, *p. 8.)*

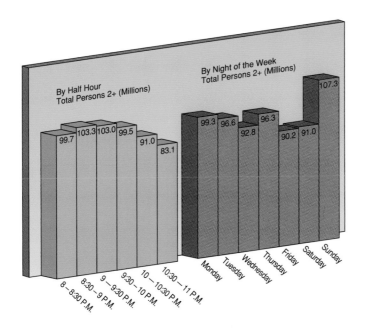

PLATE 7
Number of persons viewing prime time. *(Source: Nielsen Media Research,* Nielsen 1990 Report on Television, *p. 7.)*

PLATE 8

Comparisons between the beginning and end of the 1980s. *(Sources: Department of Commerce, Bureau of Census, Statistical Abstracts of the United States, 1990, Washington, D.C.; Broadcasting Publications, Inc., Broadcasting Yearbook, 1990; Nielsen Media Research Nielsen 1990 Report on Television; The Arbitron Ratings Company, Radio Year-Round: The Medium for All Seasons; North American Publishing, Channels 1990 Field Guide; and Neil Hickey, "Decade of Change, Decade of Choice," TV Guide, December 9, 1989, pp. 29–34. Reprinted with permission from TV Guide® Magazine. Copyright © 1989 by News America Publications Inc., Radnor, Pennsylvania.)*

	1980	1990
TV Network's Share of Prime–Time Audience	85%	67%
Average Number of Channels Available in Homes	8	28
Number of Independent TV Stations	150	444
Homes with Cable TV	22%	59%
Homes with Videocassette Recorders	1%	73%
Number of Video Stores	2,000	29,000
Homes with Color TV Sets	85%	97%
Homes with Multiple TV Sets	51%	63%
Homes with Remote Controls	18%	66%
Homes with Personal Computers	.2%	23%
Homes with Satellite Dishes	.01%	3%
Homes with Camcorders	0%	10%
Hours of TV Household Viewing Per Day	6.48	7.02
Average Number of Viewers Per TV Set	1.61	1.29
TV Network Advertising Revenues	$5.1 billion	$9.3 billion
Cable TV Advertising Revenues	$53 million	$1.5 billion
Cost of a 30-Second Commercial in the Super Bowl	$275,000	$700,000
Cost of the Rights Fees for the Summer Olympics	$87 million	$401 million
Cost of Producing a Typical One-Hour Drama	$420,000	$1 million
Cost of Producing a Typical Half-Hour Sitcom	$240,000	$575,000

Nevertheless, controversy surrounds the media's making of a president and other political figures. Are the articulate and attractive really the best people to run the country? Is the choice of a public relations firm more important to a candidate than the choice of a foreign affairs policy? Should elections go to the candidates who can afford to spend the most on political advertising?

The intertwining of politics and the media creates some strange bedfellows. Reporters are supposed to ferret out news objectively, but when they themselves harbor political ambitions or become friends with those they are covering, their judgment may not remain objective. People who cover politics *objectivity* effectively must, by nature, be independent and inquisitive. Perhaps too many of the same types of people wind up covering the political scene and, thus, inadvertently bias reports in terms of their own dispositions.

Radio and television are regulated by the very people they cover. For example, senators pass laws related to communications, city council members award cable TV franchises, and judges rule on First Amendment rights. In order to win favor with these government people, the media practitioners might be benign in their reporting. On the other hand, the media can be so powerful in creating a politician's image that the politician might be inclined to treat the media kindly. In both instances the public could be the loser.

There are many positive points to make about the media-political interrelationship, however. Politicians are much more visible now than they were before the age of telecommunications. Constituents can now see their representatives in more revealing ways than ever before and can make voting decisions based on firsthand visual observation rather than on secondhand written information.

The political process is scrutinized by the media. Innumerable changes *public scrutiny* occurred in political nominating conventions when the watchful eye of the camera made its debut. Major speeches and events were scheduled or orchestrated to coincide with prime-time hours. Delegates paid attention and were present for votes. Cameras in the halls of Congress have resulted in a bit more decorum than was present in years gone by as lawmakers realize that someone out there may be watching.

The entire media process aids in our country's dedication to peaceful *peaceful change* change. Views can be expressed through radio and television, debates can be held, issues can be discussed, and in the end, changes are made without resorting to revolution as is the case in many countries of the world.

The media serve as another balance of power by acting as a watchdog *balance of power* over the government and by reporting government corruption or wrongdoing. Sometimes, however, the media is accused of overreporting on wrongdoing and, therefore, inflicting more harm than good. Sometimes in order to obtain a story or a scoop, radio and TV reporters resort to sensationalism and blow minor infractions way out of proportion. At other times they prolong reporting on wrongdoing to the extent that it impedes the normal business of government. Some have concluded that many people qualified to run for public office do not in order to spare themselves and their families from the often inhumane scrutiny of the media.

The nation can be informed instantly of what is occurring in the political process. Sometimes this immediacy breeds danger, especially if all the facts have not been gathered, but the overall result is an informed public.

Because our country is strongly committed to freedom of the press, politics and the media will undoubtedly continue to interrelate. The balance between the two is delicate, but the forces of our society can act as a rational mediating force.

Ethical Considerations

Many of the points just made in regard to the effects of the media involve ethics. There is no law against reporting favorably about a politician in hopes that he or she will grant your station a favor. Nor is there a law that says a network can't portray women as passive or broadcast blood and gore. All of these are areas that require human judgement—a sense of ethics.

Obviously, individuals within all facets of life have a responsibility to act in an ethical manner, but the burden is often greater within the electronic media field. Part of this is because of the influence that this field has on society. An ethical slip betrays the trust people have placed in the media and leads them to wonder about the character of the country in general. This certainly happened in the 1950s when the discovery was made that quiz-show contestants that audience members admired had actually been given the answers to questions ahead of time.[19]

The temptation for unethical practices is great within the field. Hollywood has always had a reputation for living outside the bounds of morality. The entertainment business is one where a few people can get rich very fast, legitimately. But those that do often need to walk the thin line between right and wrong because that is where many other people who want to get rich fast are walking.[20] This does not justify unethical behavior, but it makes it seem more acceptable because "everyone is doing it." An aspiring actor who knows that a competitor received a part in a drama because his uncle is a friend of the producer may feel justified in telling people he once played Hamlet, even though it was just a short scene in his junior high school English class.

Then there is always the problem of determining what is right and what is wrong. Ethics, unlike law, is not codified. There are no set penalties for unethical behavior. It is just a sense of doing what is proper under the circumstances. Although some networks and other organizations have ethical codes, these are usually vague and do not give a great deal of help when individual situations arise. Some universities have ethics courses, but many do not, so people entering the field have not had the training to handle ethical dilemmas of the media.[21] Ethical decisions are not clear-cut. Deciding to do something may harm one segment of society while deciding not to do it may harm another. For example, when stations are deciding whether or not to accept condom advertisements, they are usually faced on one side by people who say if stations accept the ads they will promote promiscuity and by people on the other side who say if stations refuse the ads they are hindering progress in the control of teenage pregnancies and AIDS.

immediacy

influence

temptations

determining right and wrong

Many media-related ethical decisions, especially those related to broadcast journalism, must be made quickly and without complete information. For example, when a man holding hostages calls the news director of a station and threatens to kill the hostages if he cannot broadcast his demands immediately, the news director does not have time to read the ethics manual or hold a meeting with the top management of the station. But the news director does not know what the demands will be. Perhaps they will involve the threat to kill more people, thus making the situation worse. Giving airtime to someone of this nature leads others who are unbalanced to demand airtime. However, if the airtime is not given, the hostages may actually be killed. This is really a no-win ethical situation.

speed and lack of information

Ethical decisions can also vary depending on time and geography. The ethics involved in deciding whether or not to air a program on child pornography at 3:00 in the afternoon could be quite different than the factors that go into deciding whether or not to air that program at 11:00 at night. A news story about a man in St. Louis who has unusual ways of extracting sexual favors from women might be exploitatively titillating when aired in Denver. In fact, it might cause copycat crimes. However, if aired in St. Louis the story can help protect women by letting them know the man's procedures.

time and geography

People also need to look out for their own skins, especially in the dog-eat-dog world of the electronic media. Something may appear unethical to a certain group of people that genuinely appears ethical to the person engaged in it. For example, a salesperson may have sold a group of station ads for a particular diet. This diet is attacked by a group of health food store owners as being useless or even harmful. Both the person selling the ad and the station have a great deal to lose monetarily if the diet commercial is removed. In addition, the salesperson may have used the diet and may feel that it works very well. The health food store owners may be against the diet because it takes business away from them. So continuing to air the ads, although this may appear unethical to many, may be, at worst, confused. Ethical decisions are rarely easy. Sometimes they look easy with hindsight after all the facts are known. But because of lack of facts, peer pressure, the desire to succeed financially, and genuine differences of opinion, deciding what is right for the particular circumstance can be difficult.

self-promotion

All of this means that everyone who is in the telecommunications business must keep ethical considerations in the forefront of the brain so as not to be eased into unethical behavior. Media personnel should make every effort to abide by the truth, to treat other people equitably, to keep promises, to cause as little overall harm as possible, and to do what they truly feel is right.

Ethical Problems

The following are a few examples of situations that involve ethical decision making. Read them over, decide what you would do, and discuss them with others. The situations cover a variety of aspects of the telecommunications

ethics cases

business, because no facet of the business is exempt from ethical problems. A few of these might sound familiar because they are drawn from real situations, some of which made the headlines.

1. It's Nielsen ratings time and your TV station's nightly news programs are floundering in fifth place in your market. Your numbers must come up so you can sell your ads for more money, or there will be major layoffs in your news department. You, as the news director, hit on an idea. During your newscast you could run a series about the Nielsen ratings and the people involved. That should entice the people who are a part of the Nielsen ratings survey to watch your station. Their meters would record that they watched your station and, even though they may watch other stations the next week, they would be watching your station during the crucial time when ratings are taken. Should you run this series?

2. You are the producer for a potential new series called "The New Hit." You have talked to the network about this show and indicated several leading ladies that you feel you could obtain. The network executives responded most favorably to Betty Bigname, so you make a concerted effort to hire her. But her agent tells you there is no way; she is overcommitted. So you pursue your second choice, Mary Mediumname. She is available and eager for the part. The network gives you the go-ahead to make a pilot with Mary Mediumname, but you can tell the executives are not as excited about the series as they were when they thought Betty Bigname would star. But you sign Mary to a contract and begin planning to produce the pilot. One day Betty Bigname's agent calls you and says the major movie Betty was to star in has fallen through, so she is now available for your series. You know that legally you can buy out Mary—pay her what she would have earned but don't use her. But since she really wants the part, she will probably drive a hard bargain. Also, you would then be paying for two leading ladies. Mary really wants the part, and her public relations firm has issued a great deal of publicity saying she will be starring in this pilot. What do you do?

3. You are a salesperson for radio station KICK. Last month you sold a large package of ads to the Goody Food Restaurant. This earned you the biggest commission you have ever had. You want Goody Food to buy an even more lucrative package of ads, but the manager is stalling because she doesn't think the ads she bought did much good. You hit on an idea. You could call some of your friends and have them make reservations at Goody Food, making sure to mention that they heard an ad on KICK. Then you could call the manager and ask again if she wants the ads. Of course, your friends could later cancel their reservations, but not until you had the ad deal signed. Should you do this?

4. Your news crew is covering a story about a woman suspected of committing a series of murders, who has barricaded herself inside a house. She tells the police (and the media) that she is about to come out and surrender, so you train your cameras on the front door. The woman comes out, pours gasoline on herself, and burns herself to death. Do you show this footage on the air?

5. You rent a videotape that has several vintage TV shows on it and invite some friends over to see it. One of your friends says she is doing a term paper, and it would really help her if she could watch one of the programs over and over to analyze it. She knows you have two VCRs, so you could easily make her a copy. She is a poor, starving student with only one VCR and not enough money to rent the tape for the time she will need it. The tape has the usual copyright notices all over it, but it is not copyguarded, so you could make her a viewable copy. Should you do so?

Conclusion

Whether this field be called telecommunications, electronic media, or broadcasting, it is an area that deserves careful study. For the individual who spends four hours with TV and three hours with radio each day, it is important to know their effects in terms of education, use of time, and acceptance of role models. Individual emotional responses to TV are crucial and should be understood by those who experience them, especially in this era of possible media overload.

From a sociological point of view, telecommunications is often a unifying phenomenon that engenders a sense of communion in times of both trial and triumph. The media permeate the home and affect the family structure, particularly children, in ways that are still not totally understood despite a great deal of research using field and laboratory methodology. Schools, culture, sports, fads, and outlooks toward violence are all affected by the electronic media. The basic structure of communication, with its source, message, channel, receiver, barriers, and feedback, affect society and the media in an interactive way. The fact that TV is considered a primary source of news and information as well as entertainment makes its relationship to society all the more crucial.

Telecommunications and technology interrelate in a spiraling fashion. New developments in one cause new developments in the other. The various media also interrelate, and one medium has the potential to destroy or significantly alter another. Technology affects both the artistic quality and the communication ability of all telecommunications.

From an economic point of view, advertising seems to be the price that is paid for our telecommunications, and yet this advertising represents a small portion of the country's economic base, especially in relationship to the sociological importance of telecommunications. The American capitalistic system, based on profit, must be carefully considered when one studies both the positive and negative effects of media. The profit base can also affect the economic health of the industry, and income and outgo fluctuate.

In the realm of politics, the electronic media are the primary method by which the public is now acquainted with candidates. They are also the primary scrutiny device for overseeing political actions. The electronic media must be given credit for the ability to bring about peaceful change, but they also must be carefully watched for unavoidable and avoidable bias in reporting politics.

Because of the sociological importance of the electronic media, those who are practitioners within the industry must keep ethics foremost in mind. They must not succumb to pressure from peers or their own self-interests as they make their day-to-day decisions.

In studying the inherently interesting field of telecommunications, one should be very careful to assess criticism and understand reality. Nothing is perfect, but all elements of the media can be improved.

Thought Questions

1. Why are you interested in studying telecommunications?
2. Suggest some worthwhile research studies that ought to be conducted concerning electronic media.
3. What do you think are some of the most important ethical problems facing the electronic media at present?
4. In what specific ways and to what extent has television affected your life and the life of your family?

International
Telecommunications

Introduction

Many of the major telecommunications issues of the 1990s will involve international interrelationships of the electronic media. Broadcasting systems in most countries developed very differently than they did in the United States, so a knowledge of the history and customs of various broadcasting systems around the world will be crucial to anyone who intends to deal with foreign media. And most people who are in the business will probably have to give great considerations to international markets in the coming years, because a major trend of the present is a meshing of television technologies throughout the world.

The concept of international telecommunications encompasses a great deal. On one hand, there is the broadcasting that occurs internally within particular countries. This is intended for the national audiences, although on occasion programs are sold to other countries. In order to operate a radio-television system, countries have set up differing types of mechanisms to handle programming, distribution, finances, and ownership.

An entirely different aspect of international telecommunications involves the transmission of material by one country to another country or countries. This is often propaganda based. The intent of the transmitting country is to show itself in the best light to other countries of the world. Closely aligned with this are the negotiations that occur between countries regarding the use of spectrum space by each national entity and the uses these countries make of satellite transmission.

These various aspects of international telecommunications will be covered in this chapter. "Different strokes for different folks" is a truism that definitely applies here. Organizations vary as boundary lines are crossed, and yet there are common elements that glue the world's telecommunications systems into a unified whole.

CHAPTER
2

The orbiting satellites herald a new day in world communications. For telephone, message, data, and television, new pathways in the sky are being developed. They are sky trails to progress in commerce, business, trade, and in relationships and understanding among peoples. Understanding among peoples is a precondition for a better and more peaceful world. The objectives of the United States are to provide orbital messengers, not only of words, speech, and pictures, but of thought and hope.

Lyndon B. Johnson

Technical Standards

One of the irritating technical problems of international telecommunications is that various countries have different systems for taping and transmitting electronic information. The United States operates on a television system that scans 525 lines on the face of the TV tube and employs a color coding system for color TV that was approved by a committee of engineers called the **National Television Systems Committee.** This color coding system is referred to as **NTSC.** The Americans hoped it would be adopted throughout the world, but European engineers quipped that NTSC stood for "Never Twice the Same Color" and developed their own system called **SECAM (Sequence Couleur a Memoire),** which American engineers in turn labeled "Something Essentially Contrary to the American Method." Another system, **PAL (Phase Alternate Line),** was developed in hope that it would yield "peace at last." This was not the case. At present approximately half the world uses PAL and the other half is split almost evenly between NTSC and SECAM. Most of the SECAM and PAL systems scan at 625 lines, making the United States among the minority that scan at 525.[1]

One of the many problems with having so many different standards is that when programs are exchanged they must be dubbed from one system to another, which causes loss of picture quality.[2] For awhile it appeared that a universal standard would be accepted for a new form of television, **high definition TV (HDTV).** This system, which was developed by the Japanese, would have scanned at 1125 lines, giving everyone a better quality picture than presently available. But politics intervened, and a worldwide standard was not accepted. At present it appears that different parts of the world will each develop their own high definition standard. These will all have more lines than any of the present systems, but they will not have the same number of lines as each other. Therefore, dubbing will still be needed. Another option when exchanging programs is to distribute them on film, which does have a worldwide standard in both 16mm and 35mm. However, many programs are produced on videotape, and there is no efficient way to transfer them to film.[3]

Program Exchange

Many countries are involved in the exchange of programs. Some only buy from other countries, and some both buy and sell. The amount of foreign programming shown in a particular country often becomes very controversial. On one hand, countries want to maintain their national identity, but on the other hand, they want quality programming without high cost. Radio was inexpensive enough that most countries could handle their own programming, even when they first began the service. Television was another matter, however, and programming available through international syndication could be obtained with much less expense than similar programming produced locally. As a result, a large market built up for the exchange of foreign programming.

color system

HDTV

local versus imported

Many countries sell or exchange their programs throughout the world. Mexico provides much of the programming for South America, and the Soviet Union does likewise for the countries in Eastern Europe. Britain sells a great deal of its programming, primarily to countries that used to be part of its commonwealth. Canada, France, Germany, Brazil, and Australia all make international sales. The granddaddy of them all, however, is the United States, which sells more programming than any other country.[4]

American programming

American programming in foreign countries, especially the developing nations, has been controversial. Politically, this programming is often unwelcome because many of the cultural modes it displays are different from those of the nation in which the program is aired. Entertainment programming is criticized for its decadent morals, and news and information programming supplied by American radio and TV is often accused of having a Western bias.

CNN

CNN, however, has managed to withstand much of the criticism against American news and has been accepted by many foreign countries. It made a conscious effort to market itself as an international service, and it can now be seen in more than 100 countries.[5] In many instances CNN has been instrumental in creating as well as reporting the news. When Eastern Europe was in turmoil in 1989, Polish Solidarity leader Lech Walesa kept up on events by watching CNN. When the Soviets wanted to denounce the American invasion of Panama, they did so by making the complaint to CNN's Moscow bureau. During the war with Iraq, CNN was the only service allowed to remain in Baghdad, partly because Saddam Hussein was a regular viewer. The broadcasting systems of many countries send video reports to CNN for inclusion in the "CNN World Report." In exchange for this, the countries can use anything else on the program. This has brought about a healthy exchange of international news.[6]

quotas

Usually as countries develop their TV facilities, they buy less foreign programming and depend more on their own production. Many of them set **quotas** regarding the amount of foreign programming that can be shown. Different countries have differing attitudes about this, however. New Zealand, which feels isolated from the rest of the world, will program up to 60 percent foreign-produced shows so it can feel in touch with the world. Canada, which is always fearful of being absorbed by American culture, is constantly fighting to keep down the amount of American programming that crosses the border.[7] In the radio area, countries may produce their own programs, but the music aired on a large number of the stations represents America's ubiquitous contribution to the world.

increased outlets

Another element of foreign exchange of programming involves the number of broadcasting outlets in a country. If a country has just one or two networks, it may not purchase a great deal of outside material. But as countries increase the number of outlets, the demand for product goes up. So the cutback in the purchase of foreign programming, which most countries go through as they become developed, is in some ways being offset when they increase the number of broadcasting facilities.[8]

This is welcome news for the countries that sell programming, because they have come to depend on the foreign market. Many United States program producers do not turn a profit on productions until they are sold overseas. Increasingly, this is becoming true for British programs.

selling practices

The international market is filled with financial vicissitudes. What one nation is willing to purchase for $40,000, another may only be willing to purchase for $400. The annual program market, held in Cannes, France, is where most international buying takes place; selling patterns can vary widely from year to year. For awhile only dramas sold well overseas, but later comedies hit their stride. Once the programs are sold, there are often difficulties in collecting the money, particularly from some of the more volatile Latin American countries. In addition, **piracy** has become a big problem because many nations are not sympathetic to the United States copyright laws and see nothing wrong with dubbing copies of programs to use in a variety of ways.[9]

maintaining identity

Many of the new electronic media facilities that are springing up worldwide are trying new methods of maintaining their nationality. Although they need a quantity source of programming (such as that produced by America), they are trying to be creative about their own programming. Many countries are engaging in **co-productions,** sometimes with American companies and sometimes with other countries, where each country pays part of the cost of producing the program or series. In this way countries can have at least some say about the creative elements of the programs because they have a financial interest.[10] In addition, a number of foreign companies are buying American film and TV production companies. In this way, they may be able to have some say about what is produced.[11]

American imports

Through all of this, the country that imports the least programming of all is the United States. Generally it imports only 2 percent of programming, most of which is British material seen or heard on public broadcasting.[12]

Original Forms of Foreign Broadcasting

When most of the world's broadcasting stations and networks were started, they were organized under one of four systems—authoritarian, paternalistic, permissive, or pluralistic.[13] (These forms are often also referred to as totalitarian, public, private, or mixed.)

authoritarian

The authoritarian or totalitarian systems were most common in countries with dictators or strong central governments—for example, the Soviet Union and its Eastern European satellites. The governments owned and controlled all broadcasting. The programming was created within the government, usually by a department of education or propaganda, and material was often censored by a government committee. Many of the programs espoused the virtues of the government. In fact, radio and TV were essentially public relations arms of the government designed to make sure government policies were not questioned. The emphasis was on educational, informative programming; very little of it even tried to be entertaining. The broadcasting systems were financed by general government funds collected through taxes.

For the paternalistic (or public) systems, corporations actually owned the broadcast facilities, but the corporations were closely tied to government. The government in each country issued a **charter** outlining provisions that the corporation had to follow. Britain set the pattern for the charters and then exported this form of broadcasting to many of the countries in its Commonwealth. Most of Western Europe also adopted the public-charter form of broadcasting. Usually the government department overseeing broadcasting was the same one that oversaw the mail, telegraph, and telephone. Because of this there have been some historically close ties between the telephone and radio/TV. Programming was produced by the corporations, but the content was implicitly or explicitly scrutinized by the government. The emphasis was on information and culture, but the programming was of a more entertaining nature than that of totalitarian countries. Financing came from license fees collected from people who owned radio and TV sets. These fees were earmarked to be used for the production and distribution of programs.

paternalistic

The permissive (or private) systems were ones organized like the United States. Private companies, responsible to stockholders, owned the broadcasting companies. Programming was created by networks, stations, or production companies, all privately owned. Although some of the programming was informative, the main emphasis was placed on entertainment. Funding came primarily from advertising.

permissive

The pluralistic or mixed systems were ones that employed two different types of systems, usually because the system originally chosen did not work adequately. For example, Canada started with radio stations privately owned and supported by advertising. These stations were almost totally in the large border cities because stations in sparsely populated areas could not attract enough advertising to make money. As a result, the government-run **Canadian Broadcasting Corporation (CBC)** was established in 1936 to provide programs to sparsely populated areas, as well as cities. The CBC resembled the British system, and when television came to Canada about 1952 both private and public (CBC) TV systems were established.

pluralistic

New and Developing Forms of Foreign Telecommunications

The four forms of broadcasting coexisted in fairly unchanging forms during the days when the only form of broadcasting was radio. But when television began, it was much more expensive to operate than radio. Many of the countries with private or totalitarian systems found they could not support television through general taxes or licenses on TV sets. As a result, advertising began to creep into systems that had previously disallowed it. The ads were usually low-keyed and clustered at times that did not interrupt programming.

commercials

At first many of the countries with public systems kept the government-chartered TV systems ad free but allowed separate stations that were privately owned to advertise. But eventually, ads were seen on many of the government systems also. In this way, the countries developed mixed systems similar in form, but not history, to the Canadian system. In some countries, such as France, the main government channels were turned over to private interests.[14]

For the most part, the totalitarian systems that allowed advertising kept a close reign on the ads and did not allow private systems to develop. However, the spirit of *glasnost,* which started in the Soviet Union in the mid-1980s, and Eastern Europe changes the events in Eastern Europe during the late 1980s led to major changes in the broadcasting forms of those totalitarian countries. The media were very instrumental in helping the democratic reforms to occur in those countries. Stations changed from government mouthpieces to open forums almost overnight. Although it is too early to say exactly what form of broadcasting these countries will settle into, they appear headed toward systems that have many characteristics of private systems.[15]

Another element that changed the telecommunications structure of much of the world, including the United States, was the emergence of other forms of electronic media besides radio and broadcast television. The primary ones to have an influence were cable television, direct broadcast satellite, and videocassette recorders.

cable
Some countries, such as Canada and Germany, are heavily wired to receive cable TV. Others, such as England, Australia, and most of the Orient and Africa, have very little cable TV. The main reason cable TV has not been instituted in many countries is economic. Laying cable and providing the programming for it is very expensive, and Third World countries and others with weak economies cannot afford it. Sometimes, however, the reasons are political. Australia, for example, voted down cable, largely because it could not fill all the channels with Australian-produced programming, and it did not want a huge influx of American programming.[16]

DBS
Direct broadcast satellite (DBS) involves beaming programming from a satellite to receivers on homes. In the United States, some people have bought satellite dishes for their backyards that they use to receive programming that is intended primarily for cable TV operators. But in some countries, programming from the satellite is intended to be received by small dishes or plates mounted on houses. Japan is leading the way with DBS, broadcasting one channel throughout the day. Several hours of the day are devoted to the DBS transmission of programs in high definition TV. At present the HDTV part of the DBS broadcast is being viewed mainly in public gathering places because not many Japanese homes have sets that can receive HDTV.[17]

Europe is also experimenting with DBS. A satellite called Astra is broadcasting sixteen channels of programming intended for homes. Because Europe has so many diverse cultures and languages, many of the channels are aimed toward small subcultures, such as the Scandinavians. However, some programming, such as music videos or sports, can be viewed by a broad audience.[18]

VCRs
Videocassette recorders (VCRs) are rapidly becoming popular around the world, even in Third World countries. In India, for example, VCRs have become staple items in dowries of brides.[19]

The result of all of this—new private channels, cable TV, DBS, and VCRs—is a much greater proliferation of TV programming sources throughout the world than ever before. This means more programming needs to be produced to fill all the channels. Exactly who will profit from this, if anyone, is not known yet, but the whole area of producing for the international market is one that will be heavily addressed during the 1990s.[20]

The establishment and evolution of broadcast systems can be seen by taking a more thorough look at the British (paternalistic) and Soviet (authoritarian) systems. Many countries of the world have followed patterns that closely resemble these two. The permissive (American) system will, of course, be the main one discussed throughout the rest of this book.

The British System

For many years all of British broadcasting was under the control of the government-sanctioned **British Broadcasting Corporation (BBC)**. However, the country now has both the BBC and an ever-evolving form of independent broadcasting.

The British Broadcasting Company was formed in 1922 under a license from the Post Office. Five years later, in 1927, this organization became the British Broadcasting Corporation under a Royal Charter.[21]

early BBC radio

This charter gave the BBC a monopoly on all radio broadcasting. It created a Board of Governors consisting of twelve members appointed by the monarch for five-year terms. This was to be the policy-deciding group and the day-to-day operations of the BBC were to be carried out by a Director-General and his/her staff. The charter also stated that there could be no advertisements on the BBC, that it had to broadcast daily impartial accounts of the proceedings in Parliament, and that it could not give its own opinion on current affairs. Most of the provisions of this charter are still in effect. The major changes are that the charter now encompasses television as well as radio, and the BBC no longer has a monopoly.[22]

charter

Because there were no advertisements, the funding for the BBC came from money people paid to own a radio. This money was collected annually by the Post Office as a license fee. When TV was added, people paid licenses on both radios and TV sets. Eventually the license fee was dropped on radios, so now people pay a license only on their TV sets and this supports both BBC radio and BBC TV.[23]

license fees

Originally the programming on BBC radio was very paternalistic. Designed to upgrade tastes, it was often referred to as programming from "Auntie BBC." The original organization consisted of three program services, each designed to lead the listener to the next for a higher cultural level. At the lowest level was the Light Programme, which consisted of quiz shows, audience participation programs, light music, children's adventure, and serials. Level two was the Home Service, which included modern music and drama, school broadcasts, information about government, and news. The highest level, called the Third Programme, was classical music, literature, talk, drama, and poetry. All these services were national and were the only radio in Britain.

three program services

This programming was an easy target for outside competition. When Radio Luxembourg began broadcasting popular music of the 1950s and 1960s, many of the British people began tuning in; British companies even began buying ads on Luxembourg stations.

"pirate radio"

In 1963 **pirate ships** anchored off the coast of England began broadcasting rock music on frequencies that could be picked up with ordinary radios. This programming became so popular that when the government planned to suppress it, there was such public outcry that the BBC had to change its programming before these pirate ships could be eliminated.[24]

four program services

The result was a new, renamed, four-level radio setup. Radio 1 programs rock and pop music. Radio 2 resembles the old Light Programme with panel and quiz games, comedy shows, and light music. Radio 3 is similar to the old Third Programme, broadcasting primarily serious music. Radio 4 is the main talk service with news, current affairs, parliamentary reports, and other programs similar to the old Home Service. In addition, the BBC established local radio stations. More than thirty of these stations are now on the air throughout the country, programming primarily information of interest to local listeners.[25]

early BBC TV

BBC television began in 1936 and in 1937 televised the coronation of George VI—quite a feat for that time. The operation was stopped in 1939 because of the war and then resumed in 1946, utilizing a 405-line scanning system. In 1953 it telecast the coronation of Queen Elizabeth II, an event that greatly increased the sale of television sets and brought television to the attention of the British people. When color was introduced in 1967, the BBC converted to a 625-line PAL system.[26]

two program services

In 1964 the second BBC TV network was established, and the two were called BBC I and BBC II. Unlike radio, the overall program content of both services was similar, but the individual programming hours were different. For example, at 8:00 P.M., BBC I might have sports and BBC II a drama; then at 9:00 P.M., BBC I might have a drama and BBC II a documentary. The original intent of this programming was to upgrade appreciation.

formation of IBA

This paternalistic programming, like its radio counterpart, came under fire, and the result was the establishment of the **Independent Television Authority (ITA).** This was formed in 1954 after a heated debate in Parliament regarding the quality and role of television. In 1972 its responsibilities were extended to include the setting up of independent radio stations, more than five hundred of which went on the air. Along with these extended responsi-

IBA radio

bilities, its name was changed to Independent Broadcasting Authority (IBA).[27]

The organization of the independent radio stations was fairly simple. These were local stations similar to the BBC local stations except that they received their money from advertisements rather than government allocation.

IBA TV

The organization of independent television became quite complex, however, and very different from the BBC. Although the IBA was overseen by an eleven-member Board appointed by the Home Secretary and led by a Director-General, it was actually a regionally based national network. It did not produce any programs itself. All programming came from fifteen independent

FIGURE 2.1
The London complex
that serves as
headquarters and
production facility for
BBC TV. *(Courtesy of
Barbara and Dave Huemer)*

companies, called **ITVs,** selected and appointed by the IBA. Each of these companies served a different area of the country and programed local material, as well as the national fare. Each contributed a certain number of programs to the national feed. The companies in large cities generally contributed more than the companies in outlying areas. The closest parallel to this in the United States is public broadcasting, where local stations produce programs for PBS.[28]

IBA and the independent companies received almost all their funding from commercials. The ITVs sold commercial time within the programs they produced. Regulations for these ads were set up by the IBA in accordance with the Broadcasting Act that established it. These regulations included provisions that limited ads to an average of six minutes an hour, prohibited ads on programs aimed at children, stated ads had to come at natural program breaks, and prohibited advertisers from sponsoring program content (i.e., they could not buy programs, they could only buy commercial spots).[29]

The independent companies received the money from the ads, but they then paid IBA to rent the national television transmitters. In this way the IBA received the money it needed for its operation.

To make matters a little more confusing, there was also a shared network, **ITN,** which provided news to all the regional companies. Since 1983 there has also been a nationwide breakfast program produced by a separate company called TV-am.[30]

In 1982 a new independent network, **Channel 4 (C4),** was authorized. It was funded by additional subscription fees paid by the ITVs, but they, in turn, collected the advertising for C4 and provided it with many of its programs. Other programs came from other countries and production companies that are not affiliated with either the ITVs or the BBC. C4 paid IBA for the use of the transmitters in much the same way that the ITVs did.[31]

structure

Channel 4

FIGURE 2.2
The IBA headquarters
in London.

In 1990, a broadcasting bill was passed that revamped independent radio and TV. The IBA was dissolved, and two separate commissions were formed. One, the **Radio Authority,** oversees all radio stations in much the same way that the IBA did. The other, the **Independent Television Commission (ITC),** oversees not only independent broadcast TV but also cable TV and satellite broadcasting. Channel 4 is to be separated from the rest of the independent system and sell its own advertising. The ITVs are to be reorganized and possibly sold.[32]

When independent TV started, BBC I lightened its programming so that it could maintain its audience. BBC II appeals to narrower interests, but it, too, has some popular programs. By American standards, all the British programming services are heavy on educational programming. During the day there are many programs, both radio and TV, for children in schools. At other times there are continuing education programs. One unique partnership in which the BBC participates is the **Open University.** Established in 1969, this department produces courses for university credit that are funded by the government through its Department of Education and Science, rather than being funded by the license fees.[33]

Independent TV has a larger budget than the BBC. This is because advertising generates more money than people are willing to contribute through license fees. In fact, the hue and cry from the public against the ever-increasing license fees and the complaining from the BBC about shrinking budgets led the House of Commons to introduce a bill that would allow advertising on the BBC. This bill caused a high degree of emotion and was defeated in 1985. This makes the BBC one of the few private charter systems left in the world that does not accept commercials.[34]

bill for change

Open University

controversies

FIGURE 2.3

A production scene from "Coronation Street," one of IBA's most popular programs. Its cast of characters often interacts in the pub, Rovers Return Inn. *(Courtesy of Granada Television)*

One of the services presented by British television that is rarely seen in the United States is **teletext.** This involves using part of the television signal not used for the picture to transmit words and graphics. These include captions for the hearing impaired, as well as news items, sports scores, stock market prices, and similar information. The BBC's system is called **Ceefax** (see facts) and the independent TV's system is called **Oracle** (an acronym for Optical Reception of Announcements by Coded Line Electronics). People have to adapt their TV sets with decoders and keypads if they wish to receive the teletext at all times, but the pages are also shown when regular TV is not being broadcast.[35]

teletext

Cable TV was not introduced into Britain until 1982 and has not been very successful, in part due to regulations that state all cable must be buried under the ground. This is much more expensive than hanging the cable on telephone poles as is often done in the United States. VCRs are very popular in Britain with a majority of homes having them.[36]

cable TV

Direct broadcast satellite is operational in Britain in a limited fashion. Two companies, Sky Television and British Satellite Broadcasting, have been beaming signals, but very few homes are equipped to receive them. Because neither company is doing very well economically, they have plans to merge.[37]

DBS

The Soviet System

Russian radio began experimentally in the early 1900s. Some of the first radio stations were run by labor unions and educational organizations, and some were run by the government. Gradually the government took over the unions and educators, and broadcasting was, in reality, government-run.

early radio

Vladimir Lenin saw radio as a significant force to enlighten the illiterate masses and made radio development one of his primary goals after the 1917 revolution. Programming was primarily educational, including such information as sanitation techniques and medical care. Because the Soviet Union is such a large country, the transmitters in Moscow were high powered (40 kilowatts in 1927). Shortwave was also used to transmit radio. A fair portion of the nation was wired, with radio programs being delivered to loud speakers situated in the main squares of large cities.

national and local

By 1936 there were five state-run radio networks and a number of local stations to serve regional needs. All of them were supported by general government funds. During World War II, 1200 kilowatt stations emanated from Moscow and Leningrad, mainly broadcasting information for the troops.

program services

After the war, radio went back to an organization that included central and local state-operated systems. The five networks programmed in Moscow included: the First Program, which was the most important network devoted half to socio-political messages, including news, and half to the arts; the Second Program, which consisted of music and information bulletins; the Third Program, which was literary and musical, including such material as operas and plays; the Fourth Program, which was devoted to symphony music; and the Fifth Program, which was music and information. In 1949 radio was placed under the Council of Ministers of the Communist Party, the highest party organization in the USSR.

early TV

Television started in the 1950s, using the 625-line SECAM system. In 1957 the operation of both radio and television was given to a newly created committee, the State Committee for Radio and TV (**Gostelradio**). The people on the committee were all Communist Party members and, as a group, reported to the Council of Ministers.

production facility

The development of television became a national objective beginning in the 1950s. A large production facility was built in the outskirts of Moscow. One of the big televised events was the World Youth and Student Festival held in Moscow in 1957. In 1980 a new television facility, built to handle the Summer Olympics in Moscow, became the primary broadcasting facility.

One of the government's major problems was the delivery of programming to all parts of the country, which includes eleven time zones and more than sixty languages and dialects. The solution was satellite transmission.

satellites

In 1965 the government launched its first communications satellite and delivered the programming at many different times of day, dubbed into different languages.

networks and stations

Four TV networks and a sprinkling of local stations were developed. First Channel contained news, culture, movies, and concerts and was particularly known for its 9:00 P.M. news "Vremya" ("Time"), usually viewed by more than 100 million Soviets. Second Channel was primarily for Moscow and featured information and art. Third Channel was educational programming material for school children, college courses, and programs to teach the Russian

FIGURE 2.4
The Soviet
broadcasting center in
Moscow.

language and other languages. Fourth Channel consisted of sports, arts, and politics. Commercials were not aired, but there were programs to tell consumers the features of various goods. The programming was intended to serve the needs of the state. News that the government did not want the people to know was not broadcast. Most of the programming was ponderous and educational with very little on the light, humorous side.[38]

Cable TV was started in the USSR, but the government decided not to make it a priority and it withered. Of late, videocassette recorders have become popular.

With *peristroika* (restructuring), Soviet broadcasting is undergoing many changes. Independent stations, not affiliated with the government, have been proposed and, in a few instances, started. Many of these are being run by political parties that oppose the Communist Party. This is possible because of a decree from Mikhail Gorbachev that all parties, not just the Communists, can have access to state-run or other broadcasting facilities. Gostelradio is being restructured to reduce its power, and a proposal has been made to shake up the state television monopoly by creating four separate companies to run each of the four channels. Programming, too, is less propagandistic than it used to be; however, it is still very controlled by American standards.[39]

Future directions of Soviet broadcasting will be very heavily influenced by future directions of the entire political and economic situation in the Soviet Union.

cable TV

restructuring

Other Systems

Most of the world's broadcasting systems have borrowed in some ways from the American, British, or Soviet systems. Each country has its own peculiarities, however, which give it color and interest.

Australia has an expansive geography, so it has only one nationwide service, the **Australian Broadcasting Corporation (ABC)** run by the government. There are several private networks, but they operate only in populated areas. The country used to have a quota system requiring at least 50 percent of programming to be nationally produced. Now, however, they have gone to a point system whereby homegrown products receive more points than foreign imports. The point system also governs types of programs. For example, a drama gets more points than a game show. To keep licenses, stations are supposed to average at least one point per hour.[40]

Japan

Japan has an advanced system of broadcasting that includes the national service **NHK,** which is similar to the BBC but has more autonomy. There are also private radio and TV stations that are more or less local in nature. The production of programs (and commercials) is very high tech.[41]

China

Chinese radio started in the 1920s in the American mold and then switched to the Soviet model in the 1950s because the Russians helped the Chinese to establish TV. When the two countries severed relations, the Russian advisors left. During the Cultural Revolution of the 1960s, much of the broadcasting structure was closed down and what remained programmed endless excerpts from Mao Tse-tung's *Little Red Book* and several operas chosen by Mao's wife. When China opened up to the west, it adopted more modern broadcasting that included commercials and some American programming. The broadcasting structure and programming content, however, varies greatly depending on the political situation within the country.[42]

Cuba

At one point, Cuba was the bastion of commercialization. Radio stations would program one-minute programs that consisted of thirty seconds of news, twenty-five seconds of commercials, and a five-second time check. The commercials were loud, hard sell, and repetitive. When Castro came into power this stopped, and the broadcasting system took on characteristics similar to those of the USSR.[43]

Brazil

Brazil has five television networks, but one of them, **Rede Globo,** dominates. It claims to be the fourth largest network in the world, chosen each evening by 70 percent of the Brazilian TV viewers. Its most popular programs are novellas, which are closely akin to soap operas. Many of these are sold to other countries.[44]

Africa

Most of the African countries have systems patterned after Britain or France because they were colonies of those countries. When they obtained their independence, many Africans closed down radio and TV for awhile, but then opened them up again in a style similar to what they had been. The Union of South Africa was one of the last countries in the world to initiate television and, at one time, had separate services for blacks and whites.[45]

FIGURE 2.5

A set used for the taping of religious programs in Malaysia.

Saudi Arabia now has very modern broadcasting, but for many years an organization of Muslim religious leaders prevented its introduction. Radio won its way into the country because a demonstration transmission of readings from the Holy Quran went smoothly. The devil cannot read the Quran, so the religious leaders accepted that radio was not the work of the devil. TV had a harder time because the Quran forbids creating images of living people. King Faisal thought TV was needed for nationalistic purposes, however, and had a system built that went on the air in 1965.[46]

In Guatemala, two television systems have developed—one for the rich and one for the poor. This was not intentional, but many rich people have bought satellite dishes for their rooftops. These have been humorously referred to as "the national dish of Guatemala." These rich people watch American satellite programming such as CNN and HBO, while the poorer people watch the Guatemalan stations that play primarily imported Mexican soap operas.[47]

Malaysia modeled its broadcasting after the BBC during the years when it was a British colony. It began with two state-run networks and then added a private network, owned primarily by the leading newspaper and controlled rather closely by the government. Malaysia is primarily a Muslim country, so broadcasting has a strong religious content, including daily readings of the Quran. The Prime Minister and cabinet are somewhat authoritarian, so the government has a fairly strong hand in what is programmed.[48]

Saudi Arabia

Guatemala

Malaysia

International Broadcasting

purpose

Another entirely different type of broadcasting that many countries engage in involves programming that is not intended for the national audience, but for other countries. Countries generally do this out of nationalistic interests. They want to make sure that people in other parts of the world are exposed to events as they view them. They also do it out of altruism—a desire to see that less fortunate people are exposed to culture through radio.

countries

The leaders in this type of programming include China, Germany, Great Britain, the USA, and the USSR, but many smaller countries such as Malaysia, North Korea, and Egypt also engage in international broadcasting. Obviously, each of these services broadcasts many programs simultaneously in many different languages—forty-two for the United States, eighty for the Soviet Union, and thirty-seven for Great Britain.[49]

VOA

The main American international broadcasting is prepared by the **Voice of America (VOA),** which is part of the United States Information Agency. It is headquartered in Washington, D.C., where it has thirty-three studios. The programs are transmitted around the world by a complicated system that includes satellites, microwave, shortwave relay, telephone land lines, and transmitters.

VOA started during World War II and was part of the Office of War Information. In 1948 Congress enacted legislation "to promote better understanding of the United States in other countries" and made VOA a permanent part of foreign policy housed in the State Department. In 1953 the United States Information Agency was formed and the VOA became part of it.[50]

The programming emphasizes carefully prepared newscasts, but music, sports, and features are also broadcast.[51] American music is generally the lure that gets people to listen. Sometimes programs are shipped on tape to regular radio stations in various countries and aired that way. Most people, however, hear VOA over shortwave. It is impossible to know how many people listen, but the VOA has received over half a million letters a year from overseas, and the BBC overseas service estimates that its regular audience is 120 million.[52]

controversies

International broadcasting is not without its conflicts. Countries often try to jam other countries' signals.[53] When the VOA started Radio Marti to broadcast to Cuba in 1985, Florida radio stations were fearful that Fidel Castro would retaliate by using supertransmitters to interfere with their transmissions, but this did not happen.[54] However, when a television service to Cuba, TV Marti, was started in 1990, similar fears were expressed by TV stations in the Florida area.[55]

During periods when the international situation appears to be somewhat peaceful, funds are often cut from this type of transmission.[56] The need for and effects of international broadcasting can most assuredly be debated, but it seems to be something that the major powers feel is necessary.

FIGURE 2.6
Voice of America's
master control board.
Two technicians on
around-the-clock duty
feed programs to
VOA's U.S.
transmitters and switch
all channels at every
station break. *(Courtesy
of United States Information
Agency)*

Armed Forces Radio and Television Service

Another type of overseas broadcasting is undertaken by the American armed forces for service people who are in locations where English language broadcasting is not available.

This, too, was started during World War II by some enterprising servicemen in Alaska who set up a transmitter and wrote to some Hollywood stars asking for radio programs. The stars couldn't send the programs because of security regulations, so the servicemen contacted the War Department in Washington. The result was the establishment of the **Armed Forces Radio Service (AFRS).** Its mission was to give servicemen a touch of home and to combat Axis Sally and Tokyo Rose, dulcet-voiced girls broadcasting appeals intended to demoralize American soldiers. AFRS

Some of the broadcasting was done by shortwave and some was done by troop-operated stations. While most of the stations were in fixed locations where the troops were stationed, a number of them actually moved along with the advancing armies. Studios were set up in Los Angeles so that big name stars could perform on programs such as "Command Performance" and "Mail Call," which were sent to the AFRS facilities.

In 1945 there were about three hundred AFRS stations around the world, but this shrunk to sixty after the war. With the advent of the Korean War and the Vietnam conflict, the service once again expanded, this time including television. In 1954 AFRS became the **Armed Forces Radio and Television Service (AFRTS).** AFRTS

FIGURE 2.7
Arthur Shields (*left*)
and Gary Cooper
perform for AFRS's
"Mail Call" about
1943. *(Courtesy of True
Boardman)*

At present, AFRTS is under the Department of Defense and operates over seven hundred outlets on land and on ships at sea. It programs some material specially for the military, but most of the programming is regular commercial and public radio, and TV with the commercials deleted. Sometimes citizens in the countries that have AFRTS stations listen in to the American programming.[57]

International Telecommunication Union

Radio waves do not stop at national boundaries. If Mexico were free to allocate radio spectrum in any way it wanted, its stations might very well interfere with those of the United States. Even more to the point, just about any country in Europe can affect the broadcasting of any other European country. An agency is needed to establish international guidelines and resolve disputes.

ITU

That body is the **International Telecommunication Union (ITU)**, an arm of the United Nations. This organization periodically arranges a **WARC (World Administrative Radio Conference)** or a **RARC (Regional Administrative Radio Conference)**. Numerous countries, including the United States, send delegates to these conferences to determine how the electromagnetic spectrum will be used.[58]

ITU delegates have dealt with problems such as increasing the size of the AM band and shrinking the bandwidth of individual radio stations.[59] One of the thorniest problems they have had to deal with is the allocation of satellite orbital slots to various countries. The Third World nations want to make sure that slots are available to them when and if they develop the technology to launch their own satellites. The United States and other developed countries feel the precious slots should not lie fallow and should be used, at least temporarily, by the countries that can make use of them. A compromise has been worked out in part that would guarantee each country one orbital slot, but this whole situation is not totally stabilized.[60]

COMSAT and INTELSAT

COMSAT and INTELSAT are both organizations that provide satellite services of an international nature. Their organization, duties, and interrelationships are complex and confusing and subject to constant change.

COMSAT (Communications Satellite Corporation) was the first to be formed. In the early 1960s the United States Congress felt the need to have an organization that would represent the United States in matters dealing with satellites. As a result, it passed the Communications Satellite Act of 1962, which set up COMSAT. The provisions of the act called for COMSAT to be a private company regulated by the FCC. Half the stock was to be held by the public at large and half was to be held by companies such as AT&T and RCA, which might be engaged in satellite business. At first, the fact that the communication companies that would use COMSAT were its owners looked a bit like the inmates running the asylum. Over the years, these companies sold most of their stock to the general public and now own less than 1 percent of COMSAT.[61]

INTELSAT (International Telecommunications Satellite Organization) was formed in 1964, largely at the urging of the United States. It was set up as an organization of member nations to deal with international satellite transmission. At first COMSAT was the manager of INTELSAT on behalf of these member nations. In 1973 INTELSAT became an independent organization owned by a consortium of countries, each of which pays a yearly fee based on its use of satellite communications. COMSAT represents the United States in INTELSAT and owns the biggest share of it (about 25 percent). More than 100 different countries are part of INTELSAT, which is headquartered in Washington, D.C.

INTELSAT has launched many satellites over the years until today; it owns and operates more satellites than any other entity in the world. It has sixteen satellites, which carry two-thirds of international telephone traffic and almost all international television feeds. When a nation wants to use a satellite for some event, such as telecasting the Olympics or a World Soccer Championship, it rents satellite time from INTELSAT, which then makes all the arrangements for the satellite feed.[62]

When INTELSAT was first established, it had what amounted to a monopoly on providing satellite services for the world at large. However, after much debate at both the national and international levels, it was decided in the late 1980s that other companies should be allowed to compete with INTELSAT. As a result other companies now own international satellites and

FIGURE 2.8
The lobby of the
INTELSAT building
in Washington, D.C.,
displaying some of the
satellites INTELSAT
has launched.

still others are in the business of uplinking and downlinking to INTELSAT satellites.[63]

The whole field of international telecommunications is one that involves a great deal of gamesmanship as countries jockey for world power and try to maintain their national heritage.

Conclusion

International telecommunications consists of broadcasting within countries and without. Internally, most countries started out under authoritarian, paternalistic, or permissive systems. The Soviet, British, and American systems are good examples of each respectively. Some of the systems evolved rather quickly into pluralistic systems, such as that of Canada. Of late, the various forms of broadcasting are merging and systems throughout the world are becoming more similar. Most accept commercials to some degree, and most have undertaken at least experiments with cable TV, DBS, and VCRs.

Unfortunately different countries have different technical standards (PAL, SECAM, NTSC), and this means dubbing is necessary for program exchange. For a time HDTV seemed poised to solve this problem, but now different countries are developing their own HDTV systems. Program exchange does occur, however, with the United States being the leader in exporting programs. This may change as nations develop more of their own programming and become involved in more co-productions.

The British system of broadcasting consists of two main entities, the BBC and independent radio and TV. BBC has four radio networks and two TV networks, none of which have commercials. Independent broadcasting has evolved through an Independent Television Authority to the Independent Broadcasting Authority, which interacted with radio, ITVs, ITN, and Channel 4. Now independent broadcasting is overseen by the Radio Authority and the Independent Television Commission. The Soviet system, which consists of five radio networks and four TV networks, has traditionally been controlled and financed by the government.

External international telecommunications is usually somewhat propagandistic and is undertaken by many countries, including the United States with its VOA. Programming for service people is also beamed abroad, as evidenced by the Armed Forces Radio and Television Service. Because radio waves do not obey national boundaries, conferences sponsored by the International Telecommunication Union of the United Nations are often held to plan spectrum use. Satellites have become so essential to most countries that INTELSAT and its competitors have flourished. COMSAT handles satellite business for the United States.

The international sphere is a fascinating one that needs constant fine-tuning as world conditions change.

Thought Questions

1. What are the major advantages and disadvantages of the permissive, paternalistic, authoritarian, and pluralistic systems of broadcasting?
2. Should the networks in the United States show more foreign programming than they do? Why or why not?
3. Should the United States place more or less emphasis than it does on Voice of America broadcasts? Defend your answer.
4. If you could start broadcasting from scratch in some newly formed country, how would you organize it?

Electronic Media Forms

A wide variety of electronic media exists for the dissemination of entertainment and information. The oldest, of course, is radio, which came to the fore in the 1920s. Within less than the average lifetime, these media have proliferated into more than a dozen forms. While some wonder what the market can bear, others marvel at what the market does bear. The various media both complement and compete with each other, experiencing both the slings and security of the free enterprise system. As these media develop, change is inevitable, brought about both by external and internal forces. Although all the media forms are young, they are already rich in history and adaptation.

PART 2

Commercial Radio

Introduction

Radio, the oldest of the telecommunications media, formed most of the models of entertainment and information that are common to the media today. Its beginnings are veiled in dispute. The early inventors lived in various countries and, in some instances, devised virtually the same inventions. Ironically, this was partly due to the fact that no communication system was available for people to learn what others were inventing. This led to innumerable rivalries, claims, counterclaims, and patent suits.

The earliest inventions crucial to the field were not even intended to be used for radio broadcasting. When radio waves were first discovered, there was consternation over the fact that they were so public. Many experimenters were involved in devising methods to make the airwaves private so that messages could be sent confidentially. Only a few visionaries foresaw the use of radio broadcasting as we know it today, and none of them could have predicted the trials and tribulations, successes and failures that radio has undergone.

It is inconceivable that we should allow so great a possibility for public service as broadcasting to be drowned in advertising chatter.

Herbert Hoover
while serving as secretary of
commerce

FIGURE 3.1
James Clerk Maxwell.
*(Smithsonian Institution,
Photo No. 56859)*

Early Inventions

Maxwell

Many people believe that radio originated in 1873 when James Clerk **Maxwell,** a physics professor at Cambridge University, England, published his theory of electromagnetism. His treatise predicted the existence of **radio waves** and how they should behave based on his observations of how light waves behave.[1]

Hertz

Experiments to prove Maxwell's theory were undertaken by the German physics professor Heinrich **Hertz** during the 1880s. Hertz actually generated at one end of his laboratory and transmitted to the other end the radio energy that Maxwell had theorized. He thus proved that variations in electrical current could be projected into space as radio waves similar to light waves. In 1888 he published a paper that served as a basis for the theory of modern radio transmission.[2]

Marconi

Guglielmo **Marconi,** often referred to as the "Father of Radio," expanded upon radio principles. Marconi, the son of a wealthy Italian father and an Irish mother, was scientifically inclined from an early age. Fortunately, he had the leisure and wealth to pursue his interests. Shortly after he heard of Hertz's ideas, he began working fanatically in his workshop, finally reaching a point where he could actually ring a bell with radio waves.

Marconi then incorporated the Morse key into his system with the goal of transmitting Morse code by radio waves. Until this time, the transmission of Morse code had required the laying or stringing of wires from one reception point to another. To set his radio waves in motion, Marconi used Hertz's method, which was to generate a spark that leaped across a gap. To receive the signal, he placed metal filings in a glass tube. When the radio wave contacted the

FIGURE 3.2
Heinrich Hertz.
*(Smithsonian Institution,
Photo No. 66606)*

FIGURE 3.3
Guglielmo Marconi,
shown here with
wireless apparatus
about 1902. *(Smithsonian
Institution, Photo No. 52202)*

metal filings, they cohered and the glass tube then had to be tapped to loosen the filings to receive the next impulse. Marconi's first crude but effective system thus consisted of a Morse key, a spark, a coherer, and a tapper.

After he tested his invention outside his workshop by successfully transmitting throughout his estate and beyond, Marconi wrote to the Italian government in an attempt to interest them in his project. They replied in the negative. His determined Irish mother decided that he should take his invention to England. There, in 1897, he received a patent and the financial backing

FIGURE 3.4

Lee De Forest, shown
here with wireless
apparatus about 1920.
*(Smithsonian Institution,
Photo No. 52216)*

to set up the **Marconi Wireless Telegraph Company,** Ltd. Under the auspices of this company Marconi continued to improve on wireless and began to supply equipment to ships. In 1899 he formed a subsidiary company in the United States, the Marconi Wireless Company of America.[3]

Although Marconi maintained a dominant international position in wireless communication, many other people were experimenting and securing patents in Russia, Germany, France, and the United States. Until this time the primary use of wireless had been as a means of Morse code communication by ships at sea. Now some people were becoming intrigued with the idea of voice transmission.

A significant step in this direction was taken by John **Fleming** of Britain in 1904. He developed the vacuum tube, which led the way to voice transmission. It was later developed further by others, particularly Reginald Aubrey Fessenden and Lee De Forest.[4]

Fessenden, a Canadian-born professor who worked at the University of Pittsburgh, proposed that radio waves not be sent out in bursts—which accommodated the dots and dashes of Morse code—but rather as a continuous wave on which voice could be superimposed. He succeeded in obtaining financial backing from two Pittsburgh financiers and on Christmas Eve of 1906 broadcast to ships at sea his own violin solo, a few verses from the Book of Luke in the Bible, and a phonograph recording of Handel's "Largo."[5]

De Forest is known primarily for the 1907 invention of the audion tube, an improvement on Fleming's vacuum tube. It contained three electrodes instead of two and was capable of amplifying sound to a much greater degree than was previously possible. This tube was the most crucial key to voice transmission.

Like Marconi, De Forest was fascinated with electronics at an early age and later secured financial backing to form his own company. However,

Fleming

Fessenden

De Forest

De Forest experienced management and financial problems that frequently rendered him penniless and led him eventually to sell his patent rights to Marconi's company and to American Telephone and Telegraph (AT&T).

De Forest was farsighted in his views of radio wave utilization, and he strongly advocated voice transmission for entertainment purposes. In 1910 he broadcast the singing of Enrico Caruso from the New York Metropolitan Opera House. Several years later he started a radio station of sorts in the Columbia Gramophone Building, playing Columbia records in hopes of increasing their sales. He was also hopeful of increasing the sale of wireless sets so that more of his audion tubes would be sold.[6]

Therefore, by 1910 radio waves had been theorized by James Maxwell, proven to exist by Heinrich Hertz, put to use with Morse code by Guglielmo Marconi, and developed for voice transmission by John Fleming, Reginald Fessenden, and Lee De Forest.

Early Control

During these early stages radio grew virtually without government control. The first congressional act to mention radio was the Wireless Ship Act of 1910, which required all ships holding more than fifty passengers to carry radios for safety purposes.

Wireless Ship Act of 1910

However, the rules concerning this safety requirement were not very stringent, as was proven with the 1912 sinking of the *Titanic*. As the "unsinkable" *Titanic* sped through the night on its maiden voyage, radio operators on other ships warned it of icebergs in the area, but the *Titanic's* radio operator, concerned with transmitting the messages of the many famous passengers to friends in other parts of the world, passed the warnings on to the captain who disregarded them. When the *Titanic* did strike the fatal iceberg at about midnight on April 14, the wireless operator transmitted SOS signals, but none of the nearby ships, which could have helped to save some of the one thousand passengers who died, heard the distress calls because their wireless operators had signed off for the night. The fact that 700 people were saved can be attributed to the fact that distant ships heard the SOS calls and steamed many miles to the rescue.

sinking of the *Titanic*

The wireless operators who had been receiving the passengers' messages also heard the distress calls, so for the first time in history people knew of a distant tragedy as it was happening. The wireless operator decoding the messages in New York was David **Sarnoff,** who later became president of RCA. He stayed at his post for seventy-two hours, relaying information about the rescue efforts to anxious friends and relatives and to the newspapers. This brought wireless communication to the attention of the general public for the first time.[7]

Congress then passed the **Radio Act of 1912,** which emphasized safety and required everyone who transmitted on radio waves to obtain a license from the secretary of commerce. The secretary could not refuse a license, but could assign particular wavelengths to particular transmitters. Thus, ship transmissions were kept separate from amateur transmissions, which were, in turn,

Radio Act of 1912

FIGURE 3.5

David Sarnoff working at his radio station position atop the Wanamaker store in New York. It was here he heard the *Titanic's* distress call, causing him to stay at his post for about seventy-two hours to report the disaster. *(Courtesy of RCA)*

separate from government transmissions. All this was done without any thought of broadcasting as we know it today.

World War I

At the beginning of World War I the government took over all radio operation. Ship-to-shore stations were operated by the Navy, and many ham radio operators were sent overseas to operate radio equipment. Perhaps even more important, patent disputes were set aside for the good of the country. Marconi's company, still the leader in wireless, had aroused the concern of **American Telephone and Telegraph (AT & T)** by suggesting the possibility of entering the wireless phone business. AT&T, in an effort to maintain its supremacy in the telephone business, had acquired some wireless patents, primarily those of Lee De Forest. The stalemate that grew out of the refusal of Marconi's company, AT&T, and several smaller companies to allow one another to interchange patents had virtually stifled the technical growth of radio communications.

patent problems

With the onset of the war these disputes were set aside so that the government could develop the transmitters and receivers needed. World War I was also responsible for ushering into the radio field two other large companies, **General Electric (GE)** and **Westinghouse.** Both were concerned with electrical energy and were established manufacturers of light bulbs. Because both light bulbs and radio tubes require a vacuum, GE and Westinghouse assumed responsibilities for manufacturing tubes. GE had also been involved in the development of Ernst F. W. **Alexanderson's** construction of the **alternator** to improve long-distance wireless. During the war this alternator was perfected.

the alternator

After the war the patent problem returned and, as a result, GE began negotiating with the Marconi Company to sell the rights to its Alexanderson alternator. The Navy, which had controlled radio during the war, feared that this sale would enable Marconi, a primarily British company, to achieve a monopoly on radio communication. The Navy intervened and convinced GE president Owen **Young** to renege on the Marconi deal. This cancellation left GE sitting with an expensive patent from which it could not profit because GE

government intervention

did not control other patents necessary for its utilization. But the patent placed GE in an excellent negotiating position because of its value for long-distance transmission.

The Founding of RCA

What ensued from this situation was a series of discussions among American Marconi, AT&T, GE, and Westinghouse that culminated in the formation of **Radio Corporation of America (RCA)** in 1919. American Marconi, realizing with reluctance that it would not receive Navy contracts as long as it was associated with British Marconi, transferred its assets to RCA. The U.S. government was not directly involved in the negotiations carried on by Owen Young, but it was obvious to the Marconi people that their British ownership was an incurable liability. Marconi ousted

Individual stockholders received RCA shares for American Marconi shares, and GE purchased the shares of American Marconi held by British Marconi. AT&T, GE, and Westinghouse also bought blocks of RCA stock and agreed to make patents available to one another, thus averting the patent problem and allowing radio to grow. GE, Westinghouse, and AT&T involvement

Again, this was not undertaken with entertainment broadcasting in mind, but with emphasis on ship-to-shore transmission. The growth of radio occurred during a postwar era when the government had very little direct control over radio and when the primary companies involved—British Marconi, American Marconi, AT&T, GE, Westinghouse, and RCA—were leery of one another and insecure about their own futures.[8]

Owen Young put this complicated international agreement together, but it remained for David Sarnoff to convert it into a successful company operation. Sarnoff, a Russian immigrant who had supported his family from a very young age, was among the RCA employees who came from American Marconi. At the age of fifteen he had become an employee of Marconi and, at age twenty-one, received the distress messages from the *Titanic*. Sarnoff

Additional laws were passed that required radio equipment on ships, and these helped American Marconi's business to boom and Sarnoff's career to escalate. In 1915, at the age of twenty-four, he wrote a memo to Marconi management suggesting entertainment radio. It read in part as follows:

> I have in mind a plan of development which would make radio a "household utility" in the same sense as the piano or phonograph. The idea is to bring music into the home by wireless. . . . The problem of transmitting music has already been solved in principle and therefore all the receivers attuned to the transmitting wavelength should be capable of receiving such music. The receiver can be designed in the form of a simple "Radio Music Box" and arranged for several different wavelengths, which should be changeable with the throwing of a single switch or pressing of a single button.
>
> The "Radio Music Box" can be supplied with amplifying tubes and a loudspeaker telephone, all of which can be neatly mounted in one box. The box can be placed on a table in the parlor or living room, the switch set accordingly and the music received. . . .

The same principle can be extended to numerous other fields as, for example, receiving lectures at home which can be made perfectly audible; also, events of national importance can be simultaneously announced and received. Baseball scores can be transmitted in the air by the use of one set installed at the Polo Grounds. The same would be true of other cities. This proposition would be especially interesting to farmers and others in outlying districts removed from cities. By purchase of a "Radio Music Box," they could enjoy concerts, lectures, music recitals, etc., which may be going on in the nearest city within their radius. . . .

It is not possible to estimate the total amount of business obtainable with this plan until it has been developed and actually tried out; but there are about 15 million families in the United States alone, and if only one million or 7 percent of the total families thought well of the idea, it would, at the figure mentioned, ($75 per outfit) mean a gross business of about $75 million, which would yield considerable revenue.[9]

This idea was filed away as a harebrained notion, and Sarnoff had to wait for a more propitious time.

Early Radio Stations

Conrad and KDKA

Meanwhile, with restrictions lifted, many of the amateur radio enthusiasts began to experiment again. One of these was Frank **Conrad,** a physicist and an employee of Westinghouse in Pittsburgh. He resumed his amateur activities in his garage, programming music and talk during his spare time. A local department store began selling wireless reception sets and placed an ad for these in a local newspaper, mentioning that these sets could receive Conrad's concerts. One of Conrad's superiors at Westinghouse saw the ad and envisioned a market. Up until this time both radio transmission and reception had been for the technical-minded who could assemble their own sets. But it was obvious that sets could be preassembled for everyone who wished to listen to what was being transmitted.

Conrad was asked to build a stronger transmitter at the Westinghouse plant, one capable of broadcasting on a regular schedule so that people purchasing receivers would be assured listening fare. Thus, in 1920, Westinghouse became the first company to apply to the Department of Commerce for a special type of license to begin a broadcasting service. The station was given the call letters **KDKA** and was authorized to use a frequency away from amateur interference. KDKA launched its programming schedule with the Harding-Cox election returns, interspersed with music, and then continued with regular broadcasting hours. Public reaction could be measured by the long lines at department stores where radio receivers were sold.[10]

KDKA's success spurred others to enter broadcasting. Foremost among these was David Sarnoff, who could now dust off his old memo and receive more acceptance for his idea. He convinced RCA management to invest $2,000

Dempsey-Carpentier fight

to cover the Jack Dempsey-Georges Carpentier fight on July 2, 1921, and a temporary transmitter was set up in New Jersey for the fight. Fortunately, Dempsey knocked Carpentier out in the fourth round, for shortly after that

FIGURE 3.6
Frank Conrad, who
worked with
experimental
equipment and
supervised the
construction of KDKA.
*(Courtesy of KDKA,
Pittsburgh)*

FIGURE 3.7
A tent atop a
Westinghouse building
in East Pittsburgh
served as KDKA's first
studio. It caused some
of early radio's unusual
moments—such as the
whistle of a passing
freight train heard
nightly at 8:30, and a
tenor's aria abruptly
concluded when an
insect flew into his
mouth. *(Courtesy of
KDKA, Pittsburgh)*

the overheated transmitter became a molten mass. This fight, however, helped
to popularize radio, and both radio stations and sets multiplied rapidly.

By 1923 radio licenses had been issued to more than six hundred stations, and receiving sets were in nearly one million homes.[11] The stations had
low power (usually ten to fifteen watts) and were owned and operated primarily by those who wanted to sell sets (Westinghouse, GE, RCA), and retail

1923 status

Commercial Radio   57

department stores, as well as radio repair shops, newspapers wanting to publicize themselves, and college physics departments wishing to experiment. One licensing aspect that led to later problems was that all stations were on the same frequency—**360 meters.** Stations in the same reception area worked out voluntary arrangements whereby they could share the frequency by broadcasting at different times of the day.

Early Programming

Programming was no problem in the early days. People were mainly interested in the novelty of picking up any signal on their battery-operated crystal headphone receivers. Programs consisted primarily of phonograph record music, call letter announcements, and performances by endless free talent who wandered in the door eager to display their virtuosity on this new medium. For a time all stations in some cities would go off the air one night a week so that listeners could pick up signals from distant places. This was usually referred to as "black night."

early anecdotes

It is only natural then that early radio would produce some humorous anecdotes. For example, one early station, WJZ, operated from a shack on the roof of a building in Newark, New Jersey. It was accessible only by an iron ladder and hatchway and programmed mainly time signals, weather, and selections played on the Edison phonograph. On one occasion a woman was invited up to read a story, but after climbing the ladder and being pushed through the hatchway, she fainted.[12]

A woman who was a strong, speech-making advocate of birth control asked to speak on radio. The people at the station were nervous about what she might say, but when she assured them that she only wanted to recite some nursery rhymes, they allowed her into the studio. She then broadcast, "There was an old woman who lived in a shoe/She had so many children because she didn't know what to do."[13] She was not invited back.[13]

A young man in New Jersey wanted to let his mother know how he sounded over the air, so he dropped in at WOR, which had just opened a studio near the music department of a store. The singer the studio was expecting had not arrived yet, so this young man was put on the air before he even had time to notify his mother. He sang to piano accompaniment for over an hour as a messenger rushed sheet music from the music counter to the studio.[14]

A Chicago man wanted to discuss Americanism over a Chicago station and even submitted a script ahead of time. When he appeared at the station, he was with a group of bodyguards who covered the station premises to make sure that no buttons were pushed to take him off the air. It turned out that he was a potentate of the Ku Klux Klan, and, digressing from the script, he extolled the virtues of white supremacy.[15]

music

The primary programming of the era was dubbed **potted palm music**—the kind played at teatime by hotel orchestras. Sometimes a vocalist was featured and sometimes a pianist or small instrumental group played. Sopranos outnumbered all other "potted palm" performers.

FIGURE 3.8
Los Angeles's first station, KFI, began broadcasting in 1922. Its studio shows that although only the audio was received by the audience, special attention was given to the decor, including the "potted palms."
(Courtesy of KFI, Los Angeles)

Drama was also attempted even though engineers at first insisted that men and women needed to use separate microphones placed some distance from each other. Performers found it difficult to play love scenes this way. Finally it was "discovered" that men and women could share a mike.

From time to time radio excelled in the public affairs area. Political conventions and presidential speeches were broadcast, as well as the funeral service for Woodrow Wilson. When the six-year-old son of Ernst F. W. Alexanderson, the builder of the alternator, was kidnapped, a radio report of the child's description was responsible for his recovery.

drama

public affairs

The Rise of Advertising

As the novelty of radio wore off, people were less eager to perform and some means of financing programming had to be found. Many different ideas were proposed, including donations from citizens, tax levies on radio sets, and manufacturer and distributor payment for operating stations. Commercials came about largely by accident.

AT&T was involved mainly in the telephone business and was unwilling to see radio grow because the demand for wired services might be diminished. Therefore, one of its broadcasting entries was closely akin to phone philosophy. It established station WEAF in New York as what it termed a **toll station.** AT&T stated that it would provide no programming, but anyone who wished to broadcast a message could pay a "toll" to AT&T and then air the message publicly in much the same way as private messages were communicated by dropping money in pay telephones. In fact, the original studio was about the size of a phone booth. The idea did not take hold. People willing to pay to broadcast messages to the world did not materialize.

WEAF "toll" experiment

AT&T realized that before people would pay to be heard, they wanted to be sure that someone out there was listening. As a result, the nonprogramming idea was abandoned, and WEAF began broadcasting entertainment material, drawing mainly on amateur talent found among the employees. Still there were no long lines of people willing to pay to have messages broadcast.

Queensboro ad

Finally, on August 22, 1922, WEAF aired its first income-producing program—a ten-minute message from the Queensboro Corporation, a Long Island real estate company, which paid $50 for the time. The commercial was just a simple courtesy announcement because AT&T ruled out direct advertising as poor taste and an invasion of privacy. (There was also strong sentiment against advertising toothpaste because it was considered an intimate product.)[16] Many people of the era said that advertising on radio would never sell products. In fact, every dollar of income that WEAF obtained was a painful struggle.

WEAF frequency change

What eventually made the station succeed was the fact that AT&T was able to convince the Department of Commerce that WEAF should have a different frequency. The argument was that other broadcasters were using their stations for their own purposes, while WEAF was for everyone and therefore should have special standing and not be made to broadcast on 360 meters like everyone else. As a result, WEAF and a few other stations were assigned to the 400-meter wavelength. This meant less interference and more broadcast time. The phone booth was abandoned, a new studio was erected, and showmanship took hold.[17]

The Formation of Networks

Because AT&T was still predominantly in the phone business, it began using phone lines for remote broadcasts. It aired descriptions of football games, which came over long-distance lines, from Chicago and Harvard. It also established "toll" stations in other cities and interconnected them by phone lines—in effect, establishing a network.

AT&T's network

During this time AT&T did not allow other radio stations to use phone lines and also claimed sole rights to sell radio "toll" time. At first, other stations were not bothered because they were not considering selling ads. In fact, there was an antiadvertising sentiment in the early 1920s. Herbert Hoover, then secretary of commerce, stated that it was inconceivable that a service with so much potential for news, entertainment, and education should be drowned in advertising chatter.

However, as the AT&T toll network emerged and began to prosper, other stations became discontent with a second-class status. The fires of this flame were further fanned by a Federal Trade Commission inquiry that accused AT&T, RCA, GE, and Westinghouse of creating a monopoly in the radio business.

formation of NBC

A series of "behind-closed-doors" hearings was held by the major radio companies. The result of these complicated negotiations was the formation in 1926 of the **National Broadcasting Company (NBC)**—owned by RCA, GE, and Westinghouse—that was to handle broadcasting activities for the radio group. AT&T agreed to withdraw from the programming area and from the radio group in exchange for a long-term contract assuring that NBC would lease AT&T wires. This agreement was to bring the phone company millions of dollars per year. NBC also purchased WEAF from AT&T for $1 million, thus embracing the concepts of both "toll" broadcasting and networking. This was frequently referred to as chain broadcasting in those days because a chain of stations was interconnected.

In November of 1926 the NBC Red Network, which consisted of WEAF and a twenty-two-station national hookup, was launched in a spectacular debut that aired the New York Symphony Orchestra from New York, Mary Garden singing "Annie Laurie" from Chicago, Will Rogers mimicking President Coolidge from Kansas City, and dance bands from various cities throughout the nation. A year later NBC's Blue Network was officially launched. This consisted of different stations that had been part of a loosely knit RCA network.

In 1932 GE and Westinghouse withdrew from RCA, largely because of a U.S. attorney general's order that the group should be dispersed and partly because David Sarnoff, president of RCA, felt his company should be an entity. Again a series of closed-door meetings resulted in a divorce settlement. RCA became the sole owner of NBC, and GE and Westinghouse received RCA bonds and some of the RCA real estate. In retrospect it appears that RCA walked off with the lion's share of value. But all this happened during the depression, and GE and Westinghouse were not overly eager to hold onto what they thought might be an expensive broadcasting liability. NBC, in spite of the depression, moved to new mid-Manhattan headquarters, dubbed Radio City.[18]

GE and Westinghouse withdraw from NBC

As can be seen, both RCA and NBC have an interesting parentage. NBC was originally owned by RCA, GE, and Westinghouse, who ousted AT&T in the process of NBC's formation. RCA was formed by GE, Westinghouse, and AT&T, who ousted Marconi during the process of RCA's formation. The exact details of all these corporate maneuvers will probably never be known.

What eventually became the **Columbia Broadcasting System (CBS)** began in 1927. Arthur Judson, disgruntled because Sarnoff had not accepted his offer to supply talent for the NBC networks, established, with several associates, the United Independent Broadcasters. The original intent of the company was to supply talent to stations not associated with NBC Red or NBC Blue.

However, in the search for capital, the company joined with the Columbia Phonograph Company to form the Columbia Phonograph Broadcasting System. Judson's group was to supply talent and programs, and the Columbia group was to sell the programming to sponsors. Unfortunately, sponsors were not eager to oblige, and the project failed, with Columbia Phonograph Company pulling out.

Several investors tried to keep the project afloat and changed the name to Columbia Broadcasting System. Finally, in 1928, the family of William S. **Paley** supplied the needed capital, and Paley became president. He built a

Paley and CBS

FIGURE 3.9
William S. Paley. *(UPI/ Bettmann Newsphotos)*

network that was similar in organization to NBC in that it consisted of a chain of stations. The network became successful, and during the 1930s Paley managed to lure much of the top radio talent from NBC to CBS.[19]

The **American Broadcasting Company (ABC)** came to the fore because of actions taken by the Federal Communications Commission (FCC) in the early 1940s. By 1940 the networks had established a power base that the FCC felt could be detrimental, so it set about to limit the power by issuing what it called **chain regulations.** One aspect of these was the **duopoly rule,** which prohibited one company from owning and operating more than one national radio network. This created a situation whereby NBC had to sell one of its networks. NBC at first formed a separate company to operate its Blue Network, then in 1943 sold that company to a group headed by Edward J. Noble, the Lifesavers' millionaire. In 1945 this group changed the name of the Blue Network to the American Broadcasting Company.[20]

A fourth radio network, **Mutual Broadcasting Company,** was formed in 1934 when four stations decided to work jointly to obtain advertising. Unlike the other networks, Mutual owned no stations. It kept enough money from the ads it sold to recover its costs and to buy programs. Then it paid the rest to the stations in its network so that they would carry the programs and the network ads. It also allowed stations to sell their own ads.

Mutual, too, was involved with the chain regulations. One of the stipulations of the NBC and CBS affiliate contracts was that the local stations could not carry programs from a different network. In 1938 Mutual gained exclusive rights to broadcast the world series, but the NBC and CBS contracts

chain regulations and ABC

Mutual

would not allow their affiliated stations to carry these games, even in cities where there were no Mutual stations. The people wanted the games, the stations wanted to carry them, and advertisers wanted to pay for the coverage. Nevertheless, many Americans did not hear the 1938 world series. The FCC determined that this type of program thwarting was not in the public interest. As part of its chain regulations it stated that no station could have an arrangement with a network that hindered that station from broadcasting programs of another network.[21]

Chaos and Government Action

The problem of broadcast frequency overcrowding continued to grow during the 1920s. Secretary of Commerce Herbert **Hoover** was besieged with requests that the broadcast frequencies be expanded and that stations be allowed to leave the 360-meter quagmire, the frequency band on which most of them were broadcasting. He made various attempts to improve the situation by altering frequencies, powers, and broadcast times, and he called four national radio conferences to discuss with broadcasters problems and solutions to the radio situation. But Hoover was unable to deal with the problem in any systematic manner because he could not convince Congress to give him the power to do so.

national radio conferences

By 1925 the situation had so deteriorated that the only remedy would have been to reassign frequencies being used for other purposes. However, under the existing law, the secretary of commerce was powerless to act in this regard. Hoover threw up his hands and told radio station operators to regulate themselves as best they could.[22]

During 1926–27 there were some two hundred new stations, most of them using any frequency or power they wished and changing at whim. The airwaves were complete chaos. To help remedy this situation, Congress passed the **Radio Act of 1927.** The act proclaimed that radio waves belonged to the people and could be used by individuals only if they had a license and were broadcasting in the public "interest, convenience, and necessity."

Radio Act of 1927

All previous licenses were revoked and applicants were allowed sixty days to apply for new licenses from the newly created **Federal Radio Commission (FRC).** The commission gave temporary licenses while it worked out the jigsaw puzzle of which frequencies should be used for what purposes. In the end it granted 620 licenses in what is now the AM band. The FRC also designated the power at which each station could broadcast.[23]

Several years after the Radio Act of 1927, Congress passed the **Communications Act of 1934,** which created the **Federal Communications Commission (FCC).** This act was passed primarily because both the Congress and President felt all regulation of communications should rest with one body. The FCC was, therefore, given power over not just radio but also over telephone and other forms of wired and wireless communications. This 1934 act still governs telecommunications today.

Communications Act of 1934

The Golden Era of Radio

equipment improvements

With the chaotic frequency situation under control, radio was now ready to enter the era of truly significant programming development—a heyday that lasted some twenty years. Improvements in radio equipment helped. **Earphones** had already been replaced by **loudspeakers** so that the whole family could listen simultaneously. The early **carbon microphones** were replaced by **ribbon microphones,** which had greater fidelity, and **single-dial tuning** replaced the **three-dial system** required on earlier receivers. Battery sets were introduced for portability and use in automobiles. (However, the first portables were cumbersome because of the size of early dry batteries.)

Radio became the primary entertainment medium during the depression. In 1930, 12 million homes were equipped with radio receivers, but by 1940 this number had jumped to 30 million. During the same period, advertising revenue rose from $40 million to $155 million. In 1930 NBC Red, NBC Blue, and CBS offered approximately sixty combined hours of sponsored programs a week. By 1940 the four networks (Mutual had been added) carried 156 hours.[24]

Amos 'n' Andy

The first program to generate nationwide enthusiasm was "Amos 'n' Andy." It was created by Freeman Fisher **Gosden** and Charles J. **Correll,** who met while working for a company that staged local vaudeville-type shows throughout the country. Gosden and Correll, who were white, worked up a blackface act for the company and later tried this on WGN radio in Chicago as "Sam 'n' Henry." When WGN did not renew their contract, they took the show to WMAQ in Chicago and changed the name to "Amos 'n' Andy" because WGN owned the title "Sam 'n' Henry."

Correll and Gosden wrote all the material themselves and played most of the characters by changing the pitch, volume, and tone of their voices. Gosden always played Amos, a simple, hardworking fellow, and Correll played Andy, a clever, conniving, and somewhat lazy individual who usually took credit for Amos's ideas. According to the scripts, Amos and Andy had come from Atlanta to Chicago to seek their fortune, but all they had amassed was a broken-down automobile, also known as the Fresh-Air Taxicab Company of America. Much of the show's humor revolved around a fraternity-type organization called the Mystic Knights of the Sea headed by a character called Kingfish, who was played by Gosden.

WMAQ allowed Correll and Gosden to syndicate the show on other stations. Its success caught the attention of the NBC Blue Network, which hired the two in 1929 at $100,000 a year. Their program, which aired from 7:00 to 7:15 P.M. eastern time, became such a nationwide hit that it affected dinner hours, plant closing times, and even, on one notable occasion, the speaking schedule of the president of the United States.[25]

comedy

Many other comedians followed in the wake of the success of Correll and Gosden—Jack Benny, Ed Wynn, Lum and Abner, George Burns and Gracie Allen, Edgar Bergen and Charlie McCarthy, Fibber McGee and Molly, and the Aldrich Family.

Electronic Media Forms

(a)

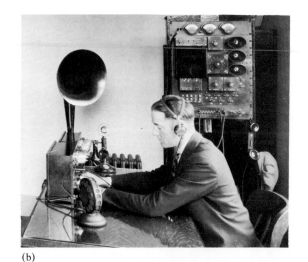

(b)

(c)

(d)

(e)

FIGURE 3.10

(*a*) A carbon microphone that was the best quality available during the formative years of radio. (*b*) An early station setup that included a carbon microphone, a multitubed audio board, and Westinghouse receivers. (*c*) A battery-operated radio receiver from about 1923. (*d*) A home radio receiver with speaker from about 1924. (*e*) Early backpack equipment for remote radio broadcasting. *(a, b, e: courtesy of KFI, Los Angeles; c, d: courtesy of RCA)*

FIGURE 3.12

Jack Benny with his
wife and costar, Mary
Livingston, in 1933.
The "Jack Benny
Show" sponsored for
many years by Jell-O
and Lucky Strike,
featured such sure-fire
laugh provokers as an
ancient Maxwell
automobile that
coughed and sputtered,
Benny's perennial age
of thirty-nine years, a
constant feud with
Fred Allen, and
Benny's horrible violin
playing. *(Courtesy of NBC)*

FIGURE 3.13

Ed Wynn, also known
as "the perfect fool."
He received this label
originally because he
performed in the first
Broadway stage show
broadcast on radio,
"The Perfect Fool."
(Courtesy of NBC)

FIGURE 3.14
Lum and Abner, played by Chester Lauck (*left*) and Norris Goff. This comedy took place in the Jot 'Em Down grocery store in the supposedly fictional town of Pine Ridge, Arkansas. In 1936 the town of Waters, Arkansas, changed its name to Pine Ridge in honor of Lum and Abner. *(Courtesy of KFI, Los Angeles)*

FIGURE 3.15
George Burns and Gracie Allen. Many jokes of this program were plays on words based on Gracie's supposed empty-headedness. At one point Gracie started searching for her "lost brother" by suddenly appearing on other shows to inquire about him. *(Courtesy of NBC)*

FIGURE 3.16
Charlie McCarthy and ventriloquist Edgar Bergen (*right*) with W. C. Fields (*left*) and Dorothy Lamour. Charlie had a running feud with W. C. and a love affair with just about all the thirties and forties beauties and even the fifties movie idol Marilyn Monroe. *(Courtesy of NBC)*

FIGURE 3.17
Marian and Jim
Jordon as Fibber
McGee and Molly. The
commercials were
integrated directly into
the program when the
announcer dropped by
the McGee home and
extolled the virtues of
Johnson's Wax.
(Courtesy of NBC)

FIGURE 3.18
Katherine Naiht,
House Jamison, Ann
Lincoln, and Ezra
Stone of "The Aldrich
Family." This
program's familar
opening was "Henry,
Henry Aldrich," to
which Ezra Stone
replied, "Coming
Mother." The program
was based on the
Broadway play *What a
Life*. *(Courtesy of NBC)*

music

Music, especially classical music, was also frequently heard. There were broadcasts of the New York Philharmonic concerts and performances from the Metropolitan Opera House. As a pet project of David Sarnoff, NBC established its own orchestra led by Arturo Toscanini. For lighter music, "Your Hit Parade," which featured the top-selling songs of the week, was introduced in 1935, and people who later became well-known singers, such as Kate Smith and Bing Crosby, took to the air. As the big bands developed, they, too, went out over the airwaves.

audience participation

One program innovation was to involve the audience. Among many amateur hours, perhaps the most famous was the one hosted by Major Bowes. Quiz shows, such as "Professor Quiz," rewarded people for responding with little-known facts. Stunt shows, such as "Truth or Consequences," which prompted people to undertake silly assignments if they answered questions incorrectly, were attracting large and faithful audiences.

Electronic Media Forms

Many programs were developed for children, including "Let's Pretend," a multisegment program that emphasized creative fantasy; "The Lone Ranger," a western; "Uncle Don's Quiz Kids," a panel of precocious children who answered questions; and "Little Orphan Annie," a drama about a child's trials and tribulations.

children's shows

During the day there were continuing dramas called "soap operas" because soap manufacturers were frequent sponsors. The segments always ended with an unresolved situation in order to entice the listener to tune in tomorrow.

soap operas

Tommy Dorsey's Band. When cigarette companies backed many of the swing bands, Raleigh-Kool sponsored the Dorsey musicians. Since the radio programs were performed before live audiences, the huge cigarette packs did make an impact. *(Courtesy of KFI, Los Angeles)*

Rudy Vallee (*right*) with Charles Butterworth (*left*), Helen Vinson, and Fred Perry. Some of the many stars who got their start on this variety format show were Bob Burns, Bob Hope, Eddie Cantor, Alice Faye, Milton Berle, and Ezra Stone. *(Courtesy of KFI, Los Angeles)*

drama

Most did. The scripts for a major portion of the soap operas were developed by a husband-wife team, Frank and Ann Hummert. They defined the basic idea for each series, wrote synopses of programs, and then farmed the actual script-writing to a bevy of writers around the country, some of whom never even met the Hummerts.

In the field of drama, the networks first tried to rebroadcast the sound of Broadway plays, but discovered that this was akin to sitting in a theater blindfolded. As a result, the networks hired writers such as Norman Corwin, True Boardman, Maxwell Anderson, and Stephen Vincent Benet to script original dramas for radio. These dramas usually employed many sound effects and were sponsored by one company that often incorporated its name into the program, such as "Lux Radio Theater" or "Collier's Hour." In 1938 Orson Welles produced "War of the Worlds," a fantasy about a Martian invasion in

FIGURE 3.23
Major Edward Bowes
of "The Original
Amateur Hour." Some
of the winners from
this amateur
competition formed a
touring Major Bowes
company that provided
talent employment
during the depression.
*(Courtesy of KFI, Los
Angeles)*

New Jersey. Upon hearing the broadcast, an estimated 1.2 million people succumbed to hysteria. They panicked in the streets, fled to the country, and seized arms to prepare to fight—despite the fact that the "Mercury Theater" program included interruptions to inform the listener that the presentation was only a drama. Some people felt that radio had become overly realistic.[26]

The depression brought about the growth of commercials. During the 1920s, advertisements were brief and tasteful, and price was not mentioned. However, as radio stations and all facets of the American economy began digging for money at any price, the commercial standards dissolved. Some advertisers felt commercials should irritate, and broadcasters, anxious for the buck, acquiesced. The commercials became long, loud, dramatic, hard-driving, and cutthroat.

Most radio programs were produced not by the networks but by **advertising agencies.** They found they could combine advertising effectiveness with human misery. Thus, a large number of personal help programs developed. Listeners would send letters to radio human relations "experts" detailing traumas, crimes, and transgressions asking for help. Often this help had commercial tie-ins. Box tops accompanying the letter qualified it for an answer; or the suggested solution might involve the sponsor's drug product; or the contentment derived from puffing on the sponsor's brand of cigarette might be recommended. By 1932 more airtime was spent on commercials than on news, education, lectures, and religion combined. The commercials brought in profits for NBC, CBS, and some individual radio stations. They also brought profits to the advertising agencies that were intimately involved in most details of programming, including selecting program ideas, overseeing scripts, selling and producing advertisements for the shows, and placing the programs on the network schedule.[27]

There were also many events that could be termed **stunt broadcasts,** such as those from heights, depths, widely separated points, gliders, and underwater locations. A four-way conversation involved participants in Chicago, New

commercials

stunt broadcasts

► THE INCOMPARABLE AMOS 'N' ANDY, returning to air via NBC Friday
Oct. 8.

RAMP, TRAMP, TRAMP.
It's NBC's parade of stars marching along to open the fall and winter season of happy listening.

For dialers, the biggest news of all is the return of Amos 'n' Andy. The two old favorites introduce a brand new show on Friday, October 8, complete with guest stars, music and the kind of laughter which made Freeman Gosden and Charles Correl famous.

TRAMP, TRAMP, TRAMP.

"The Great Gildersleeve" started the parade by huffing and puffing his way back to his fans late in August. This is Hal Peary's third season on the air with his own program, and from the way the polls were going when he went off in June, it looks like his biggest.

TRAMP, TRAMP, TRAMP.

Fanny Brice, with more antics of

▼ THE INCORRIGIBLE BABY SNOOKS and Frank Morgan on Maxwell House
show Thursday, 8:30 p. m.

▼ THE INVENTIVE ARKANSAS
TRAVELER, Bob Burns, back on air
Thursday night.

▼ THE INGENIOUS JACK BENNY,
airing with all the gang at 4
p. m., Sunday.

◄ THE INFALLIBLE H. V. KAL-
TENBORN, commentator, heard
four afternoons weekly.

Page Four
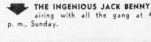

FIGURE 3.24

A wartime plug for NBC's programs. *(Courtesy
of KFI, Los Angeles)*

A Star-Bedazzled Parade Of Fast-Stepping Radio Entertainers, on March To Storm Your Listening

her inimitable Baby Snooks, and Frank Morgan, with a new batch of tall stories, followed "Gildersleeve" to NBC microphones on the first Thursday in September.

TRAMP, TRAMP, TRAMP.

Edgar Bergen and Charlie McCarthy marched back from Newfoundland, where they entertained the troops stationed there. This season they are presenting Victor Moore and William Gaxton, in addition to Ray Noble's orchestra and the songs of Dale Evans.

TRAMP, TRAMP, TRAMP.

That bad little boy, Red Skelton, was next in line, and with him were the popular members of his cast—Harriet Hilliard and Ozzie Nelson and his band.

TRAMP, TRAMP, TRAMP.

An account of Bob Hope's travels while away from his radio show for the summer sounds like a review of the war headlines—England, Bizerte, Tunis, Algiers, Sicily. Back on the air with him for the new season comes another grand trouper, Frances Langford, who also went into the battle areas with Bob. And, of course, Jerry Colonna and Vera Vague will be on hand.

TRAMP, TRAMP, TRAMP.

And so they come. Those two top

▲ THE IMPERTINENT CHARLIE McCARTHY and Bergen for Chase & Sanborn, Sunday, 5 p. m.

comedians, Jim and Marian Jordan, who have more delightful sessions with "Fibber McGee and Molly", ready for their listeners.

Eddie Cantor with another season of Wednesday night laughfests.

Jack Benny, another of radio's globe-trotters, only recently returned from the European and North African battlefronts.

And, of course, there are all the favorites who have been on NBC this summer and who will continue to make radio listening America's Number One pastime—"One Man's Family;" Bing Crosby; the Standard Symphony Hour; H. V. Kaltenborn and the other commentators who bring the world into our homes; Kay Kyser's "College of Musical Knowledge;" Ginny Simms; the Joan Davis-Jack Haley show; and the Sunday morning Westinghouse program.

Happy listening? Yes, indeed!

▲ THE INGRATIATING FATHER AND MOTHER BARBOUR on "One Man's Family," Sunday, 5:30 p. m.

▼ THE INIMITABLE FIBBER Mc-GEE AND MOLLY, back with their friends Tuesday, 6 p. m.

▼ THE INEXHAUSTIBLE LAUGH CREATOR, Bob Hope, back in his regular Tuesday spot, 7 p. m.

▼ THE IRREPRESSIBLE LITTLE KID, as played by Red Skelton, Tuesday, 7:30 p. m.

Page Five

FIGURE 3.25

The NBC radio mobile unit making contact with an airplane. This 1929 experimentation led to future possibilities for news coverage. *(Courtesy of NBC)*

FIGURE 3.26

President Franklin Delano Roosevelt delivering a "fireside chat." *(Courtesy of NBC)*

York, Washington, and a balloon. One music program featured a singer in New York accompanied by an orchestra in Buenos Aires.

public events

These stunt broadcasts paved the way for the broadcasting of legitimate public events from distant points. In 1931 nineteen separate locations around the world participated in a program dedicated to Marconi. People were able to hear the farewell address of King Edward VIII when he abdicated the British throne and the trial of the man who kidnapped Charles Lindbergh's baby.[28]

politics

Radio also figured in politics of the day. President Franklin Delano **Roosevelt** effectively used radio for his **fireside chats** to reassure the nation during the depression. Louisiana's firebrand Governor Huey Long was often heard on the airwaves, and Father Charles E. Coughlin, a Detroit priest, tried to build a political movement through radio.[29]

The Press-Radio War

News was destined to become one of radio's strongest services, but not without a struggle. At first announcers merely read newspaper headlines over the air, but gradually networks began purchasing news from the wire services. In 1932 Associated Press sold presidential election bulletins to the networks, and programs were interrupted with news flashes. Newspapers objected to all of this on the grounds that news on radio would diminish the sale of papers. From 1933 to 1935, a **press-radio war** ensued.

newspaper objections

A meeting of newspaper publishers, network executives, and wire service representatives, held at the Biltmore Hotel in New York in 1933, established the **Biltmore Agreement.** It stipulated that networks could air two five-minute newscasts a day consisting of material received from the established wire services. These newscasts had to be aired in the morning after 9:30 A.M. and in the evening after 9:00 P.M. so that they would not compete with the primary hours of newspaper sales. No "hot-off-the-wire" news was to be broadcast, and newscasts were not to have advertising support because this might detract from newspaper advertising. Newspapers were able to see that these provisions were placed in the Biltmore Agreement because they were the most numerous, most powerful, and wealthiest of the participants at the meeting.

Biltmore Agreement

But the ink on this agreement was barely dry when its intent began to be subverted. For one thing, a new company, Trans Radio Press, began selling news items to the radio world, and the networks produced sponsored newscasts with this information. The newspapers countered by forcing Trans Radio Press out of business and, in some instances, newspapers boycotted offending radio networks by refusing to carry radio program schedules in their papers.

Trans Radio Press

Commentators were also used to counteract the news restrictions. The publishers agreed that radio stations and networks could have commentators, but often these commentators became thinly disguised news reporters.

commentators

Also, NBC and CBS began their own news-gathering activities. In the case of NBC, one man, Abe Schechter, gathered news simply by making telephone calls. Sometimes he scooped newspaper reporters because just about anyone would answer a call from NBC. In addition, Schechter could reward news sources with highly-prized tickets to Rudy Vallee's show. Most of the material Schechter collected was broadcast by NBC's prime newscaster, Lowell Thomas, but an item or two usually wound up on Walter Winchell's Sunday night gossip program. CBS set up a larger news force that included **stringers**— reporters paid only for material actually used. That network's top news commentator was H. V. Kaltenborn.

network news gathering

As world tensions grew, the public became increasingly aware of news. Advertisers became interested in sponsoring news radio programs because of the growing potential listener market. At one point, two services agreed to make their news available to advertisers who would then broadcast it over radio, but they would not make it available to radio stations directly. This arrangement led to a total breakdown of broadcast news blackouts, and radio began to develop as an important news disseminator. Americans heard actual sounds of the Spanish Civil War and of Germany's march into Austria, and they heard the voices of Hitler, Chamberlain, and Mussolini.[30]

FIGURE 3.27

Lowell Thomas began broadcasting in 1929 with one of the earliest programs, called "Headline Hunters," and remained on the radio regularly until after his eightieth birthday. For many years he preceded "Amos 'n' Andy" with the news, prompting him to say of himself, "Here is the bird that everyone heard while waiting to hear 'Amos 'n' Andy.' " *(Courtesy of Lowell Thomas)*

FIGURE 3.28

Walter Winchell with the signal key he used to accent his rapid-fire speaking style. He always worked with his hat on in the studio and always began his programs, "Good evening Mr. and Mrs. North and South American and all the ships at sea. Let's go to press." *(Courtesy of NBC)*

World War II

The government did not take over broadcasting during World War II as it had during World War I. However, it did solicit the cooperation of radio for morale and public service announcements, bond purchase appeals, conservation campaigns, and civil defense instructions. Among the most famous of these solicitations were singer Kate Smith's marathon broadcasts for war bonds. Her appeals sold over $100 million worth of bonds. Many of the plays and soap operas produced during the period dealt with the war effort, and some even tried to deal with the segregation problem, which was coming to a head because of segregation in the armed forces. Several soap operas presented Negroes (the preferred term at that time) in esteemed professional roles.

war efforts

The news function greatly increased as up-to-date material was broadcast at least every hour. One of the best-known voices heard from overseas was Edward R. **Murrow** with his "This Is London" broadcasts, which detailed what was happening to the English during the war.[31]

news

One result of the war was the perfection of audiotape recorders. Events could now be recorded and played back whenever desired. Prior to the war, NBC and CBS had policies forbidding the use of recorded material for anything other than sound effects, and even most of those were executed live. This policy was abetted by the musicians' union, which insisted that all broadcast music utilize musicians rather than phonograph records. The Mutual Broadcasting Company permitted some use of recorded speech, but was considered second-rate for doing so. As a result, the live programs usually had to be performed twice—once for the East and Midwest, and once again three hours later for the West Coast.

audiotape recorders

The recording technique used before the audiotape recorder usually employed phonograph discs, for the only magnetic recording known in America prior to World War II was **wire recording.** In order to edit or splice, a knot had to be tied in the wire and then fused with heat, making it a cumbersome and essentially unusable technique. During the war, American troops entering German radio stations found them operating without any people. The broadcasting was handled by a machine that used plastic tape of higher fidelity than

FIGURE 3.30

Correspondent Edward
R. Murrow at his
typewriter in wartime
London. *(From United
Press International)*

Americans had ever heard from wire. This plastic tape could be cut with scissors and spliced with adhesive. The recorders were confiscated, sent to America, improved, and eventually revolutionized programming procedures.[32]

economic prosperity

Radio stations enjoyed great economic prosperity during the war. There were about 950 stations on the air when the war began. No more were licensed during the war, so these 950 received all the advertisements. A newsprint shortage reduced ad space in newspapers, and some of that advertising money was channeled into broadcasting. Institutional advertising became common because of high wartime taxes; companies preferred to pay for advertising rather than turn money over to the government. And with few consumer products to sell because industry was geared to the war effort, companies were happy to sponsor such prestige programs as symphony orchestra concerts. Thus radio station revenue increased from $155 million in 1940 to $310 million in 1945.[33]

Postwar Economics

Postwar radio was, above all, prosperous. Advertisers were standing in line, and the main problem was finding a way to squeeze in the commercials. To the networks, especially NBC, this boon provided the necessary capital to support the then unprofitable television development. In order to invest even more in the new baby, nonsponsored public affairs radio programs dropped by the wayside, as did some expensive entertainment. Radio fed the mouth that bit it.

station expansion

On the local level this prosperity created a demand for new radio station licenses as both entrepreneurs and large companies scrambled to cash in on the boom. The 950 wartime stations expanded in a rabbitlike fashion to well over two thousand by 1950.[34]

Part of this increase was the result of actions by the FCC and the courts. During the 1930s the FCC required that companies or individuals applying for a new station license prove that this station would cause little or no technical interference to existing local or distant radio stations. Shortly after the war, potential licensees had only to demonstrate lack of interference to local stations. This opened the door for many more stations in the frequency band.

Also during the 1930s the potential licensee had to prove that establishment of another station would not cause economic injury to other stations in the area. In other words, an applicant had to prove to the FCC's satisfaction that there was enough potential advertising revenue to go around. This thwarted many new stations, especially during the depression years. Shortly before the war, however, the U.S. Supreme Court ruled in *FCC v. Sanders Brothers Radio* that the effect on the public should be considered in designating new radio stations. It stated that the FCC was to rule in the public interest and should not be concerned with the economic well-being of particular stations. After the war, when the rash of new station applications came in, the FCC did not require that the issue of economic injury be dealt with.[35]

For several years it appeared that there was no such thing as economic injury to radio. Advertising revenues increased from $310 million in 1945 to $454 million in 1950.[36] But the bubble burst as advertisers deserted radio to try the medium that featured both sound and sight. This left radio networks as hollow shells. The two thousand local stations found that the advertising dollars remaining in radio did not stretch to keep them all in the black. In 1961 almost 40 percent of radio stations lost money.[37]

TV takeover

Postwar Format

After the war, radio networks appeared for a time to return to prewar programming—comedy, drama, soap operas, children's programs, news, and public service. But the new phenomenon was beginning to appear on the scene— the **disc jockey (DJ)**. Several conditions precipitated this emergence.

For one, the FCC altered a ruling about identification of recorded material. Previously it had been necessary to identify all recordings as they were broadcast. Such frequent announcing would have stigmatized a DJ show. But during the 1940s the FCC ruled that such announcements could be made only each half hour.

identifying records

Also, a court decision in 1940 ruled that if broadcasters purchased a record, they could then play it without further financial obligation. This ended the practice of stamping records "not licensed for radio broadcast" and added legal stature to disc jockey programs.

playing records

Then during the mid-1940s, the musician's union, which had voted to halt recording, was appeased with a musicians' welfare fund to which record companies would contribute. This opened the door to mass record production.

This mass production of records led to a symbiotic relationship between radio and the record business that is still very much in force today. The record industry became dependent on radio, and radio became dependent on records.

producing records

FIGURE 3.31

Todd Storz. *(Broadcasting Magazine)*

The beginnings of this relationship can be traced to several people, most notably Alan Freed, Gordon McLendon, and Todd Storz. During the early 1950s, **Freed,** a Cleveland DJ, began playing a new form of music he called rock and roll. The music caught the fancy of teenagers and gave radio a new primary audience and a new role in society—that of acting as a mouthpiece and a sounding board for youth.

McLendon and Storz were both station owners who began programming Top 40 music. According to radio lore, **Storz** was in a bar one night trying to drown his sorrows over the sinking income of his radio stations and noticed that the same tunes seemed to be played over and over on the juke box. After almost everyone else had left, one of the waitresses went over to the jukebox. Rather than playing something that had not been heard all evening, she inserted her nickel and played the same song that had been played most often. Storz observed this and decided to try playing the same songs over and over on his radio stations. Thus Top 40 radio was born. **McLendon** programmed the same Top 40 format and promoted it very heavily.[38]

At the same time that music was being introduced on radio, radios, themselves, were becoming more portable, and Americans were becoming more mobile. The public (especially the young) appreciated the disc jockey shows, which could be enjoyed while listeners were engaged in other activities, such as studying, going to the beach, or talking with friends.

A final important reason for the rise of the DJ is that station management appreciated the lower overhead, fewer headaches, and higher profits associated with disc jockey programming. A DJ did not need a writer, a bevy of actors, a sound effects person, an audience, or even a studio. All that was needed were records, and these were readily available from companies who would eagerly court disc jockeys in the hope that they would plug certain tunes, thus assuring sales of the records.[39]

This courtship slightly tarnished the disc jockeys' image during the late 1950s when it was discovered that a number of disc jockeys had been engaged in **payola,** the practice of accepting money or gifts in exchange for favoring

rock and roll

Top 40

mobility

lower overhead

payola

FIGURE 3.32
Gordon McLendon.
(AP/Wide World Photos)

certain records. To remedy the situation, Congress amended the Communications Act so that if station employees received money from individuals other than their employers for airing records or other material, they had to disclose that fact prior to broadcast time under penalty of fine or imprisonment. This has helped control payola, but every now and then it still rears its ugly head.[40]

As the stations began their romance with the disc jockeys and top talent left radio for TV, the stations had less and less need for the networks. The increasing number of stations also meant that more stations existed in each city, so more of them were programming independently of networks. Therefore, although the actual number of network-affiliated stations did not decrease, the percentage decreased dramatically because the new stations were independently programmed. The overall result was a slow but steady erosion of network programming that began during the 1950s. Eventually, radio stations essentially ceased to exist as conventional entertainment sources and became primarily news and music sources.

decline of networks

Frequency Modulation—FM

During the early 1930s David Sarnoff mentioned to Edwin H. **Armstrong** that someone should invent a black box to eliminate static. Armstrong did not invent just a black box, but a whole new system—**frequency modulation (FM).** He wanted RCA to back its development and promotion, but Sarnoff had committed RCA funds to television and was not interested in underwriting an entirely new radio structure despite its obviously superior clarity and fidelity.

Armstrong

Armstrong continued his interest in FM, built an experimental 50,000-watt FM station in New Jersey, and solicited the support and enthusiasm of GE for his project. During the late 1930s and early 1940s an FM bandwagon was rolling, and some 150 applications for FM stations were submitted to the FCC. As a result, the FCC altered channel 1 on the TV band and awarded spectrum space to FM. It also ruled that TV sound should be frequency modulated. Armstrong's triumphant boom seemed just around the corner, but the

FIGURE 3.33
Edwin H. Armstrong.
*(Smithsonian Institution,
Photo No. 43614)*

war intervened and commercial FM had to wait. However, FM went to war, installed on virtually every American tank and jeep.

moving FM

After the war the FCC reviewed spectrum space and decided to move FM to another part of the broadcast spectrum, ostensibly because it felt that sunspots might interfere with FM. This move was violently protested by Armstrong and other FM proponents because it rendered all prewar FM sets worthless and saddled the FM business with heavy conversion costs.

Armstrong was further infuriated by the fact that although FM sound was to be used for TV, RCA had never paid him royalties for the sets it manufactured. In 1948 he brought suit against RCA. The suit proceeded for more than a year, and the harassment and illness it caused Armstrong led him to leap from the window of his thirteenth-floor apartment to his death.[41]

slow start

FM continued to develop slowly. With television on the horizon, there was little interest in a new radio system. Many of the major AM stations acquired FM licenses as insurance in case FM replaced AM, as its proponents were predicting. AM stations simply duplicated their AM programming on FM, which naturally did not increase the public's incentive to purchase FM sets. In fact, for awhile an industry joke ran, "What do the letters FM in FM radio stand for?" The answer was, "Find me."[42]

However, as general interest in high-fidelity music grew, FM's interference-free signal became a greater asset. In 1961 the FCC authorized stereophonic sound transmission for FM, which led to increased awareness of the medium by hi-fi fans. At first classical and semiclassical music dominated the

FM success

FM airwaves. But as hi-fi equipment became inexpensive enough to be bought by teenagers, rock music became prominent FM fare. This led to an increased number of listeners followed by an increased number of advertisers.

A further aid to FM's success was a 1965 FCC ruling stating that in cities of over 100,000 population, AM and FM stations with the same ownership had to have separate programming at least 50 percent of the time. This helped FM gain a foothold because it now developed its own distinctive programming.[43]

Electronic Media Forms

AM Trials and Tribulations

During the 1970s, FM developed so successfully that it began taking the audience away from AM. In 1972, AM had 75 percent of the audience, and FM had a paltry 25 percent. By the mid-1980s those percentages had reversed, and AM stations were the ones losing money.[44]

loss of audience

The switch to FM was mostly due to the superior sound quality of the medium, including the capability for **stereo** sound. AM proponents tried to combat this by developing an AM stereo system. During the late 1970s and early 1980s, several different companies proposed stereo transmission systems that, unfortunately, were not compatible with each other. AM stations hoped the FCC would choose a standard, as it had for FM, but in 1982, the FCC refused to rule on one common standard, stating instead that the marketplace should decide. The marketplace was not quick to decide. By the 1990s, it looked like the system developed by Motorola was becoming the standard, but AM stereo sound was not as good as FM and did not seem to give AM the competitive boost it needed.[45]

AM stereo

Part of this was due to the fact that many AM stations had deserted music formats in favor of talk and news formats that did not require a superior sound quality. These formats tended to attract older audience members who were not as prized by advertisers as the younger demographic groups tuned to FM music. This added to the economic woes of AM.

change of formats

During the 1990s, the FCC began considering plans to help AM achieve a better competitive position. Most of these ideas involved ways to decrease interference on AM and mandate stereo broadcasting so that a better AM signal could be broadcast.[46]

plans for improvement

In the meantime, a new form of radio, **digital audio broadcasting (DAB)** has been proposed by several companies. Their plans would provide CD quality sound delivered by satellite directly to individual radio receivers. These proposals have made both AM and FM broadcasters nervous because they could bring competition that could obsolete the present radio station structure.[47]

Deregulation

The Reagan era at the beginning of the 1980s ushered in an overall national philosophy toward **deregulation** that greatly affected the broadcast industry and led to the FCC eliminating many of the regulations that had governed radio for many years.

Radio station licenses were granted for seven years rather than three. Radio stations were no longer limited in the amount of commercial minutes they could program and were no longer required to program news and public affairs programs. The stations did not need to keep logs, and they did not need to **ascertain** the needs of the community in order to have their licenses renewed.[48] In addition, the number of radio stations any one company could own was raised from seven AM and seven FM to twelve of each. Companies did not need to keep stations for at least three years before they could sell them as had been previously required.[49]

changes in station rules

Although many radio stations did not initiate any perceivable changes after the deregulation policies were initiated, many did greatly reduce or eliminate public affairs programs, and some totally eliminated news. Quite a few companies acquired additional stations, and the pace of station trading picked up considerably from 424 stations changing hands in 1980 to 1558 in 1985.[50]

Reemergence of Networks

Today radio audiences listen to stations rather than programs. This is because most stations have a sound that usually translates into a particular format—country and western, Top 40, easy listening, religious, jazz, all news, or talk.

However, within this station-oriented structure, there has been a re-emergence of networks. These do not serve the same function as the old radio networks that brought common programming to the entire nation. Instead, the companies owning networks program several different networks with different features and formats so that they can appeal to a wide variety of stations. They distribute their programming to stations by satellite. Some of the stations use only bits and pieces from the network, some use the network music along with local personalities, and others use a preponderance of the network material. The networks often become involved with helping their affiliate stations with promotion, marketing, and sales to advertisers, in addition to supplying programming. Also, it is not uncommon for several stations in one market to affiliate with the same network.

ABC

ABC was the first to redevelop its network. In 1968 its shell of a network was providing primarily news to a limited number of affiliates. It received a waiver from the FCC's duopoly rule so that it could, in essence, operate four networks—American Contemporary, American Information, American Entertainment, and American FM. Each network provided some combination of news, sports, and features for a different target audience (e.g., teens, young adults, adults).[51]

CBS

Shortly thereafter, CBS restructured its news network to add features and old-fashioned radio drama. The drama did not draw the audiences hoped for and was phased out.[52]

NBC

NBC experimented with News and Information Service, the intent of which was to make it economically possible for stations in small markets to afford an all-news format. This was discontinued in 1977 because of insufficient affiliates, but NBC then formed a network called The Source that was composed of news and features for young adults. It also formed Talknet, a weekend talk radio service.[53]

Westwood One

Mutual also made some minor changes in its offerings, and several other companies started radio networks, quite a few of which folded. The upstart in the network business is **Westwood One,** which purchased Mutual in 1985 and then in 1987 pulled an even bigger coup by buying the three NBC networks—NBC Radio Network, The Source, and Talknet—for $50 million. That meant the original radio network was no longer in the business, although Westwood One continued to use the term NBC News.[54]

Electronic Media Forms

As of the 1990s, the networks with the largest number of affiliates are the ABC networks. These have changed their names and organizational structure several times, but at present there are five different networks, three appealing to adults and two to young adults. ABC also has the highest rated of all radio network programs, the news and commentary presentations by Paul Harvey.[55]

ABC

Westwood One, CBS, Sheridan, Satellite Music Network, and Unistar are also relatively successful in the network business. They each have at least two networks serving differing demographics. In addition, there are specialized networks delivering such material as financial news, ethnic programming, and regional information.[56]

others

Although radio networks and radio in general have undergone some cataclysmic changes, they have rebounded nicely, and today they constitute a healthy medium.

Commercial Radio Chronology

1873   Maxwell publishes theory of electromagnetism
1888   Hertz publishes theory of radio transmission
1901   Marconi transmits letter "s" across the Atlantic
1904   Fleming develops the vacuum tube
1906   Fessenden broadcasts to ships at sea
1910   Wireless Ship Act of 1910
1910   De Forest broadcasts Caruso
1912   Sinking of the *Titanic*
1912   Radio Act of 1912
1915   Sarnoff writes radio memo
1917   Government takes over radio operations for war
1919   RCA is formed
1920   Harding-Cox presidential returns broadcast
1921   Dempsey-Carpentier bout aired
1922   Queensboro Corporation message aired
1926   NBC is formed
1927   Radio Act of 1927
1928   Paley becomes president of CBS
1929   Correll and Gosden on NBC as "Amos 'n' Andy"
1930–1950   Golden era of radio programming
1932   Westinghouse and GE withdraw from RCA
1933   Armstrong invents FM
1933–1935   Press-radio war
1934   Mutual Network formed
1934   Communications Act of 1934
1940–1945   World War II broadcasts
1941   FCC chain regulations
1945   Blue Network becomes ABC
1950s   TV creates hard times for radio

1950s Rise of the disc jockey
1958   Payola scandals
1961   FM stereo authorized
1965   FCC ruling that commonly-owned AM and FM stations must have separate programming
1968   ABC forms four networks
1981   Radio deregulated
1982   FCC refuses to set AM stereo standard
1984   Radio ownership rules changed
1985   Mutual is purchased by Westwood One
1987   Westwood One buys NBC

Conclusion

Radio has survived through periods of experimentation, glory, and trauma. Early inventors, such as Maxwell, Hertz, Marconi, Fleming, Fessenden, and De Forest, would not recognize radio in its present form. Indeed, many people who knew and loved radio during the 1930s and 1940s do not truly recognize it today. But radio has endured and along the way has chalked up an impressive list of great moments: picking up the *Titanic's* frantic distress calls, broadcasting the Harding-Cox presidential election returns, broadcasting the World War II newscasts of Edward R. Murrow, and surviving after the television takeover.

Along the way the government has interacted with radio in significant ways that illustrate the medium's growth as a broadcasting entity. The Wireless Ship Act of 1910 and the Radio Act of 1912 dealt with radio primarily as a safety medium. The fact that government took over radio during World War I but did not do so during World War II indicates that radio had grown from a private communication medium to a very public one that most Americans relied upon for information. The need for the government to step in to solve the problem of overcrowded airwaves during the late 1920s proved the popularity and prestige of radio. The ensuing Communications Act of 1934 and the various chain regulations helped solidify the government's role in broadcasting. The post-World War II rulings by the FCC led to a great increase in the number of radio stations on the air, including the late-blooming FM stations. The deregulation of radio in recent times is further evidence that radio has indeed survived.

Companies from the private enterprise sector have also played significant roles in the history of radio, starting with the Marconi Company and progressing through the founding of RCA by the still-powerful AT&T, GE, and Westinghouse. These early companies contributed a great deal in terms of technology, programming, and finance. As the networks were formed, each in its own peculiar way, they set the scene for both healthy competition, and elements of unhealthy intrigue. Intrigue also characterized the free enterprise rivalry between newspapers and radio in the prewar days, and free enterprise in its purest sense altered the format of radio when television stole its listeners. The recent formation of new networks is further proof that radio has survived.

Radio programming is indebted to early pioneers who filled the airwaves with boxing matches, "potted palm" music, and call letters, and to "Amos 'n' Andy," Jack Benny, and others who are remembered for creating the golden era of radio. Today countless disc jockeys and newscasters let us know that radio has survived.

Thought Questions

1. What are some methods of financing radio, other than advertising, that might have been successful? What would have been the advantages and disadvantages of each?
2. How might patents have been handled so that they did not impede the development of radio technology?
3. How do you think radio will change in the next ten years?

Commercial Television

Introduction

To some people it may sound presumptuous to suggest that anything as young as television even has a history. To other people it may seem as if there were no history before television. Some say television is still in its infancy, and others argue that it is past its prime. Television has been dominant only since 1952, but the years have been a blur of technological change and programming turnover.[1]

Many people are disappointed because television has not been put to general use already. They have been reading about it for so long that they are beginning to doubt that it will ever be of much importance. But those who are working with television are confident that someday television will be widely used to entertain people and to inform them.

An article in the February 13, 1939, issue of The Junior Review, *entitled "What Will be the Future of Television?"*

FIGURE 4.1

Early mechanical scanning equipment. Through a peephole, J. R. Hefele observes the image recreated through the rotating disc. The scanning disc at the other end of the shaft intervenes between an illuminated transparency and the photoelectric cell. This cell is in the box that is visible just beyond the driving shaft. *(Courtesy of AT&T Co. Phone Center)*

Early Experiments

The first experiments with television employed a **mechanical scanning** process originally invented by German Paul **Nipkow** in 1884. This process was dependent on a wheel that contained tiny holes positioned spirally. Behind the wheel was placed a small picture. As the wheel turned, each hole scanned one line of the picture.

Even though this device could scan only very small pictures, attempts were made to promote it commercially. For example, John Baird of Britain obtained a television license in 1926 and convinced the British Broadcasting Corporation to begin experimental broadcasting with a mechanical system. At GE's plant in Schenectady, New York, Ernst F. W. **Alexanderson** began experimental programming during the 1920s using a revolving scanning wheel and a three-by-four-inch image. One of his "programs," a science fiction thriller of a missile attack on New York, scanned an aerial photograph of New York that moved closer and closer and then disappeared to the sound of an explosion.[2]

While mechanical scanning was being promoted, other people were developing **electronic scanning,** the system that has since been adopted. One was Allen B. **Dumont,** who developed the oscilloscope, or cathode-ray tube, a basic electronic research tool that is similar to the TV receiver tube. Dumont was able to capitalize on this invention when the TV receiver market took hold during the 1940s.[3]

Another early electronic inventor was Philo T. **Farnsworth,** who in 1922 astounded his Idaho high school teacher with diagrams for an electronic TV system. He convinced a backer to provide him with equipment and in 1927 transmitted still pictures and bits of film. He applied for a patent and found himself battling the giant of electronic TV development, RCA. In 1930 Farnsworth, at the age of twenty-four, won his patent and later received royalties from RCA.[4]

Nipkow

Baird

Alexanderson

Dumont

Farnsworth

Electronic Media Forms

FIGURE 4.2
Early television
experimenters. Allen
B. Dumont (*right*)
giving a personalized
tour of the Dumont
Laboratories to Lee De
Forest. *(Smithsonian
Institution, Photo No.
73-11097)*

FIGURE 4.3
Philo T. Farnsworth in
his laboratory about
1934. *(Smithsonian
Institution, Photo No. 69082)*

The RCA development was headed by Vladimir K. **Zworykin,** a Russian immigrant and onetime Westinghouse employee who had patented an electronic pickup tube called the **iconoscope.** In 1930 GE, RCA, and Westinghouse merged TV research programs at RCA's lab in Camden, New Jersey, and incorporated Zworykin, Alexanderson, and other engineers into a team to further develop television.

This group systematically attacked and solved such problems as increased lines of scanning, definition, brightness, image size, and frequency of scanning. They started with a system that scanned sixty lines using a model of Felix the Cat as the star. Gradually they improved this scanning to 441 lines.

In 1932 experimental broadcasts were transmitted from the Empire State Building. Three years later, in the midst of the depression, David Sarnoff, president of RCA, announced that the company would invest millions in the further development of television. Experimental broadcasts continued with

Zworykin

experimental broadcasts

FIGURE 4.4
The television receiving screen used during the first intercity TV broadcast on April 7, 1927. Dr. Herbert E. Ives, former director of electro-optical research at Bell Telephone Laboratories, stands beside this screen, which consisted of fifty neon-filled tubes. Each tube was divided into small segments, creating a pattern of light and dark areas to form a picture. Dr. Ives is holding a photoelectric cell from the transmitter. Early scanning equipment is seen at the left. *(Courtesy of AT&T Co. Phone Center)*

FIGURE 4.5
Vladimir K. Zworykin holding an early model of the iconoscope TV tube. Zworykin and his colleagues worked to improve the quality of television. *(Courtesy of RCA)*

programs emanating from a converted radio studio, 3H, in Radio City. Because the iconoscope was not a very sensitive pickup tube, actors had to wear heavy makeup and simmer under intensive lights in order for a discernible black-and-white picture to be produced.[5] However, quality at this point was better than it had been under the mechanical scanning or the sixty-line "Felix the Cat" system.

FIGURE 4.6
Felix the Cat as he appeared on the experimental sixty-line, black-and-white TV sets in the late 1920s. This picture was transmitted from New York City all the way to Kansas. *(Courtesy of RCA)*

FIGURE 4.7
David Sarnoff dedicating the RCA pavilion at the 1939 New York world's fair. This dedication marked the first time a news event was covered by television. Sarnoff's speech, entitled "Birth of an Industry," predicted that television one day would become an important entertainment medium. *(Courtesy of RCA)*

The "Coming Out" Party

Sarnoff decided to have television displayed at the 1939 New York world's fair. President Roosevelt appeared on camera and was seen on sets with five- or seven-inch tubes.[6]

In 1939 RCA's program schedule usually included one program a day from 3H, one from a mobile unit traveling the streets of New York, and several assorted films. The studio productions included plays, bits of operas, singers, comedians, puppets, and household tips. The mobile unit consisted of two huge buses, one jammed with equipment to be set up in the field and one containing the transmitter that broadcast back to the Empire State Building. It covered such events as baseball games, wrestling, ice skating, airport interviews with dignitaries, fashion shows, and the premiere of *Gone With the Wind.* The films were usually cartoons, travelogues, or government documentaries.

"mobile" productions

Other companies established experimental stations and broadcast in New York. Sets, mostly manufactured by RCA and Dumont, had increased tube sizes to twelve inches and sold for $200 to $600. CBS began experimentation

FIGURE 4.8
The first baseball game to be televised—May 17, 1939, at Baker Field, New York—was a contest between Princeton and Columbia. *(Courtesy of NBC)*

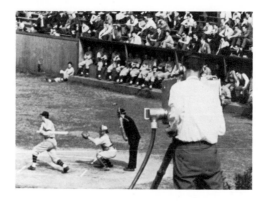

with color television, utilizing a mechanical color wheel of red, blue, and green that transferred color to the images. This color system was not compatible with the RCA-promoted system. In other words, the sets being manufactured could not receive either color or black-and-white pictures from the CBS mechanical system, and proposed CBS receivers would not be able to pick up existing black-and-white pictures.

early color TV controversy

In 1940 a group led mainly by RCA personnel tried to convince the FCC to allow the operation of the 441-line system. However, the FCC was not certain that this system had adequate technical quality so it established an industrywide committee of engineers, the **National Television System Committee (NTSC),** to recommend standards. This committee rejected the 441-line system and recommended the 525-line system, which the United States presently uses. CBS approached the committee with the idea of color television, but the committee did not think the system was of sufficient quality.

beginning authorization

In May of 1941 the FCC authorized the full operation of 525-line black-and-white television. Originally there were to be thirteen **very high frequency (VHF)** channels, but channel 1 was eliminated to allow spectrum space for FM radio. Twenty-three stations went on the air, ten thousand sets were sold, and commercials were sought. The first commercial was bought by Bulova and consisted of a shot of a Bulova clock with an announcer intoning the time.

World War II

All of this, however, was called to a halt in 1942 because of World War II. During the war only six stations remained on the air, and most sets became inoperable because spare parts were not being manufactured for civilian use. The NBC studio was used to broadcast air raid warden-training programs; volunteers traveled to their local police station to watch this instruction. So, in a very limited way, television went to war.[7]

The Emergence of Television

TV activities did not resume immediately after the war. The delay was in part due to a shortage of materials. Also, building and operating a TV station was expensive—the initial investment ranged from $75,000 to $1.5 million. And then the owners had to assume that they would operate at a loss until there were enough receivers in the area to make the station attractive to advertisers.

To add to the risk, there were rumors that all television stations might be moved to the **ultrahigh frequency** (**UHF**) band, which had been explored during the war. CBS once again raised the question of color stating that its color system was so well developed that TV station allocations should not be made until the color question was resolved. RCA represented a contrary position on the color issue and promised black-and-white sets on the market by mid-1946 and a color system shortly thereafter. The color system RCA proposed was electronic and compatible with current sets as opposed to the CBS system, which was mechanical and incompatible. In the fall of 1946 RCA did demonstrate a color system that was compatible, albeit unstable and unreliable.

UHF possibilities

color

In 1947 the FCC declared that CBS's color system would be a hardship on set owners because they would have to buy new sets. It therefore stated that television should continue as black and white and in the VHF range it had been using, channels 2 to 13.

In the next year, 1948, television emerged as a mass medium. Stations, sets, and audience all grew more than 4,000 percent within that one year. Advertisers became aware of the medium, and networks began more systematic programming.[8]

enormous 1948 growth

TV networks had already existed before 1948 as offshoots of the radio networks. As early as 1945–46 television networks had been organized by NBC, CBS, and ABC. A fourth network, Dumont, was organized by Allen B. Dumont. However, most cities had only one or two TV stations, and NBC and CBS usually recruited them as affiliates, making it difficult for ABC and Dumont to compete. ABC survived because it merged in 1953 with United Paramount Theaters and thus gained an increase in operating funds. Dumont, however, went out of business in 1955.[9]

The Freeze

Television grew so uncontrollably in 1948 that in the fall of that year the FCC imposed a **freeze** on television station authorizations because stations were beginning to interfere with each other. Mindful of the days of radio chaos, the FCC wanted to nip the problem in the bud. Additional reasons were that some of the VHF characteristics had not been predicted, there were not enough channels to meet the demand, and CBS once again raised the question of color. The freeze, which was originally predicted to last about six months, lasted until July 1, 1952. During this period 108 stations were on the air, and no more were authorized to begin operation.

What occurred between 1948 and 1952 could be termed an explosive lull. Many cities, including Austin, Denver, and Portland, had no television stations. Others, including Pittsburgh, St. Louis, and Milwaukee, had only one. Twenty-four cities boasted between two and six stations. New York and Los Angeles were the only cities with seven.

Although TV networking still could not be considered truly national, the number of sets, audience size, advertising, and programming continued to grow.

FIGURE 4.9
Milton Berle in one of
his outlandish
costumes. His "Texaco
Star Theater" was
TV's first big hit.
(Courtesy of NBC)

continued growth

By 1952 there were sets in 15 million homes. The largest ones had twenty-inch tubes and sold for about $350. TV advertising revenues reached $324 million. In TV cities, movie attendance, radio listening, sports event attendance, and restaurant dining were all down—especially on Tuesday night, which was Milton **Berle** night.

Milton Berle

People with TV sets stayed home and often invited their friends over (or allowed their friends to invite themselves over) to watch this ex-vaudevillian's show, which included outrageous costumes, slapstick comedy, lavish productions, and a host of guest stars. "Uncle Miltie," on his "Texaco Star Theater," became a national phenomenon and the reason many people bought their first television set.[10]

Early Programming

sports

During 1948 and 1949, 30 percent of sponsored evening programs were sports—basketball, boxing, bowling, wrestling, and roller skating. The wrestlers, both men and women, competed to outdo one another in costumes, hairdos, and mannerisms. The emphasis on sports was due, at least in part, to the fact that a large number of the first TV sets were in bars and taverns. But during the 1949–50 season, sports comprised less than 5 percent of evening programming, and children's programming was tops, indicating that TV had moved to the home.

comedy and variety

Other broadcasting fare of the freeze era included comedy team Sid Caesar and Imogene Coca in "Your Show of Shows," and Ed Sullivan hosting a variety of acts for "Toast of the Town." "Amos 'n' Andy" was brought to TV and featured black actors trained by Gosden and Correll. The program was condemned as an insult by the National Association for the Advancement of Colored People at its 1951 convention, much to the astonishment of the lily-white broadcasting fraternity.[11]

FIGURE 4.10
Marionette Howdy Doody and his real-life friend Buffalo Bob Smith. The show also featured Clarabell, the clown who never spoke, and an audience full of children. Each program began with "What time is it?" followed by the audience's response, "It's Howdy Doody Time." *(Courtesy of NBC)*

FIGURE 4.11
Kukla, Burr Tillstrom, Fran Allison, and Ollie. Burr operated and spoke for each one of the large cast of puppets while Fran conversed with them from in front of the puppet stage. The "Kukla, Fran, and Ollie" show was the first program to be televised coast to coast in color. *(Courtesy of NBC)*

other shows

For children there was Buffalo Bob Smith and his puppet Howdy Doody, and for children of all ages, Burr Tillstrom's puppets starred with Fran Allison in "Kukla, Fran, and Ollie." Drama was initiated with "Kraft Television Theater," "Philco Playhouse," and "Studio One." An early soap opera, "A Woman to Remember," was forgotten after its first year, but others followed successfully in its wake. For example, "Love of Life," which began in 1951, lasted until the late 1970s with one of the original members still in the cast.

In 1951 "I Love Lucy" began as a maverick of the TV world because it was filmed ahead of time, while other shows were aired live. There was a stigma against film at the time, partly because it added to the cost of the show and partly because the TV networks inherited the live tradition from radio and assumed that all shows should be produced live. But the film aspect was particularly useful and dramatic when Lucille **Ball** became pregnant and, hence,

"I Love Lucy"

FIGURE 4.12

Lucy tries to make a hit playing the saxophone. The "I Love Lucy" series, starring Lucille Ball, captured America's heart and can still be seen in reruns. *(Photo/ Viacom, Hlwd.)*

the story line dealt with Lucy's pregnancy. The episode involving the birth of Lucy's baby was filmed ahead of time, and Lucille's real baby was born the same day the filmed episode aired—to an audience that comprised 68.8 percent of the American public. Eventually, the filming more than paid for itself because the program became the first international hit. Copies of the film were made, dubbed into numerous languages, and sold overseas.

news

    TV newscasts of the 1950s developed slowly. Networks found it easy to obtain news and voices, but pictures were another matter. At first they contracted with the companies who supplied the newsreels then shown in theaters. This did not exactly fill television's bill because much of it was shot for in-depth stories, not news of the day. However, this system could not begin to cover all the news. Networks set up their own film crews, but limited budgets and bulky film equipment meant that camera operators could attend only planned events, such as press conferences, coronations, political conventions, and ribbon-cutting grand openings. The fifteen-minute newscasts tended to be reports on events that had been filmed earlier, which limited their timeliness.

public affairs

    Interview-type news shows were a further development. "Meet the Press" began its long run of probing interviews with prominent people. "See It Now" started in 1951 as a news documentary series featuring Edward R. **Murrow** and was produced by Fred **Friendly.** A historical feature of their first program was showing for the first time both the Atlantic and Pacific Oceans live on TV.

politics

    The importance of TV to the political process was already becoming evident as the 1952 political nominating conventions were covered by the TV networks. NBC's coverage was sponsored by Westinghouse and brought fame to Betty Furness as she demonstrated refrigerators over and over in live commercials, including one never-to-be-forgotten spot when the "easy-to-open" refrigerator door refused to open at all.[12]

At this point television could be proud of its achievements. Sets had increased in size and quality, TV programming had increased in hours and variety, the public was fascinated with the new medium, and advertisers were providing increasingly strong financial backing.

Lifting the Freeze

In April of 1952 the FCC issued a report that ended the freeze on new stations that had begun four years earlier. This report included a station allocation that was worked out by FCC engineers on the basis of various parameters. For AM radio, the FCC had allocated stations on a first-come, first-served allocation tables basis with the burden resting on the applicant to prove that the proposed station would not cause interference with other established stations. What the FCC discovered with this process was that first-come stations were generally sought in the more populous and hence more economically viable areas, and that by the time people were interested in establishing stations in small towns or rural areas, appropriately cleared frequencies were difficult to get. This meant residents of nonpopulous areas were not as well served as were residents of populous areas.

The FCC decided this injustice should not be repeated with TV, and as a result it allocated specific stations to particular geographic areas. These stations were to be reserved for such areas until someone was interested in establishing the stations.

Another fact apparent to the FCC as it undertook its work between 1948 and 1952 was that the twelve VHF channels being used would not be sufficient to meet demand; additional channels needed to be found. The engineers determined that there was not adequate room left in the VHF band, and they opted for adding seventy stations in the UHF band, for a total of eighty-two channels. UHF was at a much higher frequency than VHF, and very little was known about the technical characteristics of UHF at that time. However, the FCC engineers felt that by increasing the power and tower height of UHF stations, they would be equal in coverage to VHF stations.

Another item that had to be provided for in the allocation table was reservation for educational channels. Hearings on this subject were held during the freeze period, and the FCC decided to set aside 242 station assignments for noncommercial educational purposes. The allocation table that the FCC published listed 2,053 station allocations in 1,291 communities. In general, cities had both VHF and UHF stations assigned to them.

The freeze lift led to an enormous rush to obtain stations. Within six rush for stations months, 600 applications had been received and 175 new stations had been authorized. By 1954, 377 stations had begun broadcasting, and TV could be considered truly national.[13]

Blacklisting

Red Channels

To this fledgling industry came some of the country's best-known talent. They came from radio, Broadway, and film—all of which were experiencing downturns as television was burgeoning.

Unfortunately, many of these people became caught in the **blacklisting** mania of the 1950s led by Senator Joseph R. **McCarthy,** a Republican from Wisconsin. A 215-page publication called *Red Channels: The Report of Communist Influence in Radio and Television* listed 151 people, many of whom were among the top names in show business, most of them writers, directors, and performers. Listed after their names were citations against them. Some of these charges were proved to be totally false, such as associating people with "leftist" organizations to which they had never belonged. Other allegations were true, but were "leftist" only by definitions of the perpetrators of *Red Channels.* These included such "wrongdoings" as belonging to an End Jim Crow in Baseball Committee and signing a cablegram of congratulations to the Moscow Art Theater on its fiftieth birthday.

Although many network and advertising executives did not believe these people were Communists or in any way un-American, they were unwilling to hire them, in part because of the controversy involved and in part because sponsors received phone calls that threatened to boycott their products if programs employed these people.

For the better part of a decade some well-established writers found that all the scripts they wrote were "not quite right," and certain actors were told they were "not exactly the type for the part." Many of these people did not even know they were on one of the "lists" because these were circulated clandestinely among executives.

In time the blacklist situation eased. Ironically, broadcasting was influential in exposing the excesses of the Communist witch-hunt, which had spread beyond the entertainment industry. Edward R. Murrow and Fred Friendly presented a "See It Now" program entitled "The Case Against Milo Radulovich, A0569839." Radulovich had been asked to resign his Air Force commission because his sister and father had been anonymously accused of radical leanings. The program exposed some of the guilt-by-association tactics and was instrumental in helping Radulovich retain his commission. Later Murrow and Friendly prepared several programs on Senator McCarthy, who had alarmed the country by saying that he had a list of hundreds of Communists in the State Department. The Murrow-Friendly telecasts helped reveal this as a falsity.

McCarthy's downfall

Later there was network coverage of the 1954 hearings of McCarthy's dispute with the Army. As the nation watched, McCarthy and his aides harassed and bullied witnesses. Public resentment built up against McCarthy, and the Senate voted 67 to 22 to censure him.[14]

Electronic Media Forms

Programming of the 1950s was predominantly live. "I Love Lucy" continued to be filmed, and several other programs jumped on the film bandwagon as foreign countries began developing broadcasting systems. Americans could envision the reuse of their products in other countries and, hence, the possibility to recoup film costs. **Reruns** in the United States were as yet unthought of, although some programs were **kinescoped** so they could be shown at various times. These kinescopes were low-quality, grainy-film representations of the video picture. However, most of the popular series of the day originated in New York and were telecast as they were being shot.

kinescopes

This live aspect created problems for drama writers, actors, and technicians. Costume changes needed to be virtually nonexistent. The number of story locales was governed by the number of sets that could fit into the studio. Timing was sometimes an immense problem. In radio, scripts could be fairly accurately timed by the number of pages, but television programs contained much action, the time of which often fluctuated wildly in rehearsals. One writing solution was to plan a search scene near the end of the play. If the program was running long, the actor could find what he was looking for right away. If the program appeared to be moving too quickly, the actor could search the room for as long as necessary.[15]

The programming of the early 1950s is often looked back upon as the "golden age of television." This is mainly because of the live dramas that were produced during this period. One of the most outstanding plays was Rod Serling's *Requiem for a Heavyweight,* the psychological study of a broken-down fighter. Another was Paddy Chayefsky's *Marty,* the heartwarming study of a short, stocky, small-town butcher who develops a sensitive romantic relationship with a homely schoolteacher.[16]

dramas

All of these productions made abundant use of close-ups—the real emotion and action of the plays took place on actors' faces. Cameras were usually placed in the center of the studio with sets arranged in a circle on the periphery so the cameras could have easy access to each new scene. Framing shots was a little more difficult in early TV because cameras had **fixed lenses** rather than **zoom lenses.** The fixed lenses could only frame shots at one specific point. Usually four or five of these fixed lenses were mounted on a turning device called a **turret.** When the camera picture was not on the air, the camera operator could switch from one lens to another thus obtaining a closer or longer shot for the next on-air picture.

Besides conventional drama, some innovative formats were tried, many of them the brainchild of Sylvester L. "Pat" **Weaver,** president of NBC from 1953 to 1955. One of his ideas was the spectacular, a show that was not part of the regular schedule but designed to expand the horizons of creativity. The most outstanding of the spectaculars was probably "Peter Pan," starring Mary Martin, which was viewed by some 165 million Americans.

Weaver's innovations

FIGURE 4.13

A "Playhouse 90"
show entitled "The
Country Husband."
Here Barbara Hale
(*left*) can't understand
why Frank Lovejoy
insists on personally
escorting Felicia Farr
home. *(Courtesy of
Columbia Pictures
Television)*

FIGURE 4.14

A 1952 camera with
turret lenses. *(Courtesy of
RCA)*

other programs

Weaver was also involved in developing the "Today" and "Tonight" shows. "Today" started in Chicago as "Garroway at Large," an informal variety series. Its host, Dave Garroway, became host on the daily morning "Today" show, and the format changed somewhat to include both news and variety. Another star of the show was J. Fred Muggs, a baby chimpanzee. The

FIGURE 4.15
Dave Garroway with chimps J. Fred Muggs (*left*) and Phoebe B. Beebe. Muggs was discovered in classic show business style by one of "Today's" producers, who spotted him in an elevator. *(Courtesy of NBC)*

FIGURE 4.16
"Dragnet," starring Jack Webb (*right*) as Sergeant Friday and Harry Morgan as Officer Cannon. In this episode, called "The Gun," a Japanese widow has been murdered and Friday and Cannon are assigned to find the assailant. *(Courtesy of NBC)*

late-night show "Tonight" originally starred Steve Allen, and then picked up steam as Jack Paar became its controversial host.

Weaver felt that programs should be network controlled rather than advertising agency controlled. Most of the early TV and golden era radio program content had been controlled by the advertising agencies. Weaver developed a **magazine concept** whereby advertisers bought insertions in programs, and the program content was supervised and produced by the networks. This was used for "Today," "Tonight," and many of the spectaculars, somewhat to the chagrin of the advertisers.[17]

Other highlights of early programming were "Victory at Sea," a documentary series that recreated the naval battles of World War II; Elvis Presley performing on the Ed Sullivan show—from the waist up; the introduction of the crime-drama "Dragnet," which was the first show to bump Lucy from the top of the ratings; and Jackie Gleason's comedy sketches, the most famous of which was "The Honeymooners."[18]

Color TV Approval

CBS system accepted

The problems connected with color TV were not resolved until 1953. In 1950 the FCC finally accepted the CBS system of color instead of the RCA system even though the CBS system was incompatible with existing black-and-white receivers. This surprised many engineers who felt an incompatible system would not be chosen. However, the FCC was pressured by Congress and the companies involved to make a decision and felt the CBS system provided higher quality color pictures.

A great deal besides compatibility, consumer cost, and inconvenience was at stake. The company that won the color battle stood to gain enormous financial benefits through the building and licensing of TV sets. RCA continued to fight the CBS system by refusing to program in color and by gaining allies among other set manufacturing companies and TV stations.

RCA objections

end of CBS system

There were valid reasons why stations did not want to purchase CBS color cameras and other color gear. If a station were to transmit with the CBS color system, it could not simultaneously transmit with the present black-and-white system. This meant that a station could not program to people with old sets and people with new color sets at the same time.

CBS also ran into difficulty manufacturing its color sets. Certain metals that were needed for the sets were also needed for the Korean War effort, and CBS had trouble obtaining them. As a result, the CBS color system project essentially came to a halt, and a general state of confusion concerning color reigned in the TV industry.[19]

NTSC recommendation

To help solve the problem, the National Television System Committee (NTSC), the same committee of engineers that had decided on the 525-line system, volunteered to study the situation. Because this committee included more members who favored the RCA system than the CBS system, very few people were surprised when it recommended the compatible system. However, significant improvements had been made in the RCA system, some of which the NTSC initiated, and its color picture was now much better than it had been when RCA originally demonstrated it to the FCC. The FCC took the NTSC recommendation and in 1953 sanctioned RCA's electronic compatible system, which is still used in the United States today. At the time, even CBS supported the adoption.

slow growth

However, for a long time RCA-NBC was the only company actively promoting color. NBC constructed new color facilities and began programming in color, but both CBS and ABC dragged their heels, and most local stations did not have the capital needed to convert to color equipment. Even more important, consumers were reluctant to purchase color TV sets because they cost twice as much as black-and-white sets, and the limited color programming did not merit this investment. The vicious cycle was eventually reversed. As more color sets were sold, the prices were lowered and programming in color increased, causing even more sets to be sold. But not until the late 1960s were all networks and most stations producing color programs.

FIGURE 4.17
The CBS color TV set. The rotating color wheel disc was built right behind the outside of the picture tube. *(Courtesy of Ed Reitan)*

Prerecorded Programming

The days of live programming, other than news and special events, began to disappear in the mid-1950s for several reasons. One was the introduction of videotape in 1956. The expense of the equipment prevented it from taking hold quickly, but once its foot was in the door, videotape revolutionized TV production techniques. Programs could now be performed at convenient times for later airings. As the equipment became more sophisticated, stops could be made to allow for costume changes, scene changes, and the like. As the equipment became even more sophisticated, mistakes could be corrected through editing procedures. Scenes could even be taped out of sequence and assembled in order at a later time, in much the same manner as film.

videotape

However, the live era had begun to yield to film even before tape took hold. Film companies had originally been antagonistic toward TV because it stole away much of the audience that had attended movie theaters. Some film production companies would not allow their stars to appear on TV and would not even allow TV sets to appear as props in movies. But the 1953 merger of United Paramount Theaters and ABC opened the door for Hollywood film companies and the TV establishment, which was primarily located in New York, to begin a dialogue. The first result of this dialogue was a one-hour weekly series, "Disneyland," produced for ABC by the Walt Disney Studios. This was a big hit, and soon several other major film companies were producing film series for TV.

film

Some of the early filmed TV series were "Cheyenne," "Death Valley Days," "Wagon Train," and "December Bride." As the list demonstrates, westerns predominated. By 1959 thirty-two western series were on prime-time TV. The one with the greatest longevity was "Gunsmoke," which revolved around Dodge City's Matt Dillon, Chester Goode, Doc, and Miss Kitty.[20]

westerns

FIGURE 4.18
Ward Bond starring as Major Adams in "Wagon Train." This series dealt with the trials and tribulations of settlers as they moved westward. After Bond's death in 1961, John McIntire took over the leading role.
(From MCA Publishing)

FIGURE 4.19

"American Bandstand," hosted by Dick Clark (*center*). Clark was originally a Philadelphia disc jockey and became nationally prominent with his program of dancing teenagers.
(Courtesy of Dick Clark Teleshows, Inc.)

other programs

Other types of programs gained popularity also, again mainly on film or tape. The old crime-mystery formula of radio surfaced on TV in such programs as "77 Sunset Strip," and "Perry Mason." Dancer Fred Astaire presented a widely acclaimed 1958 special, and Dick Clark appealed to the teenagers with "American Bandstand." For the younger audience there was such fare as "Captain Kangaroo" and "The Mickey Mouse Club."[21]

advertiser success

The TV boom continued in the late 1950s—more TV sets, more viewers, more stations, more advertising dollars. But to this rising euphoria came a dark hour.

The Quiz Scandals

Quiz programs on which contestants won minimal amounts of money or company-donated merchandise had existed on both radio and television. However, in 1955 a new idea emerged in the form of "The $64,000 Question." If

FIGURE 4.20
M.C. of "The $64,000
Question" was Hal
March (*right*). Gino
Prato, an Italian-born
New York shoemaker,
mops his brow while
listening to a four-part
question about opera.
*(From United Press
International)*

contestants beat out challengers over a number of weeks, they could win huge
cash prizes. Sales of Revlon products, the company that sponsored "The
$64,000 Question," zoomed to such heights that some were sold-out nation-
wide. The sales success and high audience ratings spawned many imitators.
Contestants locked in soundproof booths pondered, perspired, and caught the
fancy of the nation.

From time to time there were rumors that some quiz programs had been
fixed. Then in 1958 a contestant from the daytime quiz show "Dotto" claimed
to have evidence that the show was rigged, and a contestant from "Twenty-
One" described that program's irregularities. The networks and advertising
agencies denied the charges, as did Charles Van Doren, a Yale professor who   Charles Van Doren
was the most famous of the "Twenty-One" winners. A grand jury and a House
of Representatives subcommittee conducted hearings and found discrepancies
in the testimony.

In the fall of 1959 Van Doren appeared before the House subcommittee
and read a long statement describing how he had been convinced in the name
of entertainment to accept help with answers to questions in order to defeat
a current champion who was unpopular with the public. Van Doren was also
coached on methods of building suspense and when he did win, he became a
national hero and a leader of intellectual life. After several months on the
program, he asked to be released and finally was allowed to lose. He initially
lied, he said, so that he would not betray the people who had invested faith
and hope in him. Other witnesses from various programs followed Van Doren
to testify to the means by which dull contestants were disposed of so that lively
personalities could continue to win and hence hold audience attention.

In retrospect the **quiz scandals'** negative effect on TV was short-lived.
The medium was simply too pervasive a force to be permanently afflicted by
such an incident. Congress did amend the Communications Act to make it
unlawful to give help to a contestant, but for the most part the networks rec-
tified the errors by canceling the quiz shows and reinstating a higher percentage of public service programs.

Networks also took charge of their programming to a much greater extent. The presidents of all three networks decreed that from then on most program content would be decided, controlled, and scheduled by networks, which would then negotiate sales time with advertisers. This was a further extension of Weaver's "magazine concept." Beginning in 1960, most program suppliers contracted with the networks rather than advertising agencies. This made life more profitable for the TV networks, too, because they established profit participation plans with the suppliers.[22]

The UHF Problem

When the FCC ended its freeze and established stations in the UHF band, it intended that these stations would be equal with VHF stations. In reality, they became second-class stations. UHF was technically not as effective as expected. Its weaker signal was supposed to be compensated for by higher towers, but this did not prove to work in practice.

People did not have sets that could receive UHF, so in order to tune in UHF stations, they needed to buy converters. Many were unwilling to do this, and UHF found itself in a vicious circle. In order for people to buy UHF converters, UHF had to offer interesting programming material; in order to finance interesting program material, UHF stations had to prove to advertisers that they had an adequate audience.

Because VHF stations were established first, they were the first to obtain network affiliations and hence capture the best programming. Consequently, UHF stations had to depend on syndicated material and local talent.

The FCC tried to help the fledgling UHF stations in a number of ways. In 1954 it changed the number of TV stations that one company could own from five to seven, provided that no more than five of those seven were VHF. This meant that organizations such as the three networks were free to buy two UHF stations each. And this is what NBC and CBS did—each purchased UHF stations and placed their network programs on them. The theory was that with network programming on UHF, people would be willing to buy converters. In reality people did not buy the converters, and UHF penetration increased so slightly in the markets where NBC and CBS had their stations that both networks abandoned the UHF stations after several years.

In 1957 the FCC proposed a procedure known as **deintermixture.** The intent was to make some markets all UHF and some all VHF so that in the all-UHF markets people would be forced to buy converters if they wanted to receive any television. This plan did not succeed either, however, mainly because established VHF stations fought any efforts to convert them to UHF.

A third attempt to give UHF a boost was the passage of the **all-channel receiver bill** that Congress authorized in 1962. By amending the Communication Act, the FCC was given the authority to require both a UHF and a VHF tuner on all TV sets. All sets manufactured since 1964 include both UHF and VHF, but most UHF stations still play second fiddle to VHF stations.[23]

FIGURE 4.21
A Huntley-Brinkley
newscast. Chet
Huntley broadcast
from New York, while
David Brinkley, seen
on the television
screen, broadcast from
Washington, D.C.
(Courtesy of NBC)

From the 1950s to the 1960s

The 1950s was a decade of great growth and experimentation for television. Its technical base was devised by the FCC during the freeze period, the color television problem was resolved, and attempts had been made to resolve the UHF problem. Networks and stations established programming concepts and relationships with advertisers. The videotape recorder added more flexibility to programming, and film and TV began a cautious marriage that further enhanced programming possibilities.

 The TV industry had the strength to survive both the external attack of blacklisting and the internal scandal of quiz program rigging. By the end of the 1950s even the greatest of skeptics were conceding that TV was not just a passing fad. TV had not established itself as a primary news and information medium as yet, but that accomplishment awaited the turn of the decade.

Reflections of Upheaval

Television journalism gathered force and prestige during the 1960s—the decade of civil rights revolts, the election of John F. Kennedy, space shots, satellite communications, assassinations, Vietnam, and student unrest.

 The networks encouraged documentaries and increased their nightly news from fifteen to thirty minutes in 1963, thereby assigning increased importance to their news departments. Anchoring on camera for NBC were Chet Huntley and David Brinkley and for CBS, Walter Cronkite, the person who became known as the most trusted man in America. Network news departments scored points over their print counterparts when President John F. Kennedy agreed to have news conferences televised live.

 The quiz scandals had helped precipitate a rise in documentaries. To atone for their sins, networks increased their investigative fare. Documentaries were now easier to execute because technical advances included 16mm film to replace the bulky 35mm. Film and sound could now be synchronized

success

status

news

documentaries

FIGURE 4.22

Filming preparations for "The Tunnel." NBC news correspondent Piers Anderton (*left*) reviews the building of this tunnel under the Berlin Wall with student Dominico Sesta (*center*) and cameraman Peter Dehmel. The 450-foot passage began in this basement of a West Berlin home. *(Courtesy of NBC)*

without an umbilical cord between two pieces of equipment; and wireless microphones were becoming reliable, enabling speakers to wander freely without having to stay within range of a mike cord. When portable video cameras were developed at the end of the decade, news gathering flexibility became even greater.

As ongoing documentaries, ABC established "Close-Up," and the other networks offered "CBS Reports" and "NBC White Paper." Some of the notable documentaries of the era included "Biography of a Bookie Joint," for which concealed cameras oversaw the operation of a Boston "key shop"; "The Real West," an authoritative report narrated by Gary Cooper that countered the westerns; a 1962 airing of "The Tunnel," for which NBC secured footage of an actual tunnel being constructed by young Berliners to bring refugees from East to West Berlin; "The Louvre," for which cameras were allowed inside the famous building for the first time; and "D-Day Plus 20," a World War II reminiscence between Eisenhower and Cronkite filmed on Normandy beaches.

Many documentaries reported on racial problems. "Crisis: Behind a Presidential Commitment" chronicled the events surrounding Governor George Wallace's attempted barring of the schoolhouse door to prevent blacks from attending the University of Alabama; "The Children Are Watching," dealt with the feelings of a six-year-old black child attending the first integrated school in New Orleans. Civil rights actions were covered as news—including the riots in the Watts section of Los Angeles and the death of Martin Luther King.

civil rights

Television reacted to the civil rights movement in another way—it began hiring blacks. Radio had also been lily-white, but not so visibly. Both media began hiring blacks in the 1960s. Jackie Gleason's chorus included a black dancer. Scriptwriters began including stories about blacks, which often were not aired in the South. Networks and stations hired black on-camera newspeople. A black family moved to the nighttime soap opera "Peyton Place," and Diahann Carroll in "Julia" became the first black heroine.[24]

FIGURE 4.23
One of the so-called "Great Debates." John F. Kennedy has his turn speaking while Richard Nixon and moderator Howard K. Smith look on. In the foreground are the newspeople who asked the questions. *(From United Press International)*

Television has been credited, through the **Great Debates** between John F. Kennedy and Richard M. Nixon, with having a primary influence on the 1960 election results. Kennedy and Nixon met for the first debate at the studios of Chicago's CBS station. Kennedy, tanned from campaigning in California, refused the offer of makeup. Nixon, although he was recovering from a brief illness, did likewise. Some of Nixon's aides, concerned about how he looked on TV, applied Lazy-Shave, a product to cover five o'clock shadow. Some people believe that the fact that Kennedy appeared to "win" the first debate had little to do with what he said. People who heard the program on radio felt Nixon held his own, but those watching TV, especially the reaction shots of Kennedy and Nixon listening to each other, could see a confident, attentive Kennedy and a haggard, weary-looking Nixon whose perspiration streaked the Lazy-Shave. Three more debates were held, and Nixon's makeup and demeanor were well handled, but the small margin of the Kennedy victory at the polls has often been attributed to the undecided vote swung to Kennedy during the first debate.[25]

Three years later television devoted itself to the coverage of the assassination of John F. Kennedy. From Friday, November 22, to Monday, November 25, 1963, there were times when 90 percent of the American people were watching television. As a New York critic wrote, "This was not viewing. This was total involvement." From shortly after the shots were fired in Dallas until President Kennedy was laid to rest in Arlington Cemetery, television kept the vigil, including the first "live murder" ever seen on TV as Jack Ruby shot alleged assassin Lee Harvey Oswald. Many praised television for its controlled, almost flawless coverage. Some felt TV would have made it impossible for Lee Harvey Oswald to receive a fair trial and that it was the presence of the media that enabled the Oswald shooting to take place.[26]

Through the medium of television, Americans were able to witness the nation's race into space. In 1961 Alan Shepard was launched into outer space within the view of network television cameras. Subsequently, televised pictures were sent from outer space and the moon, and television viewers witnessed Neil Armstrong's first step onto the moon.[27]

Great Debates

Kennedy assassination

space launches

Vietnam

For the first time in history, war came to the American dinner table. As the troop buildup began in Vietnam during the mid-1960s, the networks established correspondents in Saigon. Reports of the war appeared almost nightly on the evening news programs. In 1968, amid rising controversy over the war, Walter Cronkite decided to travel to Vietnam to see for himself and returned feeling that the United States would have to accept a stalemate in that country.

student unrest

Much of the controversy surrounding the war originated on the country's campuses, where students were becoming increasingly dissident regarding various issues. This, too, was covered by the media, as was the 1968 Democratic convention in Chicago, where youths outside the convention hall protested the steamrolling nomination of Hubert Humphrey. The media became embroiled in the controversy and were accused of inciting the riot conditions. The demonstrators seemed to be trying to attract media coverage. On the other hand, many people inside the convention hall learned of the protest by seeing it on a TV monitor and might not otherwise have known of this show of discontent.[28]

The role of the electronic media in the creation of news was debated frequently in the 1960s and beyond as their image as a source of information increased.

A Vast Wasteland?

Minow's speech

In 1961 Newton Minow, Kennedy's appointee as chairman of the Federal Communications Commission, spoke before the annual convention of the National Association of Broadcasters. During the course of his speech, he stated the following to the broadcasting executives:

> I invite you to sit down in front of your television set when your station goes on the air and stay there without a book, magazine, newspaper, profit and loss sheet, or rating book to distract you—and keep your eyes glued to that set until the station signs off. I can assure you that you will observe a vast wasteland.[29]

The term **vast wasteland** caught on as a metaphor for television programming. Needless to say, the executives were not happy with Minow's phrase.

violence and antiviolence

During the early 1960s the dominant fare was violence. "The Untouchables," "Route 66," "The Roaring 20s,"—all featured murders, jailbreaks, robberies, thefts, kidnappings, torture, blackmail, sluggings, forced confessions, and dynamiting. Saturday morning children's programming was also replete with violent cartoons.

A surge against violence, aided by Minow's challenge, brought forth a change to doctors' shows, such as "Ben Casey" and "Dr. Kildare" and more comedies, such as "The Dick Van Dyke Show," "Gilligan's Island," and the "Beverly Hillbillies." The last, a series about an Ozark family that struck oil and moved to Beverly Hills without changing their mountain character, headed the ratings but was heralded as the supreme example of the vast wasteland.

Spy shows, such as "Mission: Impossible" and "I Spy," were plentiful for several years and brought with them a returning encroachment of violence. A reaction to this violence brought a new wave of medical shows, comedies and variety shows—"Marcus Welby M.D.," "Here's Lucy," and "Laugh-In." This latter show was a fast-paced, free-form series that featured many one-liners. It was the first highly edited program, but the editing was done by physically cutting the videotape, a very painstaking process that is no longer used.[30]

FIGURE 4.26

George Maharis in "Route 66." In this series, two adventurers, George and costar Martin Milner, tour the country (in an automobile produced by the show's sponsor). *(Courtesy of Columbia Pictures Television)*

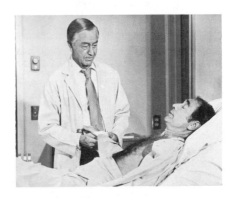

FIGURE 4.27

Robert Young as Dr. Marcus Welby. In this episode, guest star Gary Merrill plays a lawyer who refuses to step down from an important case after he has been told he has a terminal disease. *(From MCA Publishing)*

movies

During the 1960s old Hollywood movies established themselves on TV. In 1961 "Saturday Night at the Movies" began a prime-time movie trend that, by 1968, saw movies on all seven nights of the week. This rapidly depleted Hollywood's supply of old films, and some of the low-quality films that made it onto the airwaves enhanced the vast wasteland theory.

In 1966, NBC made a deal with Universal to provide movies that were specially made for television and would not appear in theaters first. This concept caught on, and by 1969, all three networks had **made-fors** on a regular basis.[31]

commercialism

The decade of the 1960s was a period of great prosperity for television. Despite objections from TV's critics, the commercialism of the medium grew. When Herbert Hoover, who as secretary of commerce had decried advertising in the 1920s, died in 1964, NBC broadcast a tribute that was followed by a beer commercial, a political commercial, and a cigarette commercial.[32]

Government Actions and Reactions

cigarette ads barred

During the 1970s government regulators and broadcasters played a cat-and-mouse game that often engendered hard feelings. Cigarette commercials disappeared in 1972, the result of a congressional bill that barred the advertising of cigarettes on the broadcast media. This action was motivated by the surgeon general's report that linked cigarettes and cancer.[33]

Two experiments of the 1970s involving the government were the **prime-time access rule** and the family hour. Prime-time access was promoted primarily by Westinghouse Broadcasting. Executives of this company pointed out that the networks monopolized prime time by programming between 7:00 and 11:00 P.M. and suggested that some of this time should be programmed by the stations themselves in order to meet community needs, as well as to allow more room for syndicated programs. In 1971 this idea was adopted by the FCC, which ruled that networks would be allowed to program only three hours a night in stations in the top fifty markets, leaving the other hour to the local stations. During this hour there were to be no old network programs or old movies; all programs had to be newly created. The rule was modified, however, so that in actuality, prime-time access became 7:30 to 8:00 P.M. Monday through Saturday. Although prime-time access was established to allow stations to broadcast significant local programming and independently produce programs of high quality, this did not happen to any significant degree. The stations filled the time with the cheapest thing they could find—game shows and remakes of old formats, although some developed local magazine shows.[34]

prime-time access

Family hour began in 1975 as an attempt to curtail sex and violence before 9:00 P.M., the time when children are usually in the audience. The family-hour concept originated with the code of the National Association of Broadcasters (NAB), an internal broadcasting organization. However, there were those who believed it came about through subtle pressure by the FCC and, hence, represented an abhorrent attempt by the commission to regulate program content. The family-hour idea was widely opposed by writers, producers, and directors, who took the concept to court. Here it was decided that the FCC had overstepped its powers and that the family-hour restrictions should be removed from the NAB code. However, most of the networks and stations continued to program that slot with material of a family nature.[35]

family hour

Another government action that engendered controversy was the **financial interest-domestic syndication** rule, often referred to as **fin-syn.** When the networks reacted to the quiz scandals by exerting a stronger hand over their programming, they often acquired a financial interest in the programs supplied to them by independent production companies. The networks would contribute money to the production of the program and when the program was syndicated to local TV stations, would reap some of this syndication profit. In 1970, the FCC adopted a rule barring networks from acquiring subsidiary interests in independently produced programs and from engaging in domestic syndication of these programs. This was done primarily because the FCC felt the networks were too powerful in the programming business and that financial interest and domestic syndication increased this power. The networks disagreed but during the 1970s accepted the ruling somewhat complacently.[36]

financial interest

The Equal Employment Opportunity Commission kept a watchful eye on broadcasters during the 1970s. The stations were still hiring and programming for blacks and Mexican-Americans. In addition, spurred on by the women's movement of the 1970s, they added female newscasters and stories about women in responsible positions.

A major dispute developed between broadcasting and the Nixon administration, arising primarily as a result of the media's disapproval of the administration's Vietnam policies. What particularly annoyed the administration was the **instant analysis** that network commentators launched after presidential speeches. After one particular November 1969 speech in which Nixon vowed to hold firm on Vietnam, some commentators expressed their displeasure and supported disengagement from the war. Shortly thereafter, Vice-President Spiro Agnew addressed a fund-raising dinner and blasted the media for its hostility toward the administration and its biased point of view.[37]

The next several years were ones of undeclared war between the media and the administration. Very little of a positive nature was reported to the citizenry concerning Nixon or his appointees. In return, administration sources accused the media news personnel of being eastern-establishment elitist liberals who were out of touch with the "silent majority" of Americans who, the administration claimed, supported Nixon and his policies.

The administration-media war came to an end with the Watergate hearings of the mid-1970s. The media expressed little sympathy for Nixon and other administration personnel who were accused of masterminding or covering up the break-in at the Democratic party offices in the Watergate complex. Eventually most of these accused people resigned, and the media could claim an undeclared victory.

Mergers, Acquisitions, and Start-Ups

The 1980s were not particularly good times financially for over-the-air commercial television. Stations and networks were faced with competition from cable TV and other newer media, which splintered the audience and the advertising dollar. Profits were down almost universally, and some stations even went bankrupt.[38]

This downturn in broadcasting in general, combined with a national phenomena of company takeovers, led to major changes in the management and ownership of all three networks and was instrumental in creating a fourth network.

The first network affected was ABC, which was purchased in 1985 for $3.5 billion by **Capital Cities Communications,** a company one-third the size of ABC. Capital Cities, known for its lean economic policies, did not hesitate long in pruning back the number of employees at ABC. Although the takeover was termed "friendly," it was the first time in history that a major network had been purchased, and it was referred to as "the minnow swallowing the whale."[39]

In less than a year, a second major network, NBC, was purchased as part of an overall takeover of its parent company RCA for $6.28 billion.[40] This acquisition brought with it an irony of history because the purchaser was General Electric, a company that had been involved in the formation of NBC and then eased out of ownership by RCA. This, again, was called a "friendly takeover." The GE employee who was put in charge of NBC was Robert Wright. He became the president and chief executive officer when Grant Tinker, who had run the network since 1981, resigned.[41] Tinker had replaced Fred Silverman, who had been dubbed "the man with the golden gut" because he had been largely responsible for putting CBS and then ABC at the top of the ratings. His magic did not work at NBC, however.[42] When Tinker took over NBC, he was able to lead the network from third place to first place.[43]

Things were not so "friendly" in the CBS executive suites. In 1985 Ted Turner of Turner Broadcasting attempted an unfriendly coup of the network that CBS was able to thwart only by buying back much of its own stock.[44] This greatly damaged CBS' financial footing, and it began laying off employees. At one point the chairman of the board, Thomas Wyman, tried to persuade Coca-Cola to buy CBS without notifying the other members of the board. This cost Wyman his job and brought back William Paley, who had retired from CBS in 1983. Paley formed an alliance with Lawrence Tisch, who had also been nibbling at the edges of CBS in what some thought was a takeover bid. Tisch was appointed chief operating officer and continued layoffs, which became particularly controversial within the news division. The division felt that its broadcast journalism tradition was being severely compromised.[45]

Even though the three major networks were suffering from losses of both money and morale, a brand-new network, **Fox Broadcasting Company (FBC),** appeared on the scene. This was promulgated primarily by Rupert Murdoch, an Australian media baron who had recently obtained U.S. citizenship. In 1985, Murdoch purchased 50 percent of Twentieth Century-Fox, which was owned primarily by oil magnate Marvin Davis. Davis and Murdoch then purchased six TV stations owned by Metromedia.[46] Shortly thereafter, Davis sold out his share of the businesses to Murdoch who proceeded to build a network using his six stations and other independent stations throughout the country.[47] The first show offered by the network was "The Late Show Starring Joan Rivers." This caused quite a stir because Rivers had been substitute hosting for Johnny Carson and left him for a show that directly competed with him. Several months later Rivers was "relieved" from the show by Fox. FBC built up the number of programs offered to the affiliated stations slowly but in a manner that was very similar to the three traditional networks. By the 1990s it had become quite successful in terms of both audience size and advertising money received.[48]

Many production companies and stations were also affected by mergers and acquisitions. Ted Turner, rebuffed by CBS, bought MGM and then turned around and sold all of it except the film library.[49] Sony bought Columbia Pictures,[50] and Matsushita bought Universal Studios.[51] Warner bought Lorimar and then merged with Time.[52]

This merger mania, which affected virtually all aspects of American business, was particularly visible within the media area. The media, by their very nature, were able to give their trials and tribulations a great deal of publicity.[53]

Deregulation

station rules

During the 1980s, the government attitude toward the industry changed. Over-the-air TV, like radio, felt the impact of the deregulatory mood. TV station licenses were lengthened from three years to five, and TV stations in small communities no longer needed to keep logs or ascertain community needs. In addition, restrictions governing the amount of commercial time stations could program were lifted.[54] The number of TV stations one entity could own was increased from seven to twelve, provided the twelve stations were not in markets that collectively contained more than 25 percent of the nation's TV homes. UHF stations were assessed for just half the homes in their reach, another attempt to give them parity with VHF stations. Group broadcasters that were more than half owned by minorities could own up to fourteen stations and reach 30 percent of TV homes.[55]

low-power TV

The FCC also authorized a new type of over-the-air TV called **low-power TV (LPTV)**. Beginning in 1980, the FCC started accepting applications for ten-watt VHF and one thousand-watt UHF stations that can be put on the air for as little as $50,000. These stations are sandwiched between regularly operating stations and cover only a radius of twelve to fifteen miles. For example, if a city has a channel 4 and a channel 6, a low-power station could operate on channel 5 provided it does not interfere with either of the full-power stations.

The FCC's main purpose for establishing LPTV was to enable groups not involved in TV ownership to have a voice in the community. The FCC hoped that women and minorities, in particular, would apply. Many did, but many large companies also applied. In fact, the FCC received more than five thousand applications by April 1981, when it decided to impose a freeze on new applications so it could sort out the old ones.[56] Eventually the FCC began giving out allocations, first to rural areas and then to cities.[57]

Some of the people who received authorization to build low-power stations have never activated their stations or have sold their permits to other groups. Several hundred are on the air, however. Some program movies, music videos, or religious programs. Others have been able to sell advertising and make modest profits from local programs, such as high school band concerts, farm reports, and information on hunting season.[58]

TV stereo

The FCC authorized TV stereo broadcasting in 1984, which caught on quickly. Although the cost to stations to convert from mono to stereo ran between $50,000 and $100,000, more than sixty stations did so within the first year and many more followed. Gradually networks and stations produced more and more programs in stereo, and more people purchased or installed stereo-receiving systems with their TVs. Stereo TV was a technological change that

Electronic Media Forms

FIGURE 4.28
The Huxtable family
of "The Cosby Show."
Clockwise from center:
Cliff (Bill Cosby),
Rudy (Keshia Knight
Pulliam), Clair
(Phylicia Rashad),
Theo (Malcolm-Jamal
Warner), Vanessa
(Tempest Bledsoe),
Sondra (Sabrina
LaBeauf), and Denise
(Lisa Bonet). *(Photo/
Viacom, Hlwd.)*

happened without the trauma associated with other changes, such as color TV and UHF.[59]

The financial interest-domestic syndication rule of the 1970s came under heavy fire during the 1980s and 1990s. The networks wanted the rules rescinded. They claimed they were no longer a power threat because of all the new media that were producing and programming material. With their shrinking profits, they wanted to be able to benefit from the programming investments. The independent production companies that supplied programs felt differently. They often went into debt producing the material for the networks and felt they were the ones who should maintain financial control and profit from the money earned in syndication.

In 1991, the FCC issued new fin-syn rules that allowed the networks to acquire some financial and syndication rights. Attempted as a compromise, the new ruling succeeded, instead, in irritating both sides. Complaints and appeals have followed, and the issue is still not satisfactorily resolved.[60]

financial interest

Innovations in Programming

Most of the programming forms of TV (dramas, situation comedies, audience participation shows, news, children's programs) were set early in its history. For the most part they were carried over from radio, which is where the roots of television were formed. When TV began forming alliances with the film industry, movies were added to the staple of programming forms.

Through the years, the forms of programming remained fairly stable, and in fact the ratings leader during much of the 1980s and 1990s was the sitcom "The Cosby Show." But some particular programs of the 1970s, 1980s, and 1990s represented changes in terms of either content or form.

For example, programming ventured into subject matter that had previously been taboo. In the 1970s, "All in the Family" featured Archie Bunker, an obviously bigoted lead, who enabled the series story lines to deal with areas

tried and true forms

taboo subject matter

FIGURE 4.29
The cast of "All in the Family" in their Queens home. Top row left to right: Rob Reiner as Mike Stivic, Sally Struthers as Gloria Stivic, and Mike Evans as Lionel Jefferson. Bottom row left to right: Jean Stapleton as Edith Bunker and Carroll O'Connor as Archie Bunker. *(Photo/Viacom, Hlwd.)*

of politics, ethics, and sex that were previously untapped. Made-for-TV movies dealt with homosexuality, incest, wife beating, and other controversially sensitive subjects.

Prime-time soap operas, which rose in 1978 with "Dallas" and reached their prime in the 1980s, also dealt with subjects that, in earlier years, would have been considered too steamy for TV. By the 1990s, so many previously taboo subjects were being discussed that the term **tabloid TV** was being used to describe talk shows and other programs that delved into the most private of subjects.

older demographics

An aging population led the networks to begin paying more attention to older demographics. Programs such as "The Golden Girls" (four older divorced or widowed women who share a house in Florida) and "Murder, She Wrote" (a mystery writer in her fifties who solves crimes) became quite popular.

spin-offs

Spin-offs also became a popular form of programming. Supporting characters who were well accepted in highly-rated shows were used as major characters in new shows. For example, two zany women who made occasional appearances on "Happy Days" were spun off into the series "Laverne and Shirley." In the same way "The Mary Tyler Moore Show" begat "Rhoda" and "Phyllis," and the nighttime soap "Dynasty" spun off "The Colbys."

anti-heroes

Programs starring anti-heroes also came into vogue. Certainly the scheming scoundrel J. R. Ewing of "Dallas" was one. The comedy series "Roseanne" revolved around a group of unattractive people who would probably not have been heroes and heroines in previous decades. Two very popular

Commercial Television

Members of the cast of
the "Mary Tyler
Moore Show."
Clockwise from bottom
left: Betty White as
Sue Ann Nivens,
Gavin MacLeod as
Murray Slaughter, Lou
Asner as Lou Grant,
Ted Knight as Ted
Baxter, Georgia Engel
as Georgette Baxter,
and Mary Tyler Moore
as Mary Richards.
*(© Zimmerman/F.P.G.
International Corp.)*

FIGURE 4.33

The cast of
"Roseanne"—from left
to right: Laurie
Metcalf, Lecy
Goranson, Sara
Gilbert, Roseanne
Barr, Michael
Fishman, and John
Goodman. Roseanne, a
working blue-collar
mother, uses an
irreverent brand of
humor to cope with her
tumultuous family.
(Everett Collection)

FIGURE 4.34
One of the Fox network's first series, "Married. . . with Children." Here Peggy (Katey Segal) and Al (Ed O'Neill) postpone going to a favorite restaurant so Al can assemble one of Bud's (David Faustine) toy cars. *(Courtesy of Fox Broadcasting Company)*

series from Fox centered around a couple who constantly criticized each other and the institution of marriage ("Married . . . With Children") and a cartoon family of self-proclaimed losers ("The Simpsons").[61]

A whole new form of programming, **home shopping**, was able to arise because the FCC lifted restrictions on how much advertising a station could program. Home shopping programs were essentially total advertisements in that they showed and extolled the virtues of products that people could buy over the telephone. This type of programming would have been impossible before the deregulation policies of the 1980s.[62]

home shopping

Some stations experimented with **teletext,** a radically different form of programming. Teletext is a method by which conventional TV stations can display words and graphics on a consumer's TV screen by transmitting the information on an unused portion of the TV signal. This part is called the **vertical blanking interval** and occurs when the video portion of the TV signal is momentarily stopped so that a new frame can be started. A regular TV set equipped with a special decoder and keypad can display the information programmed on the vertical blanking interval either on top of the regular picture, or it can replace the picture altogether.

teletext

The main material that has been programmed for teletext includes information such as news, weather, the stock market, airplane schedules, and theater listings. A number of stations have programmed teletext rather extensively, particularly KSL in Salt Lake City and WKRC-TV in Cincinnati. NBC and CBS have also experimented with it. However, the decoders needed to view the material have not sold well, and teletext does not seem to be a form of programming that a great many people are interested in receiving.[63]

FIGURE 4.35
The Simpsons. This cartoon-family of self-proclaimed losers gained great popularity in its prime-time spot on the Fox network. *(AP/Wide World Photos)*

FIGURE 4.36
An example of teletext. *(Courtesy of KSL, Salt Lake City)*

satellites

Another technological breakthrough, satellites, has enabled major changes in programming concepts. Satellite delivery during the past few decades has been particularly important in the news area. Some of the events shown live from around the world during the 1980s and 1990s included the wedding of Britain's Prince Charles and Lady Diana, the Tiananmen Square uprisings in China, the dismantling of the Berlin Wall, and the occurrences of the Persian Gulf War. Satellite transmission was also crucial for the 1985

"Live Aid" concert to help the starving in Africa. Described as a global juke box, this program originated primarily in Philadelphia and London but also in cities in eight other countries. The world's top musicians performed via satellite for a worldwide TV audience of 2 billion.[64]

Commercial television has undergone many structural and programming changes during its forty-year life, but it remains a dominant pastime for the American public.

Commercial TV Chronology

1884	Nipkow invents mechanical scanning
1927	Farnsworth transmits electronic pictures
1935	Sarnoff pledges millions to RCA's development of TV
1939	TV displayed at world's fair
1941	FCC authorizes black-and-white 525-line TV
1942	TV development halted because of World War II
1948	Many facets of TV grow 4000 percent in one year
1948–1952	TV "freeze"
1950	FCC accepts CBS's color system
1951	"I Love Lucy" is filmed
1953	FCC adopts RCA's color TV system
1953	ABC and United Paramount Theaters merge
1954	Televising of McCarthy-Army hearings
1954	Number of TV stations one company can own revised
1956	Videotape introduced
1957	FCC proposes deintermixture to help UHF
1959	Charles Van Doren confesses to the quiz scandals
1960	"Great Debates" between Kennedy and Nixon
1961	Newton Minow makes "vast wasteland" speech
1962	Congress passes all-channel receiver bill
1963	Networks nightly newscasts increased to thirty minutes
1963	Telecasting of John F. Kennedy funeral
1966	NBC and Universal sign an agreement for made-for-TV movies
1968	Cronkite travels to Vietnam
1968	Youths protest at Democratic convention
1970	FCC adopts financial interest-domestic syndication rule
1971	Prime-time access rule adopted
1972	Cigarette commercials taken off radio and TV
1975	Family-hour concept begun
1980	FCC authorizes low-power TV
1981	Grant Tinker replaces Fred Silverman at NBC
1983	Paley retires from CBS
1984	TV deregulated
1984	TV stereo authorized
1985	Capital Cities Communications purchases ABC
1985	"Live Aid" broadcast

1986 GE takes over RCA
1986 Paley returns to CBS and joins forces with Lawrence Tisch
1986 Fox Broadcasting Company begins programming
1991 FCC revises financial interest-domestic syndication rules

Conclusion

Commercial television, born into a fast-paced society, has been forced into the role of an early bloomer. Within sixty years it has progressed from a sixty-scan-line Felix the Cat to instantaneous worldwide communication.

In some ways television technology seems to prove that necessity is the mother of invention, but in other ways it seems to suggest that invention is the mother of necessity. Because the original mechanical scanning techniques might never have been adequate for a popular medium, the electronic techniques were indeed necessary. On the other hand, matters such as lines of resolution and color were hotly debated by various industry groups, but only after the technology was approved did the need for its existence become evident. The invention of the videotape recorder altered TV production techniques in ways not envisioned. Similarly, portable equipment developed before its full range of possibilities, particularly in terms of newsgathering, was explored.

Television's greatest boosters could not have predicted how quickly and thoroughly TV would be accepted by the American public. The medium's growth in terms of sets, programs, and advertising during the late 1940s and early 1950s is a phenomenon unto itself. Performers, such as Milton Berle and Lucille Ball, latched onto early TV and became instant celebrities. More reluctant "TV stars," such as Senator Joseph McCarthy, found that TV could also create notoriety.

The elements of government that had dealt with radio suddenly found a new medium featuring both sight and sound upon their doorstep. Taking a cue from radio, the FCC imposed a freeze to work out technical allocations. The choice of UHF to resolve the channel shortage was not the most fortuitous decision, and it is one that still engenders government patchwork. Other controversial government actions included eliminating cigarette ads, instituting prime-time access, influencing the family-hour concept, imposing the financial interest rule, and deregulating.

Private business has also been predominant in TV from the early founding of networks to the recent rash of takeovers.

Different forms of programming have surfaced over the years. The early forms, taken from radio, were enhanced by movies as the TV industry and Hollywood developed closer ties. The subject matter of programs has become much more liberal over the years.

Successful programs, such as "Texaco Star Theater," "I Love Lucy," "See It Now," "Marty," "Tonight," "Gunsmoke," "All in the Family," and "The Cosby Show" are dependent upon group efforts. Many individuals have also made significant contributions, among them: David Sarnoff, who laid the groundwork and financial base for the original development of TV; and Pat Weaver, who developed the spectacular and the magazine concept of advertising.

Television has survived dark moments, such as blacklisting and the quiz scandals, and has celebrated bright moments, such as telecasting the dismantling of the Berlin Wall. This medium has been accused of inciting riot conditions during the 1968 Democratic convention and praised for unifying a nation during the Kennedy assassination. Its future will undoubtedly change, but commercial television—which has taken the country by storm in a historically short period of time—seems bound to continue.

Thought Questions

1. Should the FCC have waited longer to approve the beginning of commercial television so that the technical quality would be better than it is?
2. What, if anything, could have been done to help TV personnel who were blacklisted?
3. Could a phenomenon such as the quiz scandals reoccur in TV today? Why or why not?
4. What do you think will be the role of over-the-air commercial TV networks in ten years?

Public Broadcasting

Introduction

The public radio and television system of the United States has been struggling beside its more glamorous commercial cousin—floundering and fluctuating, arguing and achieving, staggering and starring. Its history is marked by victory and defeat, and its present is replete with insecurity, but the system has nonetheless emerged as a viable aspect of American broadcasting.

CHAPTER 5

I think public television should be the visual counterpart of the literary essay, should arouse our dreams, satisfy our hunger for beauty, take us on journeys, enable us to participate in events, present great drama and music, explore the sea and sky and the woods and hills. It should be our Lyceum, our Chatauqua, our Minsky's, and our Camelot. It should restate and clarify the social dilemma and the political pickle.

E. B. White
writer and philosopher

Early Educational Radio

The roots of public broadcasting go back to radio and TV operations that were originally referred to as **noncommercial broadcasting** or **educational broadcasting.** In fact, broadcasting was born noncommercial, and many of the early stations were started by educational institutions. For example, as early as 1917, Professor Earle M. Terry of the University of Wisconsin transmitted voice and music to listeners in farm areas around Madison. The professor and his students experimented with various forms of music and particularly favored Hawaiian music because its twang carried well. During World War I the station broadcast weather and crop information and was the only radio station not under Navy control.[1]

early experiments

The early 1920s witnessed a radio rush by colleges and universities around the country. Seventy-four such institutions were broadcasting by the end of 1922. Colleges used radio stations primarily to aid extension activities, raise funds, and offer college credit courses that people could listen to in their homes. The main problem facing these stations was the same one that plagued other early stations—they all had to broadcast on the same frequency of **360 meters.** As more and more stations added their signals to the airwaves, home-study students often could not hear their lessons because of interference. As a result, faculty members became disillusioned with the effectiveness of radio, and many colleges ceased their broadcasts. In 1925 thirty-seven educational stations left the air and only twenty-five new ones began broadcasting.

radio rush of the 1920s

As commercial stations became more firmly entrenched, they overpowered the educational stations both in wattage and dollars. During the 1920s radio stations generally shared time with one another, so it was not uncommon for an educational and commercial station to alternate hours. If the commercial station decided it wanted a larger share of the time, it would petition the **Federal Radio Commission.** Both the commercial and the educational station were required to go to Washington for the appeal. This was an expensive and time-consuming process that the educational station could not easily afford. Usually the educational stations found themselves coming away with the short end of the stick on disputes involving time, power, and position to the extent that they were unable to use their broadcasting facilities effectively for any type of continuing programming.[2]

commercial station takeover

After 1925 the secretary of commerce strongly urged people who wished to enter broadcasting to buy an existing station rather than add one to the already overcrowded airwaves. As a result, many educational facilities were propositioned by commercial ventures interested in buying the stations. Frequently the institutions succumbed to the pressure because the financial drain of the stations outweighed the dwindling public service value.

The result was a downward spiral for educational radio. In 1928 thirty-three stations gave up, followed by thirteen more the following year.[3]

reservation threats

In 1929 the National Committee on Education by Radio was organized. It suggested that 15 percent of all radio stations be reserved for noncommercial educational use. The group tried to establish this reservation policy in the

Communication Act of 1934 but was unsuccessful, largely because commercial interests so touted their own cultural contribution as to negate the need for strictly educational stations. Despite defeat, educational organizations kept this issue alive while the number of educational stations continued to dwindle to approximately thirty.[4]

FM Educational Radio

With the advent of **FM** broadcasting, educators, through perseverance, began to taste victory. In 1945 the FCC reserved the twenty FM channels between 88.1 and 91.9 exclusively for noncommercial radio. These channels at the lower end of the FM band are reserved for educational use everywhere across the country, so any station broadcasting below 91.9 on the FM band is noncommercial.

By the end of 1945 there were six FM educational stations on the air.[5] This number grew to forty-eight by 1950 and passed fourteen hundred by 1990.[6] Part of this FM increase was due to a 1948 ruling authorizing low-powered, ten-watt educational FM stations that generally reached only a two- to five-mile radius and were easily and inexpensively installed and operated. For a short time, approximately half the FM educational stations were of the ten-watt variety, generally operated by educational institutions as training grounds for students.

In fact, the majority of applications for FM noncommercial radio stations were from educational institutions—colleges, universities, high schools, public school systems, and boards of education. Others were from state or municipal authorities or religious groups, and a few were owned by nonprofit community groups such as the Pacifica Foundation.[7]

As the number of educational radio stations increased, the need for some form of network grew. **The National Association of Educational Broadcasters (NAEB),** organized in 1934 as a spokesgroup for the educational radio field, formed the first workable duplication and distribution operation. This so-called **bicycle** network, begun in 1949, sent programs from one station to another by mail on a scheduled round-robin basis.[8]

The Advent of NPR

In 1967 a Public Broadcasting Act, which formed the Corporation for Public Broadcasting (CPB), was passed by Congress primarily to serve the needs of public television, but radio was also included. At that time, the NAEB ceded its radio networking duties to **National Public Radio (NPR),** an arm of CPB that began operations in 1970.

NPR was formed to upgrade the quality of public radio by providing excellent programming. In order to help with the upgrade, NPR set up certain qualifications that stations had to meet in order to become affiliates. For example, they had to have at least five full-time paid employees, and they had to program at least eighteen hours a day every day of the year. More than three hundred stations have become affiliated.[9]

programs

Under NPR's original organization, qualifying stations wanting to receive programming from NPR paid a fee of several thousand dollars per year and received about seventy hours of programming per week. This included news, senate hearings, concerts, dramas, self-help material, documentaries, and programming from foreign countries, particularly the British Broadcasting Corporation and the Canadian Broadcasting Corporation. Probably the most popular of the early programs was an in-depth evening news program called "All Things Considered" that dealt with issues, events, and people. Later two similar programs, "Morning Edition" and "Weekend Edition" were offered. The latter was hosted by Susan Stamberg, who had been very instrumental in the development of "All Things Considered."[10]

The fees from stations did not cover the costs of programming. About half of the budget came directly from the Corporation for Public Broadcasting, and other monies came from companies that underwrote programming.[11]

local programs

When stations were not airing NPR material, they programmed locally, primarily with classical music. Some stations also aired jazz, rock, swing, or other forms of music. Many local stations also produced their own dramas, children's programs, sports, news, and health programs.

distribution

Originally NPR material was sent to the stations over phone wires from the Washington headquarters. Although some of the programming distributed nationally was produced by local stations, most of it was produced in Washington, partly because the phone lines led from Washington to the local stations. Later NPR distributed by satellite, but most of the programming was still Washington based.[12]

unbundling

More recently, NPR has **unbundled** its programming so that stations can air certain blocks of programming (e.g., morning news only) without having to pay for the rest of the programs.[13]

Electronic Media Forms

FIGURE 5.2
The NPR satellite center. *(Courtesy of National Public Radio)*

The Formation of APR

Some of the structural elements of NPR were not popular with many of the stations. For one thing, NPR received most of its money from the federal government through CPB, and government cutbacks led to cutbacks in NPR programming, particularly in the cultural and concert areas. In addition, most of the programming was produced in Washington, and local stations felt their creative programs deserved wider dissemination.

reasons

As a result, a group of public stations joined together in 1982 to form **American Public Radio (APR),** an independent, nonprofit, private network headquartered in St. Paul, Minnesota. Its qualifications for affiliates are similar to those of NPR. The purpose of the network is to acquire radio programming from various local stations and disseminate it to other stations. Part of what made this possible is satellites that can pick up programs from any station and beam them to any other station.[14]

structure

American Public Radio does not receive direct federal funding. It is supported by fees paid by its members and based, to some degree, on the size of their market. These fees support the national office and the satellite distribution system. Programs are donated by many of the local stations. Usually these have been funded locally by underwriting or contributions from listeners, or they have been funded by a special program fund administered by APR that is designed to stimulate development of new ideas.

finances

The most popular personality on APR has been Garrison Keillor, first with his program "A Prairie Home Companion" and later with "American Company of the Air." Keillor's programs consist of a combination of music, interviews, and homespun philosophy.[15]

programs

In general, APR programs more cultural material and classical music than NPR. However, both services are branching out in terms of programming in order to try to reach larger audiences.[16] Recently, a number of programs

FIGURE 5.3

A feature called "Buster The Show Dog" was a regular part of "A Prairie Home Companion." Host Garrison Keillor (*left*) performs with Kate MacKenzie, Dan Rowles, Stevie Beck, and Tom Keith.
(Courtesy of American Public Radio/Rob Levine, photographer)

have switched from APR to NPR and vice versa, trying to gain better airtime, a better economic situation, or other benefits.[17]

More than three hundred stations have affiliated with APR, but most of those are also affiliated with NPR. This dual affiliation is an arrangement that is possible within public broadcasting but not commercial broadcasting.[18]

Nonaffiliated Stations

That leaves close to a thousand stations that are not affiliated with either NPR or APR. Most of these are small stations operated by colleges to train students in broadcasting. Many are not interested in network programs because they want to produce their own programming for training purposes. In general, they do not qualify for NPR affiliation because they do not have five full-time paid employees, and they do not broadcast every day of the year.

At first most of these stations were ten-watts. However, during the mid-1970s, NPR instituted a campaign to increase its number of affiliates by trying to get these stations to upgrade in power, hours, and personnel. One of the results was that in 1978, the FCC adopted new policies for the **ten-watters** stipulating that they upgrade to at least one hundred watts (Class A) or switch to a frequency somewhere in the commercial band or to a newly created frequency of 87.9, which was not on most radios. Also, all public radio stations were required to be on the air on a full-time basis or be subject to sharing their frequency with other interested entities.

Most of the stations opted to increase power; very few increased hours of operation, mainly because they did not have requests to share the frequency; and most remained uninterested in national network affiliation. NPR lost its interest in most of these stations, and the whole issue of ten-watt station upgrade simmered down or was treated with benign neglect.[19]

ten-watters

present state

Financial Stress

None of the public radio entities are rich; all are constantly scraping for money. The 1980s were particularly difficult because the government cut back support for public broadcasting. Educational institutions also had their own financial problems that led to decreased support for the stations they owned.

For a short time NPR teetered on the brink of bankruptcy and was saved only by a 1983 loan from the Corporation for Public Broadcasting.[20] In 1985 NPR voted to rearrange its financial structure so that federal funds would go directly to the stations instead of NPR. As a result of this reorganization, stations pay more dues to NPR.[21]

Public radio stations have had to rely increasingly on support from listeners and from companies who were willing to underwrite or supply grant money. Both stations and networks have undertaken some commercial fund-raising ventures, such as leasing satellite time to other users, establishing a paging system, and transmitting digital data. Aside from leasing satellite time, most of these commercial ventures have been unsuccessful.[22]

However, public radio continues to survive and provide excellent programming. Those involved with it find it worthwhile in terms of recognition, prestige, and service. One report has indicated that 13 percent of the adult population is aware of public radio, and 4 percent listens regularly. This listening audience is skewed toward people with high incomes and extensive education.[23]

Overall, the development of public radio has been slow but steady. Conflicts arise because of its diverse purposes and because it does not receive the prestige given public television, but it does make a substantial contribution to the media structure.

Early Educational Television

Early television was dominated by commercial interests with a little activity from educational institutions. Several universities offered college credit courses over local commercial channels, and in 1950 Iowa State University built a TV station that it ran as a partially commercial venture.[24]

Educators, while recognizing the value of television, also realized that they would face a losing battle akin to the early radio experience if they attempted to compete with commercial companies for stations. As a result, the same groups that fought for the allocation of noncommercial FM radio stations attempted to convince the FCC that there should be **channel reservations** for noncommercial television stations. These groups formed an organization called the **Joint Committee for Educational Television (JCET),** hired a lawyer, and raised money to lobby for their cause.

JCET was aided in its cause by FCC commissioner Freida **Hennock,** the first woman appointed to the FCC, who did not hesitate to say that she favored educational stations. When the freeze was lifted in 1952, the FCC had authorized the reservation of 242 noncommercial channels—80 VHF and 162

FIGURE 5.4
People being taken on
a tour of KUHT,
Houston, shortly after
it went on the air.
*(Courtesy of KUHT,
Houston)*

UHF. Unlike FM, these channels were scattered around the dial, as channels were reserved in different cities. For example, channel 13 was reserved in Pittsburgh, channel 11 was reserved in Chicago, and channel 28 was reserved in Los Angeles.[25]

Obtaining the channels and activating the stations were two different matters, mainly because of the huge sums of money that television stations demand. Fortunately for educational television, the **Ford Foundation** became interested in its cause and provided much of the money for the early facilities and programs. In 1953 the nation's first educational TV license, KUHT, was granted to the University of Houston in Texas. By 1955 there were nine stations on the air and by 1960, forty-four stations.[26]

The Ford Foundation also became involved in programming by helping to establish the National Educational Television and Radio Center in Ann Arbor, Michigan, which acted as a distribution bicycle network center. This helped stations acquire enough programming to fill the hours, but because tapes were mailed to stations, nothing of a timely nature could be exchanged. To compound this problem, the early programs were reproduced on kinescopes, which gave them a very grainy quality. Eventually this center in Ann Arbor dropped some of its functions (including radio), changed its name to **National Educational Television (NET),** started producing programs, moved to New York, and allied itself with the New York public TV station for production.

The Ford Foundation also funded general programming concepts, such as **Chicago TV College,** a fully accredited set of televised courses that enabled students to earn two-year college degrees through at-home viewing or a combination of at-home viewing and on-campus class attendance.[27] Another Ford-funded experiment of the late 1950s was the **Midwest Program on Airborne Television Instruction (MPATI),** by which programs for schoolchildren were broadcast onto two UHF channels by an airplane that circled two states.[28]

Despite its largess, the Ford Foundation could not underwrite all of educational television, so educators turned to the government for the additional funds to build stations. In 1962, after a year of debate, Congress passed the

Ford Foundation

Chicago TV College and
MPATI

Educational Broadcasting Facilities Act, which authorized $32 million for five years. These monies were to be made available on a matching-fund basis to states to assist in the construction of educational television facilities. At this point, government funds could be used only for facilities and not programs— a satisfactory arrangement to the educators, who were leery of strings that the government might attach to programming money. This act led to an increased interest by many groups in establishing educational TV stations. Because of the many requests, the FCC in 1966 revised its assignment table upward and set aside 604 channels in 559 communities. By the end of 1965, the number of educational stations on the air had doubled from the 1960 number of forty-four.[29]

Educational Broadcasting
Facilities Act

station ownership

The ownership of these educational television stations was somewhat akin to radio station ownership in that educational institutions owned many stations. However, many more TV stations than radio stations were owned by states and community groups. State ownership became common, particularly in the South. Alabama, the first state to own and operate TV stations, built nine stations that cover the state in a mini-network. About one-fourth of educational television stations were licensed to nonprofit community corporations. These corporations were usually formed by community leaders and supported by the communities they served.[30]

early programs

Educational TV was not without its growing pains. In many cities, the channel allocation for the educational television station was in the UHF band, making it difficult for the station to gain either audience or community financial support. What is more, educational programming was, in a word, dull. Most of it was produced on a shoestring budget and consisted of local free talent discussing issues or information. As a result, educational television became known for its talking heads and did not attract large audiences, even in VHF cities. The Ford Foundation, feeling that it alone was supplying too much of the support to educational television, began withdrawing some of its financing. Innumerable educational broadcasting organizations and councils appeared and disappeared, some because of financial problems and some because of lack of clear focus or because of political infighting. The result was a system so loosely organized that it impeded the impact of the medium.

Carnegie Commission

The Carnegie Foundation therefore set up the **Carnegie Commission** on Educational Television. This group of highly respected citizens spent the better part of two years studying the technical, organizational, financial, and programming aspects of educational television. In 1967 it published its report, "Public Television: A Program for Action."

The Carnegie Commission changed the term "educational television" to "public television" to overcome the pedantic image the stations had acquired. It also recommended that "a well-financed and well-directed system, substantially larger and far more pervasive and effective than that which now exists in the United States, be brought into full being if the full needs of the American public are to be served."[31]

The Public Broadcasting Act of 1967

Congressional changes

Most of the Carnegie Commission's many recommendations were incorporated into the **Public Broadcasting Act of 1967.** However, there were two major changes between the Carnegie recommendation and congressional passage of the act. One was that radio was added to the concept. The Carnegie Commission had occupied itself only with TV, but the radio interests that led to NPR were included by Congress. The Carnegie group had recommended that public broadcasting be given permanent funding, perhaps through a tax on TV sets, but Congress opted for one year's funding of $9 million with additional funding to be voted on at a later time.

The Public Broadcasting Act of 1967 provided for the establishment and funding of the **Corporation for Public Broadcasting (CPB)** to supply national leadership for public broadcasting and to make sure that it would have maximum protection from outside interference and control.

CPB

CPB was to have a board of fifteen lay members (later changed to ten) appointed by the president with consent of the Senate. This corporation was in no way to be considered an agency or establishment of the U.S. government. The main duties of the CPB were to help new stations get on the air, to obtain grants from federal and private sources, to provide grants to stations for programming, and to establish an interconnection system for public broadcasting stations.[32]

The Corporation for Public Broadcasting was specifically forbidden from owning or operating the interconnection system. Therefore, the corporation created the **Public Broadcasting Service (PBS),** an agency to schedule, promote, and distribute programming over a wired network interconnection (the use of wires disappeared in 1978 when PBS became the first network to distribute programming totally by satellite). PBS also had a governing board consisting of station executives. This service was not to produce programs itself, but rather was to obtain them from such sources as public TV stations, production companies, and foreign countries. Hence a three-tier operation was established: (1) The stations produced the programs; (2) PBS scheduled and distributed the programs; and (3) CPB provided funds and guidance for the activities.[33]

PBS

Setting Up a Programming Process

The advent of the Corporation for Public Broadcasting and its accompanying funding allowed public television to embark upon innovative programming of high quality. The first series to arouse interest in virtually every public TV station was the successful 1970 children's series "Sesame Street," produced by a newly created and newly funded organization, the **Children's Television Workshop (CTW).** This series helped strengthen PBS as a network because it was in demand throughout the country. In the same year, the public television drama "The Andersonville Trial" won the Emmy for Best Program of the Year.

"Sesame Street"

Other early PBS series that met with sustained popularity were "The French Chef," Julia Child's cooking show produced in Boston; "Mister Rogers' Neighborhood," a children's program produced in Pittsburgh; "Black Journal," a public affairs series dealing with news and issues of importance to blacks produced in New York; and "Civilisation," a British import on the development of western culture.[34]

other programs

In 1974 PBS established a **Station Program Cooperative (SPC),** a unique method by which many of the PBS programs were selected until the 1990s, when a more centralized form of program decision-making took hold. With the SPC, PBS conducted an audience-research survey to determine national program needs; then program proposals were solicited from stations and production companies that theoretically met the needs determined by the survey.

Station Program Cooperative

FIGURE 5.7

Bert (*right*) and Ernie,
two regulars on
"Sesame Street."
*(Courtesy of © Children's
Television Workshop)*

FIGURE 5.8

Fred Rogers (*right*)
with Mr. McFeely, one
of the many characters
who visited regularly
on "Mister Rogers'
Neighborhood."
*(Courtesy of Family
Communications, Inc.)*

These proposals were catalogued and sent to the various stations, who began voting on those programs they would like to carry. The first few rounds of voting allowed stations to express their interest in certain programs, but did not bind the stations to purchase the program. After a number of proposals had been eliminated for lack of interest, actual purchase rounds were conducted and stations were required to help pay the production costs of those programs they wished to air. As a result of this process, each program produced for PBS under the Station Program Cooperative had a slightly different group of stations paying for and airing it. In some instances, the PBS stations paying for a particular program did not all air it at the same time. Unlike commercial network affiliates, stations were free to set their own schedules. Often the money that stations contributed to production costs came from the CPB community service grants by following this route: CPB to local stations to PBS to producing stations.[35]

FIGURE 5.9

"The French Chef,"
Julia Child about to
prepare an unusual
dish. *(Courtesy of WGBH
Educational Foundation/Paul
Child, photographer)*

Only about half of all PBS programs came through the station program cooperative. Others were (and still are) underwritten by foundations, corporations, or the government, and some are purchased from foreign governments. In fact, there have been those who claim that PBS stands for "primarily British shows" because of the large amount of fare produced by the BBC. Others feel PBS stands for "Petroleum Broadcasting Service" because of the large number of shows underwritten by oil companies.

other program methods

Government Conflict

Although the organization for public broadcasting set up by the Public Broadcasting Act of 1967 looked good on paper, it had problems in actual operation. A major controversy surfaced during the Nixon administration when the CPB, which had been formed to insulate public broadcasting from government, became somewhat of an arm of the government. The conflict centered around programs that were critical of the government and around the concept of localism.

The programs that caused ire were primarily nationally aired documentaries and public affairs programs, such as "Washington Week in Review," which were against Nixon and his policies. As more and more of these programs aired, members of the administration began stating that public television should emphasize local programs, not national ones. Many within public television felt that this localism policy was espoused because the administration felt local programming would not be as influential as national programming and, therefore, any criticisms of the administration that did creep in would not become significant issues. The administration's rebuttal stated that since the three commercial networks were national in scope, there was no need for a fourth network, but there was need for local programming.

Nixon furor

localism

The administration carried its localism philosophy into the budgeting area. After Congress passed a bill that would have given public broadcasting $64 million for 1972–73 and $90 million for 1973–74, Richard Nixon vetoed the bill on the grounds that the national CPB organization exerted too much control over the local stations.

With this action the chairperson and president of CPB resigned, and Nixon appointed new people to the posts who were more in line with his way of thinking. With CPB now more in tune with the administration, a schism developed between CPB and PBS over who should control the programming decisions for public broadcasting. PBS wanted to continue news and documentary programs having a national emphasis. CPB pressed for local control and an emphasis on cultural programming rather than on documentaries.

PBS board reorganization

The controversy continued but gradually abated as Watergate consumed the administration and as new political forces came into power. In addition, both PBS and CPB reorganized their boards slightly so that both laypeople and public broadcasting station people were on both boards. This improved communication between the two groups. Monitoring committees were also set up to resolve disputes whenever PBS or CPB felt that particular programs or series were not balanced or objective.[36] In addition, public TV licensees formed a trade organization, the National Association of Public Television Stations, to do some of their lobbying and planning.[37]

All of this helped settle down the CPB-PBS government conflict, but it did not thoroughly contain it. Occasional resignations and unpleasant words still fill the air from time to time.[38]

Carnegie II

Because of the constant unrest in the public broadcasting arena, the Carnegie Commission decided to review the whole subject. This review, called **Carnegie II,** started in 1977 and culminated in 1979 with a report entitled "A Public Trust: The Landmark Report of the Carnegie Commission on the Future of Public Broadcasting."

provisions

In this report the commission recommended abolishing CPB, which, despite the best of intentions, had not been insulated from political pressure. The commission also recommended that funding be more than doubled for public broadcasting and that a spectrum fee be assessed to help raise this additional funding. This spectrum fee would be levied on users of the electronic spectrum—commercial broadcasters, CBers, AT&T, doctors' call services, and the like.[39]

lack of implementation

However, this time the Carnegie recommendations did not go into wholesale effect. They came at a time when financial cutbacks were the order of the day and when a recommendation proposing double spending was not well received.[40]

The structure of public broadcasting remained essentially as it was in the 1960s and 1970s. CPB's influence was cut slightly in that decisions on specific grants were usually made by independent panels made up of program producers.[41]

Financial Problems

Like public radio, public TV underwent difficult financial times during the 1980s. Financing public television is much more complicated than financing public radio, however, because of the greater expense.

sources of funds

During the late 1970s public TV was supported as follows: 70 percent federal, state, and local governments; 8 percent foundations; 3 percent corporate underwriting; 11 percent community subscribers and auctions; and 8 percent other.[42] By 1989, government support was less than half at 48 percent. Other percentages were: 5 percent foundations; 16 percent underwriting; 21 percent subscribers; and 8 percent other.[43]

government

The federal government reduced its support to public broadcasting, and hard-pressed state and local systems followed suit, not only in terms of percentages but in hard dollars. In addition, federal funding was only authorized for two years in advance, making it difficult for public TV to plan for the future.[44]

foundations

Foundations also reduced their contributions to public broadcasting, mostly because their money became tighter. One foundation to come to public broadcasting's aid, however, was the **Annenberg Foundation** established by Walter Annenberg, the founder of *TV Guide*. The purpose of Annenberg's grant ($150 million over a ten-year period beginning in 1982) was to develop collections of materials for college-level courses that use electronic media.[45]

corporate underwriting

Some of the slack from government funding and foundations was picked up by corporate underwriting from companies such as Xerox, Mobil, and 3M. Congress sweetened the pie for corporations in 1981 by approving the airing of company logos by public stations. Prior to this the stations could merely acknowledge that "This program was made possible by a grant from XYZ Corporation." With the change in policy, stations could show corporate logos and mention the product line of the contributing company. This was referred to as **enhanced underwriting.**[46]

citizens

People from the community also tried to give public broadcasting more of a helping hand. Monies from individuals paying a yearly fee to a local public station, and in return receiving a magazine that listed the station's programs, increased percentagewise. So did money from auctions, which had become an institution, particularly at community-owned stations. Donations of unusual objects from members of the community were auctioned off over the air with people calling in to place their bids.[47] The auctions became so profuse and consumed so much airtime, however, that negative reaction set in and many public TV stations began curtailing their auctions.[48]

ownership changes

The change in finances was also reflected in changes in ownership. Government agencies, such as public schools, colleges, and states, had been the predominant owners of early TV stations, but as their fortunes waned, nonprofit community groups became the more predominant owners. By 1990, 48 percent of licenses were held by community organizations, as compared to 33 percent by colleges, 13 percent by state authorities, and 11 percent by local educational or municipal authorities.[49]

fund-raising

Because the total amount of money from traditional sources was down, public TV, like public radio, tried to undertake commercial fund-raising ventures. It considered establishing a pay-cable network, going into subscription TV, providing data services, and leasing its satellite time.[50] Except for the leasing of satellite time, most of these ideas did not materialize. The leasing of satellite time, however, enabled PBS to recover most of its own costs for operating its satellite system.

commercials

The most controversial of PBS's fund-raising proposals, however, was the airing of actual commercials. In the spring of 1982, Congress authorized ten public TV stations to experiment with advertising in order to obtain funds.[51] The experiment, which lasted about eighteen months, had ambivalent results and, at the end of the period, the stations appeared to conclude that direct advertising was not something public broadcasting should embrace. Commercial TV networks and stations were pleased with this decision because they were unhappy with the thought of public TV stations seeking ad money from the same companies they were approaching. Also, the specter of a public broadcasting system overly dependent upon the business community seemed unhealthy to some who feared that big business could dictate programming content.

enhanced underwriting

However, PBS revised its enhanced underwriting rules to allow more commercial information to be given about companies that donate for programs. It determined that logos, slogans, locations, description of product lines or services, and trade names could be included in attractively designed announcements.[52] By the late 1980s these enhanced underwriting announcements were running as long as thirty seconds and were looking very much like commercials.[53] As a result, the PBS board once again revised its guidelines. It specifically prohibited direct comparisons with competitors, disallowed lyrics in theme music, and limited underwriters to mentioning only one product per message.[54]

By the late 1990s, funding in general had improved for public television, and the government had approved $1 billion to cover 1991, 1992, and 1993.[55]

Programming Highlights

Despite financial problems, public television could be very proud of its programming record by the 1990s. It had "Frontline," which was the only regularly scheduled prime-time documentary, and "The MacNeil/Lehrer News Hour," which had begun in 1983 and was often acclaimed for its in-depth look at news issues.[56] These programs, and other public affairs shows, were not without their critics who accused them of left-leaning biases.

controversy

Two of the most controversial programs of the 1980s were "Death of a Princess" and "The Africans." The former was a docudrama, co-produced by the BBC and WGBH in Boston, which graphically depicted the execution of a Saudi Arabian princess and her common lover, both of whom had been accused of adultery. The Saudi's objected to the portrayal, complaining it was

FIGURE 5.10

Robert MacNeil (*left*) and Jim Lehrer of "The MacNeil/Lehrer News Hour," an in-depth examination of the day's top news stories produced jointly by WNET in New York and WETA in Washington. *(Everett Collection)*

an attack on the Islamic religion.[57] "The Africans" was a point-of-view mini-series about the influence (mostly negative) of western countries, such as England and the United States, on the African continent. It had been under-written in part by the National Endowment of Humanities, which labeled it an "anti-Western diatribe" and demanded that the NEH credit be removed from the series.[58]

editorializing

Public broadcasting stations won the right to editorialize during the 1980s. Previously they had been prevented from this on the premise that fed-erally funded stations could take on the appearance of government propa-ganda organs. However, in 1984 this was declared unconstitutional and public stations were given the right to editorialize.[59]

In 1990 the eleven-hour epic, "The Civil War" drew the highest audi-ence ever for PBS—approximately 9 percent of all households.[60] In the drama department, many of the BBC's finest productions came to public television, particularly on the long-running "Masterpiece Theatre." Newly written American drama was also encouraged on such series as "American Play-house," "Hollywood Television Theatre," and "Visions." "Great Perfor-mances" was the longest running performing arts series in TV history.[61]

drama

An area where public television shown most brightly was children's pro-gramming. In the dramatic series "Wonderworks," it had the only regularly scheduled prime-time show designed for children. "Sesame Street," which by now was as old as most college students, was still one of the most popular public TV offerings. It had been followed by other noteworthy soft-teaching programs such as "Electric Company," "3–2–1 Contact," and "Square One TV."[62]

children's programming

FIGURE 5.11

A scene from "Death of a Princess." *(Courtesy of WGBH Educational Foundation, Boston)*

FIGURE 5.12

Director Ken Burns editing scenes from his highly-rated, highly-acclaimed PBS series on the Civil War. *(Everett Collection)*

Frequently these children's programs, which were designed primarily for children at home, were used within the schools. Public TV also produced some programs designed more for in-school use than at-home use, such as "Newscasts from the Past" and "The Challenge of the Unknown."[63]

formal education

In the area of formal education, PBS, in 1981, created **Adult Learning Service (ALS).** This is a cooperative effort between public TV stations and more than 1500 colleges and universities to offer college credit courses that

FIGURE 5.13
A scene from one of the "Wonderworks" programs called "Mighty Pawns." Paul Winfield plays the principal of an inner-city school where student Alfonso Ribiero leads a chess team to victory.
(Courtesy of Wonderworks)

people can watch in their homes. ALS researches courses that are available and then helps stations and colleges schedule and promote the courses that fit their needs. More than 250,000 students a year enroll in these courses.[64]

The structure of programming changed in the 1990s to a more centralized form. The Station Program Cooperative was suspended, and decision making regarding which programs would be aired was turned over to a newly created position within PBS, executive vice president of national programming and promotion services. This gave PBS more authority to select and schedule programs, including some powers to make all stations schedule the same programs at the same time.[65]

structural changes

More than 93 million people (56 percent of the population) watch public television at least once a week. Although this translates to ratings that are only in the 2 to 3 percent bracket, many of the programs have more impact than those found on commercial TV, in part because of the upscale nature of the audience that watches.[66]

Public Broadcasting Chronology

1917 Professor Terry of the University of Wisconsin transmits signals
1925 Many educational radio stations taken over by commercial interests
1929 National Committee on Education by Radio asks for station reservations
1945 FCC reserves twenty FM channels for education
1949 NAEB begins bicycle network for radio

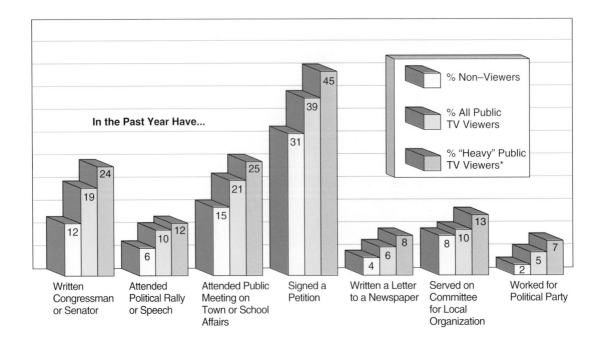

In the Past Year Have...

	% Non–Viewers
	% All Public TV Viewers
	% "Heavy" Public TV Viewers*

Written Congressman or Senator: 12, 19, 24
Attended Political Rally or Speech: 6, 10, 12
Attended Public Meeting on Town or School Affairs: 15, 21, 25
Signed a Petition: 31, 39, 45
Written a Letter to a Newspaper: 4, 6, 8
Served on Committee for Local Organization: 8, 10, 13
Worked for Political Party: 2, 5, 7

*"Heavy" viewers are defined as those who watched more than 3 hours "last week."

FIGURE 5.14

Political and social activity of public TV viewers. *(Source: PBS Research, Who Watches Public Television?, 1986 and Nielsen Media Research.)*

1950   Iowa State University operates commercial TV channel
1950s Ford Foundation funds various TV stations, NET, Chicago TV College, and MPATI
1952   FCC reserves 242 noncommercial TV channels
1953   First educational TV license granted to University of Houston
1962   Educational Broadcasting Facilities Act passed
1965   Carnegie Commission formed
1966   FCC sets aside 604 educational TV channels
1967   Public Broadcasting Act authorizes CPB
1970   "Sesame Street" begins
1970   NPR incorporated
1970s Public TV and President Nixon clash
1974   Station Program Cooperative established
1977   Second Carnegie Commission formed
1978   FCC adopts new rules for ten-watters
1981   Congress approves public broadcasting's airing of company logos
1981   ALS created
1982   First Annenberg grants awarded
1982   APR established
1982   Ten public TV stations experiment with commercials

1984　Public broadcasting given right to editorialize
1984　PBS revises underwriting rules to loosen them
1988　PBS revises underwriting rules to tighten them
1990　PBS given more programming power

Conclusion

Public broadcasting is caught between the best and worst of democracy, free enterprise, and culture. America's democratic traditions make governmental financial support of broadcasting in any form a risky undertaking. The specter of possible government influence poses a threat, and yet to date, the government seems to be the only source able to finance such a culture-oriented programming service. The free enterprise system allows the marketplace to determine what products and services should survive, but public broadcasting, if it is to maintain even a partial noncommercial nature, will probably continue to be unable to compete in this commercial marketplace. Although culture and education are supposedly highly revered in America, they cannot draw the audiences and finances necessary to ensure their survival.

All of this has influenced the history and operation of public broadcasting. The early radio stations, which were operated by colleges, were gradually overpowered by commercial interests. Through concerted effort, educators persuaded the government to reserve stations in the FM band for educational broadcasting. Most of these stations exist on modest budgets provided by educational institutions and on endowments, grants, and public donations. They do provide a hallmark of cultural programming, including classical music and some highly rated programs from National Public Radio and American Public Radio.

Public television began with channels specifically reserved for it, but educational institutions were unable to bear as large a percentage of the costs as they had for radio. Fortunately, the Ford Foundation underwrote much of the early development of educational TV, but it could not shoulder the entire weight. Therefore, the government, amid much controversy, began in 1962 to provide grant money for facilities only.

Still, educational broadcasting was not able to compete in the free enterprise system with its commercial counterparts. The Carnegie Commission spent several years studying the situation and recommended a public television system heavily supported by government funds for both facilities and programming. The Carnegie plan was quickly passed by Congress. By recommending the Corporation for Public Broadcasting to act as a buffer and to administer funds to the Public Broadcasting Service and those producing programs, the Carnegie Commission hoped to avoid the pitfall of government influence over programming. But this was not to be—both internal and external schisms over programming philosophy rocked the CPB-PBS-NPR family during the Nixon administration. Following that, the government withdrew much of its funding, leaving public broadcasting to find other methods of financing, turning more sharply to the leaders and concepts of free enterprise.

The purpose of financial recruitment, of course, is to find support for the programming services that have improved greatly since the Carnegie plan went into effect. The Station Program Cooperative programs, foreign programs, underwritten programs, and local programs have moved public broadcasting from talking heads to dynamic culture.

The triangle formed by democracy, free enterprise, and culture has created many controversies as public broadcasting tries to maintain an equilibrium among government influence, corporation influence, and audience demographics. Just where public broadcasting is headed is unclear, but this is nothing new for an industry that has spent most of its life in a state of uncertainty.

Thought Questions

1. How can public broadcasting be insulated from government influence over program content?
2. In your opinion, what is the proper balance for local versus national programming within the public broadcasting system?
3. Should the recommendations of Carnegie II be implemented? Why or why not?
4. What are some ways public broadcasting could ease its financial strain?

Cable Television and Competing Media

Introduction

Cable television is almost as old as television broadcasting. However, broadcast TV grew rapidly once it was actually introduced, while cable struggled for several decades before it received any widespread recognition. Now cable TV is well established and must look over its shoulder so that it is not overtaken by competing media.

Cable has become like air conditioning. You don't need it, but once you live with it, you can't live without it.

Lloyd Trufelman
Cable Television Advertising
Bureau

The Beginnings of Cable

There are many different stories about how cable TV actually began. One is that it was started in a little appliance shop in Pennsylvania by a man who was selling television sets. He noticed that he was selling sets only to people who lived on one side of town. Upon investigation, he found that the people on the other side of town could not obtain adequate reception, so he placed an antenna at the top of a hill, intercepted TV signals, and ran the signals through a cable down the hill to the side of town with poor reception. When people on that side of town would buy a TV set from him, he would hook their home to the cable.

Another story on the origin of cable service tells of a "ham" radio operator in Oregon who was experimenting with TV just because of his interest in the field and because his wife wanted "pictures with her radio." He placed an antenna on an eight-story building and ran cables from there to his apartment across the street. Neighbors became excited about the idea, so he wired their homes, too. The initial cable subscribers helped pay for the cost of the equipment and after that was paid off, he charged newcomers $100 for a hookup.[1]

first systems

Whether either of these stories represents the true beginning of cable TV is hard to say. During the late 1940s and early 1950s, in many remote and mountainous places, friends and neighbors built rudimentary cable TV systems. They were intent on providing television reception for themselves because they could not receive over-the-air signals.

freeze effect

One factor that helped cable TV in its beginning was the **freeze** on television station expansion from 1948 to 1952. The only way that people could receive TV if they were not within the broadcast path of one of the 108 stations on the air was to put up an antenna and, in essence, catch the signals as they were flying through the air.

early importation

As early as 1949 a cable system was established in Lansford, Pennsylvania, as a moneymaking venture. With no local signal available, the system carried the three network signals by importing them from nearby communities, a practice that became known as **distant signal importation.** Fourteen such signal importation companies were in operation by the end of 1950, and the number swelled to seventy by 1952.[2]

FCC opinions

At this point the FCC was convinced that as the number of stations increased, the need for Community Antenna Television (CATV) would diminish and gradually vanish entirely. The one factor that eluded its attention was that many communities were too small to support the expense of station operation. If the basic philosophy of "mass communication for an informed public" was to be a reality, CATV would have to grow. Signals would have to be imported from cities where stations could find support for their operations.

However, even with 65,000 subscribers and an annual revenue of $10 million in 1953,[3] cable TV was still only a minor operation. Most broadcasters felt no concern about this business that was growing on the fringes of their signal contour. Some, however, were becoming alarmed by the attitude of permanence growing in some cable systems. **Coaxial cable** was replacing the open-line wire of early days, and space was leased on telephone poles for line distribution instead of the house-to-house loops augmented here and there by a tree.

Placing the cable on telephone poles was a source of major resentment for the cable industry. The cable companies had to rent space on the telephone poles for their cable because the telephone companies naturally had a monopoly on all the telephone poles in town. The telephone companies could charge rates that the cable companies felt were unfair. The cable companies also felt that the phone companies often gave them an unnecessarily hard time about where, how, and when the cables were to be attached on the poles.

telephone pole problems

As cable TV grew, it became more sophisticated. In addition to importing signals into areas where there was no television, cable companies began importing distant signals into areas where there were a limited number of stations. For example, if a small town had one TV station, the cable system would import the signals from two TV stations in a large city several hundred miles away.

more importation

In general, during the decade of the 1950s, cable TV grew from a few very small operations run by friends and neighbors to a system of television reception for small and medium-sized areas that had poor reception or a very limited number of TV stations. Most of these systems were operated by small, locally-run organizations that charged subscribers an initial installation fee and monthly fees compatible with modest profits.[4]

Cable's Muddled Growth

The importing of distant signals brought about the first objections to cable TV. Existing TV stations in an area would find that the size of their audience had shrunk because people were watching the imported signal. In fact, sometimes the signal imported would be the same as the one showing on the local station. For example, a local station might be showing a rerun of "I Love Lucy" and find that the imported station was showing the same rerun, splitting the audience of the show in half. As a result, local stations could no longer sell their ads for as high a price as they had before the importation.

objections to importations

Some of the stations in areas affected by cable TV appealed to Congress and the FCC to help them in their plight. In 1959 the Senate Commerce Subcommittee suggested legislation to license CATV operators. The actual extent of CATV operations, even at this late date, was impossible to identify because the operators were not required to report to any governmental agency—in spite of the fact that in many areas of the country the audience served by CATV ran as high as 20 percent of the available viewers. Attempts to draft legislation were filled with arguments and debates that lasted until 1960 and ended with the defeat of a bill.[5]

legislation failure

local franchising

With the failure of federal intervention came a rash of state and local attempts to assert jurisdiction over CATV. In most areas the local city council became the agency that issued cable **franchises** and placed stipulations on how the cable system was to conduct business. Competing applicants for a cable franchise would present to the council their plans for operation of the system, including such items as the method of hookup (for example, telephone poles or underground cable), the speed with which hookups would be made, the fees to be charged to the customer for installation and for the regular monthly service, and the percentage of profit that the company was willing to give the city for the privilege of holding the franchise. Based on this information, the council would award the franchise to the one company it felt was most qualified.[6]

early 1960s figures

During the 1960s this was a very calm process; usually only one or two companies applied for a franchise in a given area. Areas with good reception did not even bother with franchising because no companies would have applied.

Although the number of cable systems doubled between 1961 and 1965, cable TV was still small business. In 1964 the average system served only 850 viewers and earned less than $100,000 annually. **Multiple system ownership**—ownership of a number of cable TV systems in different locations by one company—was less than 25 percent because of the lack of economic incentives.[7]

The FCC maintained a policy of nonintervention in cable TV matters during the early 1960s, hoping that the problems between the operators and broadcasters would be settled by court decision. Unfortunately, the situation only became more confused as court cases piled up. However, one court case,

Carter Mountain case

referred to as the *Carter Mountain* case, did set a precedent. In 1963, an appeals court ruled that the FCC could refuse to authorize additional facilities to Carter Mountain Transmission Corporation, a cable company in the Rocky Mountains, because the cable company might damage the well-being of the local broadcast stations. This gave the FCC power to restrict cable in order to protect broadcasters.[8]

The early 1960s then saw the beginnings of a rift between broadcasters and cable TV people over distant signal importation, a lack of action on the part of the federal government, and the beginning of actions on the part of local governments.

The FCC Acts

FCC rules

In April of 1965 the FCC took its obligations regarding cable TV more seriously and issued a notice that covered two main areas: (1) All CATV-linked common carriers from this time forward would be required to carry the signal of any TV station within approximately sixty miles of its system; and (2) No duplication of program material from more distant signals would be permitted fifteen days before or fifteen days after such local broadcast.[9]

The rule of local carriage, which became known as **must-carry,** caused little or no problem. Most CATVs were glad to carry the signals of local stations. However, the thirty-day provision caused bitter protest from cable operators because it limited their rights in relation to what they could show on their distant imported stations. This rule, which became known as **syndicated exclusivity,** meant that if a local station was going to show an "I Love Lucy" rerun on January 15, a cable TV operator would have to black out that rerun on a distant imported station if it were on any time during the month of January.

The cable TV industry marshalled its forces and in May of 1966 succeeded in having the thirty-day provision reduced to only one day. This, of course, angered the broadcasters.

As background to this conflict, one must visualize the frustration of broadcasters who had invested large sums of money in their stations. Many were barely able to survive on the revenue generated by their station while cable operators, who had invested far less money, were realizing profits by carrying the broadcasters' signals and importing distant signals. During this period there were both filed and pending applications for cable coverage of areas that would account for at least 85 million people. It is easy to understand the fears of station operators. Their basic desire was for the FCC to provide them with full administrative protection.

stations' apprehension

The attitude of the FCC during this time was strongly in favor of local TV stations. The second report issued by the FCC in 1966 put more restrictions on cable service. This order came when 119 cable systems were under construction, 500 had been awarded franchises, and 1,200 had applications pending. All of these systems would be required to prove that their existence would not harm any existing or proposed station in their coverage areas. By making no increase in staff to handle this load, the FCC was, in essence, freezing the growth of cable operations in the top one hundred markets. This made station owners very happy.[10]

FCC attitude

The effect of this ruling on cable operators during the remaining years of the decade was the reverse of what the FCC had intended. Cable systems that were unable to expand were sold to large corporations that could withstand the unprofitability of the freeze period. Multiple system ownership was quite prevalent by 1970.

In 1972 the FCC issued another policy on cable TV that changed the must-carry rule from all stations within sixty miles of the cable TV to all stations within thirty-five miles plus other local stations that, as shown by polls, were viewed frequently by people in the area. The policy also reconfirmed syndicated exclusivity and set up guidelines delineating how much distant signal importation a cable system could undertake, depending primarily upon the size of the market in which it was located and the number of stations already in the market.

more FCC rules

Early Programming

When cable TV first began, it was a common carrier similar to the telephone company; that is, cable companies picked up signals and brought them into homes for a hookup fee and regular monthly fees. This meant that the programming was only that which was picked up from local stations or from distant signal importation.

twelve channels

The very early systems had only three channels for the three networks. As television and its resulting technology grew, cable systems provided as many as twelve channels of programming. The cable system could use each channel from 2 through 13 because its signals were on wires that were not subject to the same interference that makes it impossible to use all twelve channels of broadcast TV in a particular area. Different cable systems placed different programming on these twelve channels, but this usually consisted of converting all the local VHF stations plus all the local UHF stations to VHF space on the dial. If there was not a large number of local stations, then the cable system would bring in stations from nearby communities.[11]

early local origination

Under this early system there was no **local origination** of programs. Gradually, however, some of the cable facilities began to undertake their own programming. The most common "programming" in the beginning involved unsophisticated weather information. Cable TV operators would place a thermometer, barometer, and other calculating devices on a disc and have a TV camera take a picture as the disc slowly rotated. This would then be shown on one of the vacant channels so that people in the area could check local weather conditions. Some systems had news of sorts. This might just involve a camera focused on a bulletin coming in over a wire-service machine or on three-by-five-inch cards with local news items typed on them. At any rate, it was a simple, inexpensive, one-camera type of local origination.

Gradually, studio-type local origination was used, usually in the form of local news programs, high school sports events, city council meetings, local concerts, and talk shows on issues important to the community. The first regularly scheduled cable local origination took place in 1967 in Reading, Pennsylvania, and shortly after that, local programming appeared in San Diego, California. The FCC became involved and gave the San Diego cable system the authorization to engage in local programming. After this, other cable systems began such programming.[12]

1970 rule

In October of 1969 the FCC issued a rule that required all cable TV systems with 3,500 or more subscribers to begin local origination no later than April of 1970. The purpose was to promote local programming in areas where it did not previously exist.

By April of 1970 many of the cable TV operators who were not engaging in local origination claimed hardship, telling the FCC that they did not have the funds to build studios, buy equipment, and hire crews. As a result, the FCC order was not enforced and later was modified to say that systems with more than 3,500 subscribers only had to make equipment and channel time available to those who wished to produce programs.

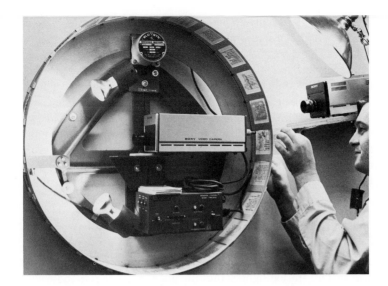

FIGURE 6.1
A simple system
whereby local
announcements are
placed on the TV
screen one after the
other as the drum
rotates. *(Courtesy of
TeleCable of Overland Park,
Kansas)*

This brought about a different type of local programming that became known as **public access.** Individuals or groups would use the equipment provided by the cable operator to produce programs that did not have the operator's input or sanction and then cablecast those programs over one of the system's channels. This differed from local origination, which was programming planned by the cable system.

early public access

In most areas public access was not a success, and the equipment provided by the cable company went unused for lack of interest on the part of the local community. In other areas, particularly where cable systems had shown an active interest in local programming, some exciting and innovative projects were undertaken in the public access area. These programs included public affairs programs and video art. Unfortunately, in several places the people who made use of the public access time were would-be stars who were using cable for vanity purposes or people from fringe groups who wished to propagate various causes or even lewd modes of behavior. Such individuals and groups made it difficult for the cable operators to present balanced opinions and also gave the concept of public access an unsavory reputation.[13]

Another form of local programming that was instituted by some cable systems was the showing of movies. A cable company would use one of its channels, or perhaps only the evening hours of one of its channels, to show movies that only subscribers who paid an extra fee could receive. These movies were leased from film companies and were shown without any commercial interruptions. Regulations prevented cable systems from showing the well-known films that broadcasters wanted to show during prime time, but because the small cable systems only had a small number of subscribers who would pay for the movies, they could not afford to pay for blockbuster movies anyway. The channels for movies were not a huge success, but they did bring extra income to some cable systems.

early movie channels

By the mid-1970s only 20 percent of the cable companies were engaging in any type of local programming, and the most popular type of such programming was local sports.[14]

promises made

However, throughout the 1960s and 1970s promises were made, broken, and remade concerning the potential services and programs that would be available through cable. Cable lived on the edge of a promise that "within the next five years" cable TV would perk your coffee, help your kids with their homework, balance your checkbook, bring you the top sports events, secure your home, do your shopping, and teach the handicapped. But except for a few isolated experiments, cable remained, until the mid-1970s, primarily a medium to bring broadcast signals to areas with poor reception.

The Copyright Controversy

producers versus cable companies

Another point of contention between broadcasters and cable operators was payment of **copyright fees.** When cable systems rebroadcast network, local station, or distant station signals, they did not pay any fees to those who owned the copyrights to the materials. In some instances, the networks or stations had paid for and created the material, so they owned the copyright. In other cases, the networks or stations had purchased or rented program material from film companies or independent producers and had paid copyright fees.

The stations, networks, film companies, and independent producers felt that cable companies should pay copyright fees for the retransmission of material because these retransmission rights were not included in the broadcast TV package. Cable operators felt that copyright had already been taken care of by the TV stations and networks, and cable systems were merely extending coverage.

United Artists's case

The basic fight over copyright began in 1960 when **United Artists Productions,** holders of license rights to a large library of feature films purchased by a network at a cost of $20 million, sued two cable operators for copyright infringement of their performance rights. The initial ruling that was handed down in 1966 favored United Artists. However, on appeal to the U.S. Supreme Court, the decision was overruled in 1968, and the controversy once again boiled over.

Further complicating the issue was the fact that the copyright law, under which the United States then operated, had been written in 1909. Congress had been trying for many years to update copyright regulations with new legislation, but the many controversies and special interests surrounding the issue, including the cable feud, made this one of the most lobbied bills in history.

1976 law

Finally, a new copyright law was passed in 1976 and went into effect in 1978. Under this law, cable TV systems paid a **compulsory license** fee to a five-member, newly created government body called the **Copyright Royalty Tribunal.** This body would then distribute the money to copyright owners. The amount of this compulsory license fee was 0.7 percent of a cable operator's revenue from basic monthly subscriptions. The copyright law gave the Tribunal authority to adjust the rate for inflation. Both cable operators and copyright holders seemed happy with this plan.[15]

However, two major problems arose. One involved the distribution of the money collected from the compulsory license. Many interests claimed a share of the money because their program material was retransmitted over cable, and it was the Tribunal's job to decide who got how much. In 1979 it decided that the $20.6 million collected should be distributed as follows: 70 percent to movie producers and program syndicators; 15 percent to sports; 5.25 percent to public television; 4.5 percent to broadcasters; and 0.25 percent to National Public Radio. Hardly any of the entities were satisfied with their pieces of the pie and most went to court in an attempt to gain increases.[16]

The other problem arose because the cable industry changed greatly after 1976, and the $20.6 million collected through compulsory licensing seemed like a paltry sum to copyright holders. Cable systems with expanded channel capacity were now retransmitting signals from stations all over the country. However, according to copyright holders, they were paying as though they were only retransmitting the local must-carries and a few distant imports.[17]

By the early 1980s there was seething discontent with the infant copyright law. Congressional lobbyists from many entities were trying to initiate bills to better their lot in life.

The Beginnings of Satellite Cable

In 1975 the stage was set for a dramatic change in cable programming when a company called **Home Box Office (HBO)** began distributing movies and special events via satellite.

Home Box Office was actually formed in 1972 by Time, Inc., as a movie/special pay service for Time's cable system in New York. The company decided to expand this service to other cable systems and to set up a traditional broadcast-style **microwave** link to a cable system in Wilkes-Barre, Pennsylvania. In November of 1972 the company sent its first programming from New York to Wilkes-Barre.

During the next several years HBO expanded its microwave system to include about fourteen cable companies in two states with more than eight thousand subscribers. This was not an overly successful venture, and it was not profitable for Time.

Then in 1975, shortly after domestic satellites were launched, Time decided to bring two of its cable systems the Ali-Frazier heavyweight championship fight from Manila by satellite transmission. The experiment was very successful, and HBO decided to distribute all of its programming by satellite.[18]

This distribution method was easier and cheaper than the system that required huge microwave towers, and it meant that as soon as HBO sent its signals to a satellite, they could be received throughout the country by any cable system that was willing and able to buy an earth station satellite receiving dish.

HBO then began marketing its service to cable systems nationwide, which was no easy chore at first. The original receiving dishes were ten meters in diameter and cost almost $150,000, a stiff price for cable systems, many of which were just managing to break even. But the technology of satellites moved quickly enough that by 1977 dishes as small as 4.5 meters sold for less than $10,000.[19]

subscription TV

Another problem HBO encountered centered on the rules that had been established, mainly for **subscription TV,** an over-the-air form of broadcasting with a scrambled signal. Programming (mostly movies or sports) was sent from a station antenna into the airwaves, but the signal was scrambled so that it was necessary to buy a special decoder in order to see a coherent picture. Subscription TV was started during the 1950s, but was stifled just as cable TV was by FCC restrictions that were pro-broadcasting. For example, no more than one STV operation was allowed in any one community, and STV systems were prevented from **siphoning**—bidding on programming, such as movies and sports events, that conventional broadcasters wanted to show. The fear was that pay-TV might simply take over, or siphon, the TV station's programming by paying a slightly higher price for the right to cablecast the material.[20] HBO and several cable TV system owners took these rules to court. In March of 1977, the court set aside the pay-cable rules restricting programming, leaving HBO free to develop as it wanted.[21]

Of course, this allowed subscription TV to develop, too, and for awhile subscription TV and the movie services of cable TV systems competed rather heavily. However, cable won because it offered many channels, and subscription TV offered only one. Subscription TV died out during the mid-1980s.[22]

marketing problems

Another problem HBO had when it first started marketing its service dealt with its financial relationship with cable systems. At first, it offered the cable system owners 10 percent of the amount they collected by charging subscribers extra for HBO programming. Approximately 40 percent of the fee was to go to HBO and 50 percent to the program producers. When cable owners indicated that they were not happy about their percentage, HBO raised it so that the systems retained about 60 percent of the money, and HBO and the program producers split the other 40 percent.[23]

With receiving dishes manageable in terms of both cost and size, with programming of an appealing nature, and with financial remuneration at a high level, cable systems began subscribing to HBO in droves. Likewise, HBO became very popular with individual cable subscribers, and by October of 1977, Time was able to announce that HBO had turned its first profit.

HBO success

Shortly after HBO beamed onto the satellite, Ted **Turner,** who owned a low-rated UHF station in Atlanta, Georgia, decided to put his station's signal on the same satellite as HBO. This meant that cable operators who had bought a receiving dish for HBO could also place Turner's station on one of their channels. This created what was referred to as a **superstation** because it could be seen nationwide. A company transmitting the station charged cable operators a dime a month per subscriber for the signal,[24] but they in turn did not charge the subscriber as they did for the HBO pay service. The economic rationale for the superstation was that the extra program service would entice more subscribers. The charge to the cable companies did not cover the station's costs, but it was able to charge a higher rate for its advertisements once the superstation had a bigger audience spread out over the entire country.

Turner's superstation

FIGURE 6.2

A "live" in-concert taping by HBO of singer Diana Ross before a regular nightclub audience at Caesar's Palace in Las Vegas. This was one of HBO's early programs. *(Courtesy of Home Box Office)*

With two successful program services on the satellite, the floodgates opened and cable TV took on an entirely new complexion.

Cable's Gold Rush

In the late 1970s and early 1980s cable TV experienced a phenomenal growth that was sparked mainly by the development of satellite-delivered services. In 1975, 469,000 homes subscribed to pay cable, but that number grew to 9 million by 1980.[25] Similarly, revenues from pay-cable increased. In 1979, pay revenues grew 85 percent over 1978 figures. The industry predicted that this was a peak that would subside to about 50 percent growth.[26] However, in 1980 pay revenues grew 95.5 percent over 1979 revenues.[27] The number of cable subscribers also increased. During the 1970s cable added subscribers at the rate of about 1.1 million per year, but in 1980 3.1 million subscribers were added.[28] Between 1975 and 1980 the percentage of households subscribing to cable jumped from 12 to 20 percent,[29] and during this same period of time, the cable TV profits grew 641 percent.[30] With figures such as these, it is no wonder that cable experienced a veritable gold rush.

One of the main areas in which this gold rush manifested itself was franchising. Cities that could not have given away franchises in earlier years because of their clear reception of TV signals suddenly became prime targets for cable and its added programming services. In general, cabling moved from rural areas into major cities such as Pittsburgh, Boston, Los Angeles, Dallas, and Cincinnati, where numerous companies applied for the franchises.

In most areas the local city council made the decision regarding which company would win the franchise, although in some areas state regulatory bodies also had a voice. Most city governments were not accustomed to dealing with such matters. Gradually, through the use of consultants, city councils established lists of minimal requirements that they wanted from the cable companies. These requirements pertained to number of channels, length of

late 1970s and early 1980s figures

franchising

Cable Television and Competing Media 161

FIGURE 6.3

Growth of cable TV.
*(Source: Department of
Commerce, Bureau of
Census,* Statistical Abstracts
of the United States, 1990,
*Washington, D.C., pp. 550,
553, and 554.)*

Year	Number of Systems	Number of Subscribers	Percent of Homes	Pay-Cable Subscribers
1955	400	150,000	.5	–
1960	640	650,000	1.4	–
1965	1,325	1,275,000	2.4	–
1970	2,490	4,500,000	7.6	–
1975	3,506	9,800,000	12.0	469,000
1980	4,225	16,000,000	20.0	9,144,000
1985	6,600	32,000,000	43.0	30,596,000
1990	10,823	54,280,000	59.0	40,100,000

time to complete the wiring job, subscriber rates, local input into programming, and similar issues.[31]

Cable companies, in their fervor to obtain franchises, usually went well beyond what the cities required. They, too, hired consultants who contacted city leaders to learn the political structure and needs of the city and to decide how the company should write its franchise proposal to ensure the best possible chance of winning the contract. Since only one company could receive a franchise for a particular area, cries of scandal sometimes accompanied these procedures as the cable companies tried to gain influence. A practice dubbed "rent-a-citizen," whereby a cable company would give blocks of its free stock to influential citizens in return for their support, came under sharp attack.[32]

The development of local programming became an important part of the franchising process. Cities usually requested that cable companies set aside a certain number of their channels for access by governmental, educational, community, and/or religious groups. In order to program these channels, cable companies promised to provide equipment and sometimes personnel.

As franchising competition became more intense, cable companies began promising cheaper rates, shorter time to lay the cable, more channels, more equipment, more local involvement, and generally more and better everything. All of this was written in a bid (usually several volumes thick) and submitted to the city council, who then evaluated all bids, usually with the help of consultants. When the winner was announced, the losers might call foul play over the procedures used by the winner, and the decision could wind up in court or back in the city council's lap. Often the winning cable company was unable to meet all the requirements stipulated in the bid, especially in regard to the speed with which the system was to be built. This led to court cases and fines for breach of contract.[33] However, cabling did move forward at a rapid rate.

As the cable companies promised more and more, they realized that they might not recover their investment for about a decade. This hastened a process that was already in full swing within the cable industry—the takeover of small "ma and pa" cable operations by large multiple system owners and then the consolidation of these MSOs with other large companies.[34] Large companies emerged in the cable industry, partly because they wanted to be part of the

margin: franchising problems

margin: MSOs

gold rush and partly because only large companies had the resources to withstand the expenses of the franchising process and the other start-up costs of laying cable, marketing, and programming.

Other groups also began to stake their claims in cable's gold rush. Advertisers who saw cable reaching close to 30 percent penetration of the nation's households in the early 1980s became interested and began placing ads on cable's programming channels, both local and national.[35]

Members of the various unions and guilds that operate in the broadcasting industry were not involved in cable programming when it first began because the cable companies did not recognize the unions or abide by union regulations. Original cable programming was written, directed, produced, and crewed by nonunion members. However, after several long, bitter strikes during the early 1980s, the unions won the right to be recognized and to receive residuals from cable TV. Thus, the same people who worked in broadcast programming began to work in cable programming.[36]

Perhaps the biggest winners in the cable gold rush were the equipment manufacturers who supplied the materials needed to build the cable systems. The suppliers of the converters that enable a regular TV set to receive the multitude of cable channels, the earth station dishes, and the cable itself found their order desks piled high. Space on a satellite became a precious commodity as more and more companies wanted to launch national programming services. Satellite transponder time that had rented for about $200 an hour or $1 million a year took on a new dimension in 1981 when RCA auctioned off seven **transponders** (parts of a satellite that carry particular program services) on its Satcom IV satellite for figures ranging from $11.2 to $14.4 million for six years. The FCC later disallowed the auction method of leasing transponders, and RCA had to settle for charging all successful bidders $13 million for seven years and nine months of use.[37]

Although large companies, advertisers, unions, and equipment manufacturers all flocked to cable during the late 1970s and early 1980s, the most noticeable type of cable growth was in the area of programming and special services.

Growth of Programming Services

Shortly after Time became successful with its HBO service, Viacom launched a competing pay-TV, **Showtime.** Viacom, like Time, owned various cable systems throughout the country and had been feeding them movies and special events through a network that involved bicycling and microwave.[38]

Following the launching of Showtime, Warner Communications, in conjunction with American Express (Warner-Amex), began **The Movie Channel,** which consisted of movies twenty-four hours a day.

Other **pay-TV** services started in fairly rapid succession: Cinemax, a Time-owned service that consisted mostly of movies that were to be programmed at times complementary to HBO; Galavision, a Spanish-language movie service; Spotlight, a Times-Mirror movie service; Bravo and The Entertainment Channel, both cultural programming services; The Disney

FIGURE 6.4

Jason Robards starring in Eugene O'Neill's *Hughie*, one of Showtime's Broadway productions. *(Courtesy of Showtime)*

Channel, a family-oriented pay service featuring Disney products; and Playboy, "adult" programming that included R-rated movies, skits, and specials. Playboy got into the cable programming business by joining forces with an already established service, Escapade.[39]

basic services The late 1970s and early 1980s also saw a proliferation of new satellite delivered programming services that became known as **basic cable.** Some of these were supported by advertising, some were supported by the institutions that programmed them, and some were supported by small amounts of money that the cable companies paid to the programmers. These were lumped together and referred to as basic cable because the subscriber did not usually pay any substantial additional fee for the services.

Most of these were born within the time span of only a few years: ESPN, all-sports programming; CBN, with Christian-oriented programming; Nickelodeon, a children's service; CNN, twenty-four hours a day of news, owned by Ted Turner; USA, a network with a variety of programming types; C-SPAN, with public service oriented programming, including live coverage of the House of Representatives; ARTS, an ABC-owned cultural service; CBS Cable, an advertising cultural service owned by CBS that featured a great deal of originally-produced American material; MTV, the creator of music videos on an around-the-clock basis; Satellite News Channel, a Westinghouse-ABC joint venture of twenty-four hour news, established to compete with CNN; Daytime, a service geared toward women; and Cable Health Network, programming about physical and mental health.

superstations The number of superstations also grew and were considered part of basic cable. Ted Turner's WTBS was first, but shortly thereafter WGN in Chicago and WWOR in New York also went on satellite. In addition, a number of audio services were developed for cable.

For several years both new pay and basic services were announced at a rapid rate—sometimes several in one day. Some of these never got off the ground, others existed only for short periods, and others showed signs of longevity.[40]

Electronic Media Forms

FIGURE 6.5

A scene from *The Greek Passion* performed by the Indiana University Opera Company and taped by Bravo. This was the first opera to be taped for cable television. *(Courtesy of Bravo)*

FIGURE 6.6

A scene from "Everything Goes," an adult quiz show on Escapade. *(Courtesy of Escapade)*

Because the franchising process placed so much emphasis on the local community in which the cable TV system originated, local programming took on an entirely new dimension in the late 1970s.

local programming

The older systems that still only had twelve-channel capacity usually allocated only one channel to local programming. But the newly franchised systems that were able to take advantage of improved technology to provide twenty, then thirty-seven, then fifty-four, then more than a hundred channels usually promised an entire complement of local channels.

At least one of these channels was generally reserved for local origination—programming that the cable system itself initiated. The others were some

FIGURE 6.7

An ESPN taping of a college basketball game. *(Courtesy of © ESPN)*

FIGURE 6.8

Host Reggie Jackson (*right*) interviews high school rodeo champ Mike Esposito for a Nickelodeon program. *(Courtesy of Warner-Amex Satellite Entertainment Company)*

access channels

combination of access channels: public access for the citizenry at large; community access for such community groups as the Girl Scouts and the United Way; government access for local officials; educational access for schools and colleges; religious access for religious groups; and leased access for businesses, newspapers, or other individuals interested in buying time on a cable channel to present their messages, sometimes with commercials included.

How these channels were to be organized and operated differed widely from community to community. Sometimes cable companies provided equipment, studio space, and professional personnel to help the various groups create their programming and then were responsible for seeing that the programs were cablecast over the system. At other times, cable companies donated equipment to a community nonprofit access organization, the city council, or

FIGURE 6.9
MTV's taping of the rock group .38 Special as they perform at Denver's Rainbow Music Hall. *(Courtesy of Warner-Amex Satellite Entertainment Company)*

a local school district, and the organization then produced its own programming with its own hired or volunteer staff. Some cable companies merely made a channel available for programming, and the local organizations or individuals interested in doing the programming used their own resources for equipment and crew.

Of course, not all the local programming planned by cable systems and local groups actually materialized, but the quality of access programming improved greatly from the early days when access consisted mainly of vanity TV.[41]

Interactive cable was highly touted in franchise applications. Once again, promises were made that cable would perk the coffee, help the kids with homework, do the shopping, protect the home, and teach the handicapped. In some areas of the country a fair amount of interactive cable was actually undertaken.

The pioneer and most publicized interactive system was **Qube,** which was operated by Warner-Amex cable. Qube was initiated in Warner's Columbus, Ohio, system in 1977 and later was incorporated into several of its other systems.[42] The basic element of the original interactive Qube was a small box with response buttons that enabled Qube subscribers to send an electronic signal to a bank of computers at the cable company that could then analyze all the responses. An announcer's voice or a written message on the screen asked audience members to make a decision about some question, such as who was most likely to be a presidential candidate, whether a particular congressional bill should be passed, what play a quarterback should call, whether an amateur act should be allowed to continue, or whether a city should proceed with a development plan. Audience members made their selection by pressing the appropriate button in multiple-choice fashion. A computer analyzed the responses and printed the percentage of each response on the participant's screen.[43]

Home security was another interactive area that cable entered. Various burglar, fire, and medical alarm devices could be connected to the cable system. A computer in a central monitoring station could send a signal to each participating household about every ten seconds to see if everything was in order.

interactive cable

Qube

home security

If any of the doors, windows, smoke detectors, or other devices hooked to the system were not as they should be, a signal was sent back to the cable company monitoring station, which then notified the police.[44]

For the most part, these interactive services did not become profitable, but proposals for them caught the attention of city councils and were often responsible, at least in part, for decisions regarding franchise awards.

Consolidation and Retrenchment

The mid-1980s saw a defoliation of the cable industry. Perhaps the promises had been too lavish and the anticipation too great, but the bloom fell off the rose. The rate of new subscriber growth leveled while the rate of disconnects from old subscribers increased.[45] Programming services consolidated and went out of business, and those that were left sported much of the same type of play-it-safe programming that had traditionally been found on ABC, CBS, and NBC. MSOs drowned in red ink as they tried to live up to the promises they had made in regard to wiring big cities. Advertisers did not respond to the cable lure as quickly or profusely as expected. Even a magazine produced by Time about cable TV programming failed.[46]

ownership changes

Companies took actions to stem financial woes. Warner-Amex, for example, sold several of its cable systems so that it could concentrate resources on a few cities. Group W sold all of its systems to five cable TV companies who then divided the systems among themselves. Storer and Times-Mirror traded several systems so that the systems each owned would be more geographically contiguous. In general, a few companies, such as Tele-Communications, Inc. (TCI) and American Television and Communications (ATC), acquired more systems, while other companies folded or retrenched. These big players in system ownership also began to acquire large shares of ownership in many of the pay and basic cable networks, leading to criticism that the cable industry was engaging in too much **vertical integration**—a process through which select group of companies had the ability to produce, distribute, and exhibit their products without input or decision making from other companies.[47]

Programming Changes

pay programming

The multitude of programming services went through a shakedown during the 1980s. In the area of pay programming, Showtime and The Movie Channel merged in 1983.[48] HBO, the perennial leader, remained profitable but suffered a slowdown that led to changes in the executive suites.[49]

Spotlight went out of business, and Galavision converted from a pay service to basic. Bravo changed owners, programming emphasis, and marketing strategy several times trying to find its niche; eventually it left the pay-TV realm to become a basic service. The Entertainment Channel turned its programming over to the basic service ARTS, which then became the Arts and Entertainment Channel.[50] The Playboy Channel hardened and softened its

FIGURE 6.10
The news room/studio
at Cable News
Network. *(AP/Wide World
Photos)*

programming several times, caught between angry citizens' groups who did not want nudity and a small but loyal group of viewers who did. This churn in philosophy angered its partner, Escapade, and the two parted company with Playboy paying a $3 million divorce settlement to Escapade.[51]

The pay services also began making exclusive deals with various movie studios so that certain films could appear only on one company's channels. These services also began making their own exclusive movies and programs.[52]

Problems occurred in the basic services also. The most highly touted failure was that of the CBS-owned cultural service, CBS Cable, which stopped programming in 1983 after losing $50 million. The service had programmed ambitious, high-quality television but did not receive sufficient financial support from either subscribers or advertisers. Because it had touted its service so aggressively, its demise was almost heralded by some of the cable companies who resented the encroachment of the broadcast networks into their cable business.[53]

basic services

Another well-publicized coup occurred when Ted Turner's Cable News Network slew the giants, ABC and Westinghouse, by buying out their Satellite News Channel. This meant less competition for CNN, which was then able to proceed on less tenuous financial footing because it did not have to compete for advertisers with Satellite News Channel.[54]

MTV ownership changed from Warner-Amex to Viacom, and its programming drew less attention than it had when the service first began. A number of imitative music video services were started, but most of them failed, including one launched by Ted Turner that lasted less than a month.[55]

FIGURE 6.11
Dr. Ruth Westheimer
in a characteristic pose
she used on her
popular sex-oriented
talk show on Lifetime.
(Courtesy of Lifetime)

On other fronts, USA Network was bought by Time, MCA, and Paramount, who did not always agree on the network's direction. Getty Oil was bought out by Texaco, which did not have any interest in maintaining the as yet unprofitable ESPN—it wound up being purchased by ABC. Daytime and Cable Health Network merged to form Lifetime. The children's service, Nickelodeon, began accepting commercials and changed ownership from Warner to Viacom. CBN changed from a religious format to a family-oriented format that included very old reruns.[56]

Although the rash of introduction of new programming services stopped, some new basic networks were introduced in the mid to late 1980s. The Discovery Channel was launched in 1985 to provide nonfiction programming about such areas as nature and history.[57] NBC launched the Consumer News and Business Channel (CNBC),[58] and Ted Turner started a new general purpose network, Turner Network Television (TNT).[59] Several comedy channels were started and soon merged because of lack of audience and lack of programming.[60] For awhile, a plethora of home shopping was touted. Many of these disappeared as soon as they began—or sooner; the market could not bear them all.[61]

local programming

Activity on the local access and local origination channels also slowed during the mid-1980s. Cable systems that had promised truckloads of production equipment to local organizations tried as best as they could to drag their feet on these costly obligations. Local programming departments that had included five or six employees dwindled down to a precious few. In some instances, the director of marketing took over programming "on the side." For some systems the only local programming became messages typed on the screen about local news and events. They had, in essence, reverted to a sophisticated

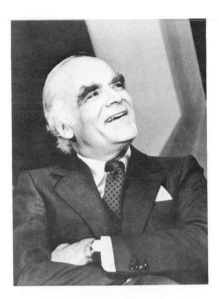

version of the early three-by-five-inch cards. Local programming did not by any means disappear, but it did not fill the multitudes of channels promised in many of the franchise agreements.[62]

In the area of interactive services, Warner-Amex killed its highly-touted Qube system by 1984, mainly because it was not profitable. No other similar interactive services rushed in to fill the void.[63] Cable essentially removed itself from the home security business, leaving it to companies who specialized in just that.

interactive

1983	1990
1. ESPN Getty Oil Co. 27.5 Million	1. ESPN Capital Cities/ABC, RJR Nabisco 61 Million
2. CBN Christian Broadcasting Advertising Network 22.2 Million	8. The Family Channel Christian Broadcasting Network 51.8 Million
3. CNN Turner Broadcasting 20.8 Million	2. CNN Turner Broadcasting, TCI, Time Warner, United Cable, and 14 other MSOs 56.5 Million
4. USA Network Time Inc., Paramount, MCA 18.5 Million	3. USA Network Paramount, MCA 53.8 Million
5. MTV Warner Amex 15 Million	7. MTV Viacom 52.4 Million
6. C-Span Cable Satellite Public Affairs Network 14 Million	10. C-Span Cable Satellite Public Affairs Network 51.5 Million Merged with Daytime To Become Lifetime
7. Cable Health Network Viacom, Dr. Arthur M. Ulene, Jeffrey Reiss 13 Million	
8. Nickelodeon Warner Amex 12.7 Million	5. Nickelodeon Viacom 52.9 Million
9. ARTS Hearst/ABC 11.8 Million	13. Arts and Entertainment Hearst, Cap Cities/ABC, NBC 48 Million No longer exists.
10. The Inspirational Network Heritage Village 10 Million	
11. The Weather Channel Landmark Communications 9.8 Million	14. The Weather Channel Landmark Communications 46 Million
12. Nashville Network Group W 9.6 Million	9. TNN Gaylord Broadcasting 51 Million
13. Daytime Hearst/ABC 9.5 Million	11. Lifetime Cap Cities/ABC, Hearst, Viacom 50 Million

1983	1990
14. Satellite Program Network Satellite Program Network, Inc. 8.7 Million	No longer exists
15. Financial News Network Financial News Network 8.5 Million	18. Financial News Network FNN Inc., Infotechnology Inc. 35.4 Million
16. Modern Satellite Network Modern Talking Picture Service, Inc. 8.2 Million	No longer exists
17. Black Entertainment Television Robert L. Johnson, TCI, Taft 4.7 Million	19. Black Entertainment Television Robert L. Johnson, TCI, HBO, Great American Communications 29.1 Million
18. AP Newscable Associated Press 4.7 Million	No longer exists
19. CNN Headline News Turner Broadcasting 4.6 Million	15. CNN Headline News Turner Broadcasting, TCI, Time Warner, and 14 other MSOs 44 Million
20. The Learning Channel Appalachian Community Service Network 3.8 Million	24. The Learning Channel American Community Service Network, Infotechnology 20.9 Million
	4. The Discovery Channel Cox Cable, Newhouse Broadcasting, United Artists, TCI 52.9 Million
	6. Nick at Night Viacom 52.9 Million
	12. Turner Network Television Turner Broadcasting 50 Million
	16. VH-1 Viacom 38.8 Million
	17. QVC Shopping Network QVC Network Inc. 35.7 Million
	20. American Movie Classics Cablevision Systems, TCI 28 Million

FIGURE 6.14

Comparison of the top twenty basic cable services (in terms of subscribers) between 1983 and 1990. Note the number of services that went out of business or merged. (Given are position, name, owner(s), and number of subscribers.)
(Source: North American Publishing, Channels 1984 Field Guide *and* Channels 1991 Field Guide.*)*

FIGURE 6.15

An ad for a pay-per-view service offered by Playboy. *(Courtesy of Playboy At Night)*

A different type of programming form was touted as cable's new frontier during much of the 1980s. This was **pay-per-view,** which enabled viewers who paid an extra one-time-only sum, to see special events such as boxing matches and first-run movies. Most pay-per-view plans asked subscribers to phone the cable company to order the event that would then be sent through the wire only to those homes requesting and paying for it. Special addressable equipment was needed in order to service part of the cable system while preventing the event from entering the homes of people who did not want the pay-per-view event. Once the equipment was developed and installed in cable systems, a proliferation of pay-per-view networks were proposed—and launched, re-organized, renamed, and discontinued. Pay-per-view was not the instant success many had hoped it would be. But as the years progressed, those companies that stayed with the business experienced a slow increase in subscribers and revenue. Going into the 1990s, pay-per-view was considered an evolutionary business that would have continued growth.[64]

pay-per-view

Deregulation and Regulatory Issues

Cable TV, like its cousin broadcasting, has been affected by the deregulatory mood of government. In 1980 the FCC abolished both the syndicated exclusivity and distant signal importation rulings, opening the way for cable systems to import as many stations as they wished and to play the programming of those stations without having to worry about whether the same programming would be on a local TV station.[65] Broadcasters, motion picture producers, and sports interests, concerned about the effects of this **deregulation** on their businesses, appealed the FCC ruling, but the Supreme Court ruled in favor of the cable interests.[66]

However, in 1988 the FCC reinstated syndicated exclusivity, once again giving broadcasters their way on this controversial subject. Now cable systems had to worry again about blacking out programs on superstations and distant stations. The superstations cooperated by trying not to show programs that might be airing on local stations until at least several weeks after the local broadcast programs had aired. Of course, the cable interests also appealed the ruling, so this issue is likely to continue to seesaw.[67]

Another major piece of deregulation legislation, the **Cable Communications Policy Act of 1984,** was passed by Congress and delineated the role of cable systems and local governments. The main positive change for cable operators was that cities with at least three broadcast stations could no longer regulate the rates that cable systems charged their customers for the basic service. Instead, these prices were to be determined by the marketplace. The most positive aspect for the cities was that they were allowed to raise their franchise fee from 3 percent to 5 percent.[68]

The deregulation of cable did lead to a sharp increase in the rates charged to subscribers—as much as 10 percent a year.[69] This led the government to take another look at its deregulation position and consider reregulating cable. During the 1990s various pieces of legislation were proposed that would reestablish some sort of overseeing functions regarding cable rates. These bills generally contained other provisions to cut down on the monopolistic powers gained by cable companies, as most are sole providers of cable programming within their communities. The bills were also intended to reduce the power and influence of the large cable companies such as TCI.[70]

The old issue of must-carries surfaced again for several reasons. Some cable practitioners began promoting the abolishment of must-carries so that cable systems could fill their channels with satellite programming rather than with local station programming. They reasoned that local stations could be received over the airwaves, so cablecasting the channels constituted an unnecessary duplication of services to the consumer. Local stations, of course, did not agree because they wanted to be carried on the cable systems.

In a series of legal maneuvers, the FCC and the courts outlawed and reinstated must-carries a number of times, each time changing the provisions slightly as to which stations needed to be carried. As of the beginning of the 1990s, must-carry was not in effect, so cable companies did not need to carry local stations. However, cable companies were supposed to sell A/B switches to customers so that they could switch from receiving cable signals by wire to receiving broadcast signals over the air. Cable systems balked at this and were given several reprieves as to when they needed to make these switches available. Most cable systems have continued to carry most broadcast TV stations; however, they often place them on upper channels, which are not as desirable as the lower numbers. Broadcasters, meanwhile, are looking to the courts and to Congress to reinstate must-carry, and broadcasters are suggesting that cable systems should have to pay for broadcast signals that they do carry.[71]

Another issue to rear its head again was copyright. The Copyright Tribunal, acting under its right to adjust fees, ordered cable systems to pay

3.75 percent (a hefty increase from the original 0.7 percent) of their gross receipts for each imported distant signal that was above the number of distant signals allowed under the 1972 rulings. This ruling had set up a complicated formula for distant signal importation that depended on the size of the cable system, the size of the market, and the number of stations in the market. The systems most affected were fairly large ones in densely populated areas, but these systems did serve about 10 million people.[72] The cable companies, of course, objected and appealed, but the courts ruled in favor of the Copyright Tribunal. This made the program producers very happy because it meant more money for them if the cable companies paid the fee. In 1983, the first year the 3.75 percent fee was in effect, cable operators paid $25 million more to the Copyright Tribunal than they did in 1982 for a total of $69.1 million. Cable companies continue to fight to have the fees reduced, so this issue, like others, still dangles.[73]

Another legal issue to surface was called **overbuild** and involved whether or not a city could give one company exclusive rights for cabling the city or a part of the city. After several trips to the courts, this issue was decided in favor of the companies that wished to overbuild—i.e., set up second systems. In other words, cities had no right to refuse to allow a company to wire a city just because it did not win the franchise. This idea is, of course, opposed by the established cable companies. In reality, it is not much of a threat to the establishment, however, because very few communities are able to support two profitable cable companies, and, therefore, hardly any companies are applying to build a second cable system.[74]

Even the old issue of pole attachments recycled. This time the Supreme Court ruled that the rates the phone companies charged the cable systems could be regulated by the FCC.[75]

Regulation, or lack thereof, has been very important in the history of cable TV and will no doubt play a large role in the future. Now that cable has matured, it may be competing for favorable regulation not only in relation to broadcasting, but also in relation to competing developing media.

overbuild

pole attachments

Satellite to Home Broadcast

One of the developing media most likely to cause cable trouble is program delivery directly from satellites to homes. This concept has been around since 1979 when the FCC issued a ruling stating that no one needed a license in order to have a receive-only satellite dish. At this point, some individuals, particularly those in rural areas where TV reception was poor, bought the same kind of large three-meter dishes that cable TV systems bought and set them up in their backyards. This became known as **backyard satellite reception.** The dishes cost about $3000, but the people who bought them could then receive all the cable programming for free. This caused a minor stir among cable TV program suppliers, but the number of backyard dishes was so small that the cable folk did not aggressively pursue the issue.[76]

beginnings of backyard reception

FIGURE 6.16

DBS configuration.
(From Gross, Lynne Schafer,
The New Television
Technologies. 3d ed. © 1990
Wm. C. Brown Publishers,
Dubuque, Iowa. All Rights
Reserved. Reprinted by
permission.)

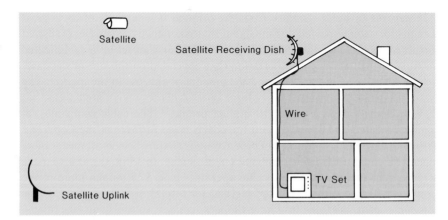

Satellite

Satellite Receiving Dish

Wire

TV Set

Satellite Uplink

DBS proposals

A more grandiose type of satellite delivery, known as **direct broadcast satellite (DBS),** was proposed by a company called Satellite Television Corporation (STC), which in 1979 informed the FCC it wanted to develop a programming service that would go directly from satellites to homes.[77] The FCC agreed with the basic idea and invited all interested parties to apply for DBS licenses. Thirteen companies applied, and in 1982, the FCC approved eight of the applications and expected that the services would be operational by 1985. The applicants planned to beam from high powered satellites to home dishes or plates that were no more than three feet in diameter. However, technical and financial difficulties overcame all the applicants and, going into the 1990s, none of them were yet operational. In fact, most of them, including STC had given up.[78]

New companies became interested in DBS periodically, however. Two particularly high-powered groups who promise DBS in the near future are Sky Cable and K Prime Partners. Sky Cable's main movers are Rupert Murdoch (who started Fox network and who is instrumental in a British DBS system that is operational); NBC; a cable TV company, Cablevision Systems Corporation; and Hughes Communications, a company that has been active in the satellite business for many years. The forces behind K Prime Partners are Time, Warner, TCI, and General Electric.[79] Two of the original DBS applicants, Dominion, and United States Satellite Broadcasting, are also still trying to develop viable systems.[80]

backyard growth

While DBS was having trouble getting off the ground, the more mundane backyard dish business was growing; hence, the companies with programming on satellites became more upset that the dish owners were receiving the programming for free. The result was that many of the services began **scrambling** their signals. At first they seemed unwilling to allow dish owners to view the programming at any price, but eventually, with a little urging from Congress, they allowed independent companies to sell descramblers and various programming services to dish owners. Having to pay for programming caused a dip in satellite dish sales during the mid-1980s, but by the 1990s the business was picking up again, and more than 2 million households had purchased home dishes.[81]

Electronic Media Forms

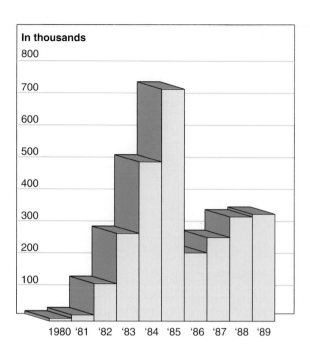

FIGURE 6.17

Consumer satellite television dish sales. *(Source: Satellite Broadcasting and Communications Assn. of America.)*

The main advantage of satellite delivery over cable TV is that wires do not need to be strung along streets and into homes. The most advanced of the satellite reception devices are small, unobtrusive mirrorlike plates that can be attached to the outside wall of a house. Only time will tell whether or not these will replace cable TV.

SMATV

Another competitor to cable, **satellite master antenna TV (SMATV),** has been in operation for more than a decade and does not appear to be a substantial threat. SMATV is a system for supplying TV programming to apartments, hotels, and private developments. It became possible because of the same 1979 FCC ruling that allowed for DBS—the one that stated that no one needed to apply for a license in order to install and use an earth station that received satellite TV signals.

Prior to this, owners of apartment complexes often had outside contractors install Master Antenna TV (MATV) systems. In its simplest form, this was a regular TV antenna installed on top of an apartment building that was wired to each apartment. In this way, all apartment dwellers could receive the regular broadcast TV signals without each having to place an antenna on the roof of the building.

After the FCC's 1979 decision, the apartment owners could contract to have satellite signals fed into the Master Antenna system so that cable TV network programming could be played on empty channels not occupied by broadcast TV. This combination of satellite and broadcast signals became known as SMATV.[82]

1979 ruling

MATV

FIGURE 6.18

SMATV configuration.
*(From Gross, Lynne Schafer,
The New Television
Technologies. 3d ed. © 1990
Wm. C. Brown Publishers,
Dubuque, Iowa. All Rights
Reserved. Reprinted by
permission.)*

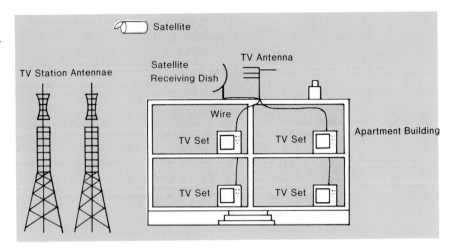

Although SMATV operations compete with cable companies that may be trying to wire apartment complexes, SMATV does not come under the FCC definition of a cable TV system. The FCC's definition specifically excludes systems that serve fewer than fifty subscribers or serve subscribers in one or more multiple-unit dwellings under common ownership, control, or management. This means SMATV is not regulated and does not need to pay fees to city governments.[83]

cable-SMATV conflicts

There is no love lost between cable and SMATV, and each takes the other to court frequently to try to solve the problem of who can go after whose customers. Cable systems try to prevent SMATV operations from taking hold in areas they have wired, and SMATV operators try to prevent cable representatives from soliciting in their apartment complexes. In addition, most companies supplying pay-cable programming try to avoid contracting with SMATV companies.

Overall, SMATV is a small but rather steady business with between 450,000 and 600,000 customers, and it seems unlikely that SMATV will compete with cable TV in any arenas except apartment complexes.[84]

MMDS

frequencies

Multichannel multipoint distribution service (MMDS) is broadcast in a manner similar to over-the-air commercial TV except that the frequencies used are two thousand megahertz higher than conventional broadcast frequencies. As a result, they cannot be received with a regular TV set. In order to receive these signals, a special antenna and downconverter system is needed. The MMDS systems also have a range of about twenty-five miles, so they cannot be used to send televised material over long distances.

The frequencies for MMDS have been available since 1971, but they laid dormant for many years, mainly because no one could figure out a way to make money using the channels. However, with the advent of HBO and other pay-TV services, several companies began to see a way of making money

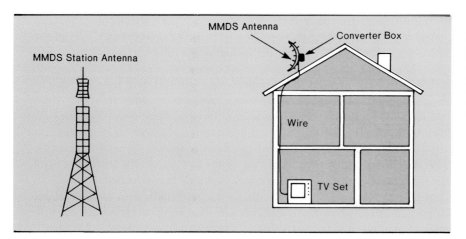

FIGURE 6.19

MMDS configuration.
*(From Gross, Lynne Schafer,
The New Television
Technologies. 3d ed. © 1990
Wm. C. Brown Publishers,
Dubuque, Iowa. All Rights
Reserved. Reprinted by
permission.)*

utilizing the frequencies to transmit pay movies to apartments, hotels, and homes.

The first MMDS operations were called MDS (multipoint distribution service) because they only used one channel. They were similar to subscription TV in that they programmed one channel with movies and charged the customer for the equipment needed in order to see the picture on a regular TV channel and for the programming itself.[85]

As the business grew, the FCC allocated more channels for MMDS so that individual companies could have four or more channels and in that way offer more programming services.[86] This multi-channel growth alarmed the cable operators somewhat because they feared MMDS would take away business from cable systems. So the cable networks withheld programming from MMDS, much as they did from SMATV. This has been a bone of contention to MMDS operators. They have asked Congress to legislate their right to the programming and MMDS operators have taken the cable networks to court. However, they have not received a clear-cut victory, so cable networks still withhold programming from them.[87]

MMDS does not appear to be a significant threat to cable. Most systems have less channels than cable systems do; its reach is not very far; and it only programs to about 350,000 homes.[88]

Telcos

The latest competitive threat to cable TV comes from the telephone companies (**telcos**). Under present law, they are specifically prohibited from bringing video services into homes. This is a remnant from the old telephone pole controversy. The phone companies were renting space on the poles to the cable companies, and if they, themselves, could deliver what cable TV was delivering, they could easily undercut the whole cable business. Also, during the 1950s and 1960s when cable TV was beginning, AT&T was a huge monopoly, and there was no desire to further its monopolistic control into the television arena.

MDS

programming conflicts

old regulations

But times have changed. AT&T has been broken up, and it is the cable companies that are acting monopolistic with their rate increases. The phone companies, which offer a vast array of audio and data services, want to add video. They point to the convenience of one wire, instead of two, coming into each house. Some of what they are proposing is interactive services, such as home shopping and home banking, which cable has abandoned.[89]

GTE waiver

In 1988 the FCC granted a waiver to General Telephone (GTE) in Cerritos, California, so that it can build and own a cable system in conjunction with (but financially separate from) the phone system it operates in that city. GTE is going to experiment with fiber optics as the delivery means and with various two-way programming services, some of them related to video-on-demand pay-per-view. In other words, customers will be able to call and ask for particular movies at particular times. GTE also plans to experiment with home shopping and with instructional subjects, such as cooking and fishing.[90]

The FCC, Congress, cable companies, and the telcos are watching the Cerritos experiment closely. Whether or not telcos should be allowed into the cable TV business is one of the major debates of the 1990s.

Cable TV Chronology

1940s First cable TV begins
1949 Distant signal importation begins
1960 Bill to regulate cable TV defeated in Congress
1960 United Artists sues cable TV for copyright infringement
1963 Carter Mountain case gives FCC right to restrict cable
1965 FCC issues rules on must-carry and syndicated exclusivity
1966 FCC publishes report to restrict cable service
1967 First regularly scheduled local origination programming starts
1970 Cable systems are ordered to have local origination
1971 Frequencies set aside for MDS
1972 Revision of must-carry rule from sixty to thirty-five miles
1975 HBO begins distributing movies by satellite
1977 Qube initiated
1977 HBO turns a profit
1977 Rules prohibiting syphoning set aside
1978 New copyright bill goes into effect
1978 Franchising boom begins
1978–1981 Numerous cable programming services started
1979 FCC says a license is not needed for a receive-only satellite dish
1979 STC proposes DBS
1979 SMATV systems started
1980 Pay-TV revenues grow 95.5 percent
1980 FCC abolishes syndicated exclusivity and distant signal importation rules
1981 RCA auctions off satellite transponders
1982 Copyright Tribunal increases distant signal copyright fees to 3.75 percent

1983   CBS Cable fails
1983   Showtime and The Movie Channel merge
1983   Daytime and Cable Health Network merge
1983   CNN buys SNC
1983   MMDS receives more channel allocations
1984   ABC buys ESPN
1984   Qube is ended
1984   Congress passes Cable Telecommunications Act deregulating cable
1984   Warner-Amex sells some cable systems
1985   Group W sells all of its cable systems
1985   Overbuild sanctioned
1985   Discovery Channel launched
1985   Must-carry rules temporarily abolished
1988   FCC reinstates syndicated exclusivity
1988   FCC grants GTE waiver to build cable system
1988   NBC begins CNBC
1988   Turner starts TNT
1990   Congress considers reregulating cable
1990   Sky Cable DBS plans announced

Conclusion

Cable TV has undergone many changes, especially in recent times. What started as one- to three-channel systems during the 1950s promised in excess of one hundred channels in the 1970s and settled for about fifty channels during the 1980s. A technology that involved only wires has now wholeheartedly embraced satellites for distance distribution.

An industry that maintained stable, modest, consistent numbers for decades suddenly boasted figures that jumped hundreds of percents in one year and then settled back to modest, slow-growing numbers just as quickly.

Ownership of cable systems changed greatly, too, from the early community-oriented, ma-and-pa owned systems to multiple system owners. These MSOs, in turn, merged and acquired more cable-related properties and then began a consolidation and withdrawal process because they were over-extended. Eventually several emerged as major powers in the business.

The whole regulatory scene has undergone evolution. Early cable TV was essentially unregulated with rules arising only as broadcasters convinced federal governmental bodies of the need to protect them from encroachment by the cable systems. The government's reaction, primarily from Congress and the FCC, was one of confusion and abstention; however, during the 1960s and early 1970s rules were enacted dealing with must-carry, distant signal importation, syndicated exclusivity, pole attachments, and copyright. As the government developed a deregulation policy during the 1980s, cable TV became even less regulated by both federal and local agencies than it had been. However, all the old issues from the 1970s continued to rear their heads, and reregulation seemed on the horizon.

The franchising of cable systems was picked up by local governments almost by default and was a sleepy process until cable boomed in the late 1970s. Then municipalities were besieged by cable companies eager to wire the cities. Franchise proposals became huge documents, and the franchising process bred large stakes, complete with a certain amount of accompanying corruption. As the wiring of major cities became complicated and expensive, the fervor for obtaining additional franchises abated, although some companies did become interested in overbuild.

More than anything else, there was an immense change in cable TV programming. Original cable systems retransmitted broadcast programming of both a local and imported nature. A few systems originated their own programming through public access, local origination, or movie channels, but the original attempts at this were largely unsuccessful. Promises, not performance, were the hallmark of early cable. But the tide turned with the late 1970's success of the HBO satellite venture, followed closely by the beginning of superstations. The floodgate opened to a raft of programming services produced exclusively for cable TV as opposed to broadcast TV. Cable programming became a glamour business with material distributed on both a national and local basis. A multitude of pay services and basic services emanated from the satellites, and the cable systems themselves revived the concept of local programming, often with the promise of numerous local channels to meet various needs within the community. The interactive services that had been long on talk and short on action began to undergo experimentation, most notably in the Qube system. The programming concepts of the late 1970s and early 1980s changed the entire face of the cable industry. A shakeout occurred during the mid-1980s and many of the national programming services merged, were bought out, or went out of business. Local programming was greatly cut back, and interactive services were virtually eliminated. Cable programming is still abundant, but experimentation and growth have stabilized. Pay-per-view is a concept that is growing slowly.

Cable and broadcasting have always considered themselves competitors, but cable has other competitors, too. One involves signals that go directly from satellites to homes. This is operational with backyard reception and is proposed by companies planning for DBS. SMATV and MMDS both have small markets that are usually in competition with cable. Telephone companies are also lobbying to enter the cable business.

The cable industry is still in a state of flux, making it an exciting phenomenon. Its future course has yet to be determined.

Thought Questions

1. What should be the policies regarding cable copyright payments, distant imports, must-carries, and syndicated exclusivity?
2. Do you foresee any changes in the types of programming that will be on cable TV in the future? If so, what?
3. What do you think will be the state of cable TV ten years from now?

Personal and Organizational Telecommunications

Introduction

Many uses are made of telecommunications by such entities as schools, corporations, hospitals, government offices, libraries, the military, and individual consumers. Although these uses may not be as "glamorous" as traditional broadcasting, they are often quite complex and quite abundant.

Much of the material produced in and for personal and business telecommunications is not intended for the mass audience. Just about anyone can turn on NBC news, listen to music on radio, or watch sports on cable TV. But only a select few can see a video about a new airplane design by Boeing, or access the TRW credit information data file, or work with a videodisc that trains fighter pilots, or even see the home videotape of Suzie's first birthday party. All of these are specialized uses of telecommunications intended for small groups who have something in common and have a specific need for the information.

The telecommunication uses currently being undertaken by companies and individuals are closely tied to advancements in technology. Much of it has been made possible by the shrinking (both in terms of size and cost) of various video equipment, particularly the camcorder and the videocassette recorder. Other uses result from the introduction of the videodisc, a significant upgrading of the telephone, and the advancements made in computer technology.

The umbrella term to describe this type of telecommunications has not stabilized. At one time, video production undertaken by businesses was referred to as "industrial TV." But the word "industrial" had a rather grimy, dingy connotation so it was abandoned, and other terms such as "business television" and "corporate video" surfaced. Educational institutions, government facilities, hospitals, and other organizations do not consider themselves industries or corporations, so they often used the terms "instructional TV," "organizational TV," or "institutional TV." Video used by individuals has been called a number of terms including "home video," "consumer video," "individual video," and "personal video." When other communications methods, such as discs, computers, and telephones, are used in conjunction with or separate from the more traditional video production, terminology becomes even more muddy. The word "telecommunications" applies, but so do other words such as "nonbroadcast communications" and "private communications."

This chapter title, "Personal and Organizational Telecommunications," represents one current way of naming this overall phenomenon. However, various other terms are used throughout the chapter, especially as they apply to particular subdivisions of telecommunications at particular times in history.

I believe in the future . . . a man in one part of the country may communicate with another distant place.

Alexander Graham Bell, 1878

Organizational Video

old equipment

The first uses of television equipment for anything other than broadcast TV began during the late 1960s and early 1970s. Before that, video equipment was so bulky and expensive that only TV stations and networks could afford and house it. Cameras that produced black-and-white pictures weighed about fifty pounds and cost in the neighborhood of $30,000. Videotape recorders stood six feet high, used bulky two-inch tape, and cost several hundred thousand dollars.

educational uses

But during the late 1960s, smaller black-and-white cameras that were not as high quality as the broadcast version, and recorders that used one-inch tape were developed. One of the first areas to become involved with the use of this equipment was education. Colleges bought cameras and other video equipment to teach students to become broadcasters. These students could learn the principles of video production using this cheaper, lower-quality equipment and then transfer this knowledge when working with the professional gear.

Schools also used this equipment to enhance instruction. Teachers would tape lessons to show in the classroom, particularly when something was difficult to show to a large class. For example, a science teacher could tape the dissection of a frog and show it on a large monitor for a whole class to see. This worked better than trying to have a whole class gather around a table to view the procedure.

industrial training tapes

Industry, too, began using video primarily to improve training of personnel. For instance, a tape could be produced to show how to perform a particular assembly line job. Then when new people were hired, they could be shown the videotape, and a supervisor would not need to show each one the various techniques. In a similar manner, the military used tapes to train recruits, and hospitals produced programs to keep doctors up to date.

Another way tape was used as a training method was to tape people performing a certain task and then replaying the tape so they could see where improvements could be made. For example, simulated sales calls were set up and taped to see how a new salesperson would handle the situation. The tape was then played back and critiqued by an experienced salesperson.[1]

During the early 1970s, even smaller, cheaper equipment was developed, primarily for network and station newsgathering, utilizing color cameras that could be carried on the shoulder and **videocassette recorders,** called **U-matic,**

three-quarter-inch tape

that used three-quarter-inch tape. Once this equipment was on the market, many more educational, industrial, government, medical, and military organizations began using it. Sometimes they produced in the studio, and sometimes they took this more portable equipment to locations. Sometimes organizations bought equipment and hired a staff to produce the tapes, and sometimes they hired outside independent production companies to produce for them on an as-needed basis.

In 1971 two groups were formed to aid communication among these various video users. These groups were called the Industrial Television Society and the National Industrial Television Association. Eventually these two groups

FIGURE 7.1
One organizational use of video involves the taping of homes on sale for in-office presentations, as is being undertaken by this Southern California realtor.
(© Spencer Grant/Photo Researchers, Inc.)

merged into an organization that is now called the **International Television Association (ITVA).**[2]

ITVA

    In addition to producing training and instructional tapes, organizations started making tapes dealing with internal communication. For example, the company president would tape a talk that would be shown to all employees. Within some businesses, regularly scheduled video newsletters were produced to keep their employees up to date on company business. Orientation tapes that covered the history of the organization, its functions, and its benefits were shown to new employees so they could better understand the organization they had joined. New company products were demonstrated on videotapes that were then sent to sales representatives around the country.[3]

other industrial programs

Closed-Circuit TV, ITFS, and Teleconferencing

As the decade progressed, several methods of distribution were established for the tapes produced by various organizations. **Closed-circuit TV** systems were set up primarily by educational institutions. This involved wiring a number of buildings or rooms so that all could see the same programming being played back from a central location. Educators used these to teach entire courses to large groups of students. For example, all students in a grade school could receive Spanish instruction at the same time from one Spanish-speaking teacher working from a central studio. Or, all college students taking American history at 9:00 A.M. could receive televised lectures from one professor who was either live in a studio or had been taped ahead of time, even though the students were scattered in ten different rooms around the campus.[4]

closed-circuit TV

    Another type of distribution used by educators was **Instructional Television Fixed Service (ITFS),** which was a special form or over-the-air broadcast

ITFS

FIGURE 7.2

Master control of an ITFS facility owned by the Long Beach Unified School District. This operation, which has been in existence since the 1960s provides programming to 84 schools and over 73,000 students. It also sends programming to a local cable TV system. *(Courtesy of Long Beach Unified School, Ed Kniss, photographer)*

authorized by the FCC in 1971. It involved giving educational institutions frequencies in the 2,500 megahertz range. This is much higher than regular broadcast frequencies, so regular TV sets cannot receive the signals unless they are attached to special downconverters. Also, the signals from ITFS could only reach about twenty-five miles, a relatively short range. Some school districts used ITFS to interconnect all the grade schools in an entire district. In this way a Spanish teacher could teach all third graders in one district at the same time. These ITFS systems were also used to retransmit public TV programs such as "Sesame Street" to be used by teachers at times other than the times the programs were broadcast by the local public TV station.

A number of universities also set up ITFS systems, some of them in cooperation with local businesses who often wanted to upgrade training for their employees (perhaps a course in marketing or economics). A professor at the university could offer the course, which was then sent through the ITFS transmitter to the antenna and downconverter located at the business. The course was scheduled so that employees could watch it during their lunch hour or right after work. Hospitals were big customers for these types of courses because busy doctors and nurses need updating on rapidly changing medical technology.[5]

When satellites became predominant during the late 1970s and early 1980s, organizations began **teleconferencing** (also known as **videoconferencing**). This involves renting satellite time and using it to transmit video information from place to place. Usually this video information consists of people

teleconferencing

FIGURE 7.3

A teleconference being held for the purpose of training teachers in new techniques.
(© Cynthia Dopkin/Photo Researchers, Inc.)

making presentations to people in other parts of the country or the world. Teleconferencing usually has an interactive telephone element so that viewers can phone questions to the speakers.

One of the early teleconferences was held in 1979 and involved the American Soybean Association. This conference linked together speakers in Tokyo, Rio de Janeiro, London, and Atlanta through a satellite conferencing setup. In this way, approximately fifteen hundred soybean growers were able to hear experts from four continents discuss market outlooks for their crops. Television cameras in the four cities were connected to uplinks that enabled the remarks made at each location to be fed to satellites that transmitted the messages to a number of meeting halls.[6]

Similar teleconferences have been held for many other organizations— medical associations, universities, and business affiliates. These are of a one-shot nature and are usually handled by companies that specialize in arranging teleconferences. These companies provide the rooms, the television equipment, and the satellite time. Several hotel chains, such as Holiday Inn and Hilton, have equipped many of their hotels with satellite dishes and rooms that are permanently prepared for teleconferencing so that conferences can be held in many cities at the same time, all using the same hotel chain.[7]

Sometimes teleconferencing is conducted on a regular basis as opposed to a one-shot conference. For example, J.C. Penney Company regularly uses teleconferencing to link its product managers with buyers at the various stores. This greatly cuts down on the amount of time required to get the product through the system. Federal Express gives daily reports on weather and air traffic to its 800 offices around the country. Automobile companies use satellites to introduce their new lines to their distributors. The Army uses teleconferencing for technical training.

Some corporations with branches in several cities arrange for their top executives to "meet" through teleconferencing. Cameras and microphones at each location pick up what executives there say and beam it to the other locations. The executives can then interact in a manner that is similar to what they would do if they were all in one room.[8]

problems and failures

These distribution methods have not been overwhelmingly successful. Closed-circuit TV has essentially disappeared. It did not give the warmth and personal attention of classroom teaching. When videotapes, which have replaced closed-circuit TV, are played back in a classroom, they are usually played on a portable cassette recorder, so elaborate wiring is no longer needed. ITFS systems still exist, but not nearly as many educational institutions wanted into the field as the FCC had anticipated, so in 1983 the FCC wound up giving some of the ITFS frequencies to the commercial **multichannel multipoint distribution service (MMDS).**[9] Some companies feel very positively toward teleconferencing and use it extensively. Others feel the method is too impersonal and either have not tried it or have been disappointed when they have tried it.[10]

By the 1980s, organizational video was off to a good start, but further improvements and uses would come as a sister function, home video, came to the fore.

Personal Video—The VCR

The development of video equipment for the home consumer started during the mid-1970s. At first this equipment was very different from that used by professional broadcasters or by industry and education. Eventually, however, the equipment used at all three levels—professional, industrial, and consumer—began to merge.

Betamax

The first videocassette machine specifically designed for the home-consumer market was the Sony **Betamax,** introduced in 1975. This machine used half-inch tapes that could record for one hour. It sold for $1,300, but the price was actually $2,300 because the recorder could only be purchased with a new Sony color TV set.[11]

VHS

The price went down and the features went up as **Sony** encountered competition from **Matsushita,** which in 1976 introduced a cassette recorder with half-inch tape that it called video home system (**VHS**) and marketed through its **Japan Victor Company (JVC).** This JVC recorder had a marked advantage in that it could record for up to two hours. People who bought the recorders were using them primarily for recording feature films off the air, so the two-hour format enabled them to record an entire feature on one cassette.[12]

features

With the advent of competition, the technological war was on. Sony altered its Beta format so that it could record longer too. Both Beta and VHS adopted such features as pause buttons that enabled consumers to eliminate commercials while they were recording, timers that allowed recordings to be made while the owner was not at home, devices that allowed one program to be recorded while another was being viewed, still frame and variable speed

FIGURE 7.4

A Sony Betamax videocassette recorder. *(Courtesy of Sony Corporation of America)*

FIGURE 7.5

A VHS videocassette recorder. *(Courtesy of JVC Industries Company)*

that allowed the owner to hold a particular picture on the screen or view the picture either in slow motion or fast speed, and, of course, taping methods that allowed for longer and longer recordings.[13]

Many other companies also began manufacturing cassette machines, with more of them adopting the VHS method than the Beta method. The two methods were not compatible, so a tape recorded on a VHS machine could not be played back on a Beta machine even though both used half-inch tapes. The main difference between the two was the method by which the tape threaded itself in the machine.[14]

Once these videocassette machines became somewhat sophisticated, their sales soared beyond expectations. More than 1 million recorders were sold in 1979,[15] and more than 3 million in 1981.[16] People used them primarily for recording off the air, but consumers also began purchasing prerecorded videocassettes that contained movies, self-help programs, and educational material.

A whole new business of video rental stores sprang up. Fotomat was one of the first, renting videocassettes from its various film developing locations. Soon, however, chain video stores were springing up on almost every corner.

success

video stores

Tapes were also being sold or rented from grocery stores, convenience stores, bookstores, and even movie theater lobbies.

copyright suit

Into this success story came a lawsuit. In 1976, a year after Sony introduced its Betamax, Universal Studios and Walt Disney Productions brought suit against Sony claiming that any device that could copy their program material was being used in violation of **copyright** and should not be manufactured. The case was dragged through the courts, and in 1979 a federal judge sided with Sony, saying that copying TV program material for use in the home comes under the "fair use" doctrine and is not an infringement of copyright.

However, two years later an appeals court sided with Universal and Disney, saying that home recording was indeed a violation of copyright and that Sony, which obviously knew its recorders were being used for this purpose, was at fault.[17] Fortunately for Sony and other VCR manufacturers, in 1984 the Supreme Court reversed this decision and ruled that home taping does not violate copyright laws.[18] This decision was fortunate for consumers, too, for by 1984 more than 13 million recorders had been sold.[19]

Betamax demise

Ironically, after Sony won the copyright battle, it lost the crucial marketing war with VHS, which became the preferred choice of consumers. During the mid-1980s, Sony began phasing out Beta, and, in 1988, it began selling its own VHS machines.[20]

Throughout the years, some of the consumers who bought VCRs also bought small cameras to go with them so that they could make "home movies." At first these cameras were fairly low quality and were units that were separate from the VCR. However, in 1984, **camcorders** were introduced to the consumer market and these all-in-one camera-recorders became quite popular.[21]

camcorders

A Merging of Professional, Industrial, and Consumer Video

Beginning in the mid-1980s, both Sony and JVC began manufacturing consumer camcorders and videocassette decks that they claimed were high enough quality for industrial, and even broadcast applications. The JVC entry was **Super-VHS,** an improvement on the VHS format that had better picture resolution and better rendering of color. It was somewhat compatible with VHS in that tapes made for VHS could play back on Super-VHS machines, but tapes made using Super-VHS could not play on regular VHS machines.

Super-VHS

Sony made some last minute improvements in its dying Beta format, but then switched its emphasis to an entirely new format, 8mm. This consisted of camcorders and decks that used tape that was approximately one-fourth-inch wide so that all the equipment could be smaller and lighter than the half-inch formats. Sony introduced **Video-8** in 1985 and followed this in 1989 with **Hi-Band 8,** a format it said could be used across the board, from the consumer level to broadcast stations.[22]

8mm

It is still too early to tell which of these formats will be adopted by which entities. Regular VHS still dominates in the home, mainly because most movies at rental stores are in this format. The 8mm format seems to be a popular

FIGURE 7.6
Bob Saget, host of
ABC's "America's
Funniest Home
Videos," the first of the
programs that featured
material shot by
consumers with their
camcorders. *(Everett
Collection)*

choice with consumers for home taping, especially the models that are about the same size as a still camera and can, therefore, be easily taken on vacations. Many companies made large investments in three-fourths-inch U-matic equipment and are reluctant to leave that format.

Meanwhile, programming at the professional, business, and personal levels is beginning to merge. Tourists who just happen to have their home video cameras at the scene of some news-making event, such as an earthquake or an airplane crash, can find that their footage is sought after by networks.[23] Several very popular network prime-time programs emerged during the early 1990s featuring humorous home video excerpts. So many people from around the country sent in cassettes that the networks had to increase the number of people they hired to screen the video tapes.[24]

merging programming

The forms of programming undertaken by businesses, educational organizations, and other institutions expanded. Although training and internal communication tapes still led as the primary purpose for corporate video, tapes were also produced for the public at large. For example, many companies that had handed out printed news releases to the press for years began making video releases for local TV stations and local origination cable channels. Nonprofit groups, such as the Heart Association and the Girl Scouts, made public service announcements for television broadcast. The medical profession produced tapes on various illnesses to show to patients waiting in doctors' offices. Some of this material was also distributed on home cassettes. Some of the instructional or student-produced material produced by educational institutions also wound up being distributed to the home market, cable, and broadcast TV.

program forms

The quality of program material was also becoming uniform throughout the professional, business, and personal realms. Corporations had access to the same fancy digital effects equipment as the networks and often used the same professional talent seen on TV. In fact, in some cases, businesses rented out their facilities for the production of local or national TV programs. During the early 1990s, low-cost editing equipment became available at the consumer level, enabling individuals to produce programming with a "professional look." In addition to taping family vacations, birthdays, and the like, some people began making tapes that could be called personal essays or video art.[25]

Organizational and personal video are important forces in the overall telecommunications scene. About 8,500 companies produce at least some programming. One difficulty faced by people engaged in corporate video, however, is that it is very dependent upon the economic health of the particular company. Many people in top management positions see the function of producing videos or holding teleconferences as expendable, and when the bottom line is not healthy, these services are likely to be cut.[26]

More than 70 percent of American homes have VCRs and 10 percent of them have camcorders.[27] The home video industry, since its very beginning, has racked up one success after another; however, it seems to be leveling off in the 1990s. People are not buying more recorders or renting more movies each year as was true for a long time. This may be due simply to the fact that the market is fairly well saturated. Or, it may signal a more serious disenchantment with all forms of video—network, cable, organizational, and personal.

Videodisc Players

Videodiscs, unlike videocassettes, have had a stormy, problem-packed existence. They started as a consumer product, but their main use, at present, is in the organizational area.

Development of the disc began in earnest during the late 1960s and was soon followed by many pronouncements of the potential efficiency, low cost, and flexibility of the disc. Announcement after announcement followed that the videodisc would be on the market "next year," but many next years came and went with the promises unfulfilled.

During these development years, there were two different disc systems. The less advanced was the **capacitance disc** developed by RCA. It had a diamond stylus that moved over grooves similar to those of a phonograph record and was marketed primarily as a device that could be used to watch movies.

The second system, the **laser disc,** had a very muddled parentage that included marriages, divorces, and partial custody by a large number of companies, including MCA, Philips, Magnavox, IBM, and Pioneer. The laser disc used a laser beam that read information embedded in a plastic disc, a technology that guaranteed the disc would not wear out and that any frame could be quickly accessed and viewed in freeze-frame almost indefinitely. Discs also

FIGURE 7.7

RCA's Selectavision
stylus videodisc player.
*(Courtesy of RCA
Corporation)*

FIGURE 7.8

A Pioneer LD-V2200
laser disc player.
(Courtesy of Pioneer)

had both picture and sound quality that was far superior to that of videocassettes. It, too, was intended primarily for viewing movies, but instructional software was also developed on disc that took advantage of its freeze-frame capability. However, unlike VCRs, both disc systems were playback-only; they could not record programs off the air. [28]

The laser disc player, then owned by MCA-Philips and called Disco-Vision, was the first on the market, selling for $700. The first models of the machine itself sold out shortly after its 1978 introduction, and many copies of movies on discs were also sold. However, the discs were so technically flawed that their sale and manufacture had to be discontinued.[29]

laser disc introduction

In 1981 RCA introduced its capacitance disc player, called Selectavision, for $500 each. RCA predicted it would sell 200,000 machines in the first year. At first sales were brisk, 26,000 players and 200,000 discs in the first five weeks,[30] but the pace quickly dropped, and RCA fell far short of its 200,000 goal for the first year. In fact, after three years only 500,000 machines had been sold, and RCA abandoned Selectavision. The company blamed videocassette recorders, which could record as well as play back, for its failure.[31]

capacitance disc introduction

capacitance disc demise

When RCA threw in the towel in 1984, the laser disc, which **Pioneer** had bought in 1982 and renamed LaserDisc, was still suffering from technical glitches and videodiscs seemed dead.[32] However, Pioneer managed to solve the manufacturing problems and began marketing the disc players to education and industry. In these markets, Pioneer emphasized the vast storage capability of the laser disc and its ability to freeze pictures and access them quickly.[33]

A number of companies did begin using disc systems, primarily for data storage, training, and consumer orientation. Some hospitals store entire medical records, including X rays, on discs, and companies can store vast amounts of past financial data on discs. Discs can easily store the equivalent of slide presentations because one disc is equal to 900 carousel trays. Any small bit of information, such as a patient X ray from November 14, can be accessed instantly. One problem with storing information on discs, however, is that it cannot be updated. Once information is written on the disc, it cannot be erased or changed. For this reason, computers, which are updatable, are much better for changing information such as stock market prices and weather information.

In the training area, the disc's main advantage is its ability to respond quickly from input by the learner. For example, Lockheed has developed a disc system for training military pilots. Potential pilots are placed in a huge dome where three-dimensional visual images, created by a disc and computer, simulate aerial combat. These images change to respond to the actions of the pilot in training. A mistake on the pilot's part results in nothing more harmful or expensive than a simulated crash.[34] Bank tellers, assembly line workers, grade school students, and repair persons can be trained in a similar but less flashy manner.[35]

The main way that disc systems are used for consumer orientation involves **kiosks.** These little booths exist in hotels, shopping malls, banks, airports, hospitals, and other places in order to give information to people visiting there. For example, a hotel may list on a video screen the organizations holding meetings there. By touching a particular part of the screen, a person could find out the meeting rooms of a particular organization and then the session occurring in a particular room. Likewise, someone in a shopping mall could touch a video screen to find out the names of shoe stores in that mall and could even be shown a map of how to walk to a certain shoe store.

The travel industry also uses video discs for customer orientation. Travelers can be shown different hotels at a resort and can then view individual rooms, such as the exercise rooms or the dining rooms of a variety of hotels. They can then make their decision on where they want to stay. Different recreational activities can be shown the same way.[36]

Although video disc technology has been somewhat successful in the organization area, its proponents are still lured by the consumer market. Many of the major video companies such as Sony, Sharp, and Philips have joined Pioneer in manufacturing disc machines for the home. The motion picture companies are very cooperative about allowing movies to be recorded on discs.

FIGURE 7.9

A computer/video directory to help visitors find their way around. *(Courtesy of Nelson and Harkins, Chicago)*

When audio **compact disc players (CDs)** became very popular during the mid-1980s, the video disc industry tried to latch on to this success by marketing machines that could play both video and audio discs. Some of these only played five-inch discs that could play either five minutes of video (primarily music videos) or twenty minutes of audio, but other machines were made so that they could play either the five-inch CD or a twelve-inch video discs capable of holding a movie.[37]

A great deal of research has also been undertaken to develop discs that can record. Discs that the consumer or a company can record on once have been developed, but the erase and replace function is still not readily available. Disc manufacturers continue to come up with improvements and gimmicks. For example, some movies on discs include information such as outtakes, script pages, and director's notes that allow film buffs to study the movie carefully.[38]

But, despite predictions to the contrary and despite excellent picture and sound quality, video discs have still not been a huge success in the consumer market.

The Telephone

The **telephone** has been used as a means of communication by industry and individuals for more than a century. Invented by Alexander Graham **Bell** in 1876, it predates the age of modern electronic media such as radio, television, and cable. But within recent years, new capabilities have been developed that have made the telephone a significant player in the organizational and personal telecommunications field.

Bell and two of his friends set up a company that eventually became the American Bell Telephone Company. From the very beginning this company

joining with CDs

recording discs

invention

AT&T formed

FIGURE 7.10
Alexander Graham
Bell demonstrating an
early telephone. *(AP/
Wide World Photos)*

leased phones to customers and charged them for phone calls. Bell first established local switching systems and then developed switching systems that linked
various cities in the east. Shortly after it began linking cities, American Bell
formed a subsidiary, **American Telephone** and **Telegraph Company (AT&T),**
to handle this long-line business. After a period of time, the child swallowed
the parent and AT&T became the main organization, owning American Bell
and other local phone companies. AT&T was set up as a monopoly, regulated
by the FCC, and for many years maintained the same basic organizational
structure.

monopoly situation

During this period of time the telephone business grew into a major national and international service, but it remained primarily a voice service connecting one individual with another. Improvements were made in switching
systems and in ease of phone installation, but because AT&T was a monopoly
making a guaranteed profit, it had little incentive to add new services for the
customer. In fact, in 1956 a consent decree barred AT&T from entering nonregulated businesses.[39]

Then during the 1970s, the Justice Department began taking a hard look
at AT&T's monopoly situation. This was spurred by several companies who
were making noises that they wanted to sell telephones directly to consumers
and several independent companies, primarily MCI and Sprint, who wanted
to enter the long-distance phone business. This was all quite controversial because AT&T's long-distance service had been subsidizing its local phone company service in order to keep local phone rates low. This meant other companies
could enter the long-distance business and offer much lower rates than AT&T,
but they could only do this because they did not need to subsidize the local
rates.

Electronic Media Forms

All of these issues somewhat coalesced, and in 1980 the Justice Department issued a **consent decree** that broke up AT&T. By order of this decree, AT&T had to divest itself of its twenty-two local phone companies. These were spun off into seven new companies serving seven sections of the country. Independent companies were allowed into the long-distance business, and customers were able to choose which long-distance company they wanted. AT&T was also required to notify customers that they could purchase phones instead of renting them from the phone company as they had always done in the past. AT&T kept its long-distance service; was still allowed to operate its research arm, Bell Labs; could still manufacture equipment; and was allowed to enter some unregulated fields, such as computer sales.[40]

break-up of AT&T

One of the fields AT&T was originally allowed into was **electronic publishing.** However, newspaper publishers, who were beginning to operate data services called **videotext,** did not like the idea of AT&T entering this field. They felt that because AT&T was so big and powerful and because it already had long lines in place, it could absorb the videotext field before others got to the gate.

The newspapers were particularly loud spoken because they were afraid videotext would replace their newspapers, so they wanted to be the ones in charge of it. They did not want the phone company, or anyone else, controlling the dissemination of news by means of the written word. Newspapers voiced powerful enough objections that they were able to get the consent decree modified to prevent AT&T from entering the electronic publishing business.[41]

newspaper opposition

With that out of the way, the newspapers began setting up videotext systems that delivered large amounts of information, similar to that in newspapers, to TV sets in consumer's homes. The information was stored in a large computer and delivered to the home by phone lines. The systems were two-way so that the consumer, by using a keypad, could ask the computer for particular information. The whole system was very similar to what eventually became computer technology, but the few data bases that were available through computers in the early 1980s were not very user friendly. Videotext, with easy-to-follow menus, was designed to be easy for the consumer to operate.

videotext

The two main newspaper chains that became involved in videotext were Knight-Ridder and Times-Mirror. Knight-Ridder set up a system called **Viewtron** in southeast Florida in 1983, and Times-Mirror set up **Gateway** in southern California in 1984. For both systems, subscribers had to buy equipment that cost about $600 and then pay about $12 a month for services. In addition to news, weather, sports, stock market quotes, and other information usually found in newspapers, the videotext services experimented with banking and home shopping. Unfortunately, both systems collapsed in 1986. Viewtron had more than 20,000 subscribers at its peak, and Gateway had less than 5,000. In general, subscribers liked the service but did not feel it was worth the cost. They continued to patronize the old-fashioned newspaper that was even more user friendly. It could be scanned more quickly than the videotext systems, and newspapers could be carried on the subway.[42]

FIGURE 7.11

A combination FAX,
answering machine,
and telephone. *(Courtesy
of Panasonic)*

Meanwhile, with competition going full force, the various phone companies began offering a new array of services. Telephone folks began talking about POTS (Plain Old Telephone Service) and PANS (Pretty Amazing New Service). Although some features such as the hold button and the provision for answering machines had been part of phone technology for awhile, the post-divestiture period of the 1980s saw a rise in many new features that were soon to be used by both businesses and individual consumers.

Telephones now have provisions for redialing numbers all by themselves, telling when another call is waiting, allowing a number of people to talk to each other all at once, and storing selected numbers. FAX machines that send written pages over the phone lines are becoming commonplace, both at work and home. There are 800 numbers that allow for toll-free calls, and, of course, there are the controversial 976 dial-a-porn numbers. Even airplanes have telephones so business people can call the office from thirty thousand feet up.[43]

All of these advances have aided the corporate-government sectors by enabling them to conduct business more efficiently. They also make life more convenient for individuals. These advances have also spawned quite a few new businesses, such as answering machine and FAX machine manufacture and sales, the design and sale of telephones themselves, and various dial-a-message services. The dial-a-porn services, which receive several million calls a day, cost about $3 million a year to operate and earn about $9 million.[44]

In the planning stages for the future is **ISDN (Integrated Services Digital Network).** This is an electronics technology for which a worldwide standard is being devised that theoretically allows two-way processing, storing, and transporting of whatever information anyone desires in simultaneous voice, data, graphics, or video.[45]

Cellular Phones

Another innovation in telephones involves **cellular phones,** mobile telephones used primarily in automobiles. A **mobile telephone service** had been available for years, but it involved relatively high-powered transmitters that covered a large area. Because of the high power, only a limited number of phones could operate in any particular area, totaling about 160,000 mobile phones in the entire country. There were always long waiting lists to obtain a mobile phone, so they did not add a great deal to business efficiency.

With the advent of cellular phones, however, an almost unlimited number of mobile phones is possible. The system operates by dividing a particular geographic area into rather small, discrete parts called "cells." Low-powered transmitters and receivers serve each cell—so low-powered, in fact, that several phones can transmit on the same frequency in the same cell without causing interference to each other. A person making a phone call from an automobile does so with a low-power transmitter whose signal is received at a central cell complex. This cell complex is tied by phone lines to the regular local and national phone system so the caller in the automobile can reach anywhere in the country. Similarly, the automobile caller can receive calls from anywhere in the country. The call goes by phone lines to the central cell location and then goes from a cell low-power transmitter to a receiver in an automobile.

The limited area of a cell might appear to be a disadvantage for cellular radio, but the real beauty of it is that as traffic moves from one cell to another, the cellular system has the ability to hand off signals from one cell to the next. This is accomplished by a switching system that determines when an in-progress call is at too low a signal level and, therefore, switches it to a closer cell.[46]

The technology for cellular radio was developed by AT&T's Bell Labs. The FCC authorized the concept in 1974 and several years after that allowed several experimental systems to be developed. One was in Chicago and was operated by what was then AT&T's Illinois Bell Telephone Company. The other was in the Baltimore-Washington area and was operated by American Radio-Telephone Service, Inc.—a company involved in mobile radio. Response to both systems from customers was overwhelmingly positive. In April of 1981 The FCC authorized cellular radio and invited applicants to apply for franchises to operate in various areas.

The structure for cellular radiotelephone was designed to foster healthy competition by following a model that allowed both phone companies and other companies to be involved. Up to two systems can operate in any area. One of

FIGURE 7.12

A drawing showing
how the cells of a
cellular radio-telephone
system overlap.

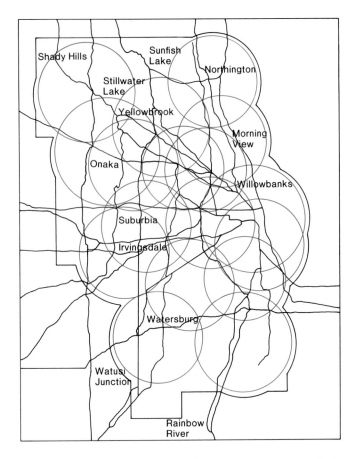

competition

these systems automatically goes to the local phone company if that entity
desires it, but the other can be operated as competition to the phone company
by some other corporation.

Cellular radio has been a good business. Once established within an area,
the sign-up rate and sale of equipment is high. In fact, the concept has been
so well received that in 1990 Motorola proposed a global cellular system designed to bring Third World countries up to state-of-the-art in their telephone
communication systems.[47]

cellular phone uses

Mobile phones are bought for both business and private use. A salesperson can make phone calls while driving between sales calls. Someone stuck
in traffic can answer the phone messages that have piled up on his or her desk.
A doctor can keep in touch with what is happening at the hospital. Individuals
can confirm (or break) a date or remind someone to pick up the laundry.[48]

Now that videotext is dead, the phone companies are tiptoeing into electronic publishing. For the most part, they are passing data bases that are owned
by other companies through their phone system. But they are also thinking of
creating banks of information that they can sell to consumers. Most of these
services revolve around another very important contribution to modern day
organizational and personal telecommunications—the computer.

FIGURE 7.13

A computerized
newsroom. *(Courtesy of
Dynatech Newstar)*

The Computer

The **computer** has revolutionized the modern office and the modern home. It has greatly affected the methods by which people obtain and manipulate information. At the simplest level, it is used for **word processing** and accounting procedures. On a more complex level, it is used for interactive services that join together large groups of people.

uses

All facets of the telecommunications industry make use of computers. Station and network newsrooms are now computerized so that stories typed into a computer by a reporter can be retrieved at another computer by a producer, who can then rewrite or integrate the story into a newscast.[49] Stations, cable systems, and corporations use computers to bill customers, keep program logs, analyze ratings, keep track of income and outgo, obtain news, and handle a myriad of other functions.

Mechanical computers, consisting of levers, gears, and odometer-like readouts existed as long ago as a hundred years, but they were slow, bulky, and expensive for the little they accomplished. Electronic computers began with a concept developed at Iowa State University during the 1930s. The resources to build the first practical **vacuum tube** computer came from the U.S. Army during World War II when it had a need to calculate ballistic/artillery trajectories quickly. As a result, the first digital computer, called ENIAC, was operating in 1946.

early development

The first time the general public became aware of computers was in 1952 when UNIVAC, a computer manufactured by Remington-Rand, was used for election coverage. With only 5 percent of the vote counted, it predicted that Dwight Eisenhower would defeat Adlai Stevenson for president.

UNIVAC

FIGURE 7.14

Dr. J. Presper Eckert (*center*), describes the functions of the UNIVAC I computer he helped develop in the 1950s, to newsman Walter Cronkite. *(AP/ Wide World Photos)*

transistors

In the early 1950s, **IBM** brought out its first electronic digital computer—a vacuum tube machine. By 1958 it introduced its first **transistor** digital computer, which sold very well to business. This computer filled a whole room but did much less than today's personal computers.

integrated circuits

During the 1960s, the **integrated circuit** was introduced for processing data. It consisted of many transistors on a single silicon chip. These chips, which are about the size of a fingernail, reduced computer size and cost, increased speed and reliability, and led to a boom in purchasing and renting of computers by the business community.

VLSI

During the 1970s, the size of computers was further reduced when very large scale integration (**VLSI**) greatly multiplied the number of electronic components that could be placed on a chip. By 1975 electronic hobby magazines were advertising computers that could be assembled from kits and used in the home. Shortly after that, the **Apple** computer was developed and marketed to the home user. A number of other companies, including IBM, entered the personal computer market, and it became a booming success.[50]

Data Banks

modems

It was during this period that the **modem** enabled the home and business computers to be connected to the telephone. Several companies began compiling **data banks** of information that users could access through phone lines by paying a fee. These data banks are essentially computers filled with information. Bits and bites stored in a computer can be sent through a modem, which, in essence, translates them into telephone language and sends them over phone lines to an individual telephone connected to a modem. This second modem

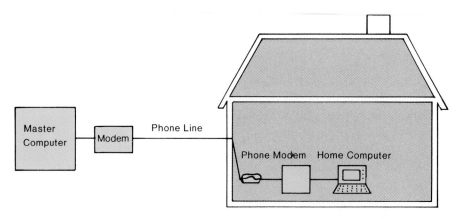

FIGURE 7.15

Telephone-computer configuration. *(From Gross, Lynne Schafer,* The New Television Technologies. *3d ed. © 1990 Wm. C. Brown Publishers, Dubuque, Iowa. All Rights Reserved. Reprinted by permission.)*

translates them back into computer language and sends them to the individual's computer, where the information can be viewed and manipulated.

Once this process became available, many companies developed data banks of information with the idea of selling this information to others. The most common setup was that a person or company paid an initial fee to join the data bank and then paid for the amount of time it spent accessing the information. For example, a data bank might charge $300 as an initial fee and for that the users would receive a special password enabling them to access the bank. Then they would be charged $6 for each hour of time they spent retrieving information. Some of the hourly fee would go to the company that had set up the data bank, and some would go to the phone company to cover line charges, in a similar manner to the way phone calls are charged.

One of the first groups to be targeted for data bank services was stockbrokers. They are in need of instant information that is constantly updated, which is an ideal use for data banks. Several companies such as Quotron Systems and Bunker Ramo set up these financial services.

Another fairly early data bank service was provided to lawyers. They need to keep up to date on the myriad of legal decisions that are made each day. Lexis set up such a service and earmarked it for law libraries and individual law offices.[51]

Data banks are also ideal for librarians. The card catalogue for the Library of Congress is on a data bank, as are most of the indexing services such as *Readers' Guide* and *Psychological Abstracts.* By using these services, someone can come up with a list of articles or books on a certain subject much more quickly than the old-fashioned way of looking through drawers of cards and stacks of books. There are also services such as Nexis (a brother of Lexis in that both are owned by Mead Data, which is a subsidiary of the Mead Paper Company), which offer on-line access to entire articles from hundreds of publications.[52]

Throughout the years thousands of data banks have been established; at present there are close to twenty-five hundred.[53] Many have come and gone because they have not been able to attract enough users to cover costs; however, there are large industries that can support numerous banks.

data bank procedures

stockbrokers

lawyers

librarians

other uses

The medical industry, for example, needs current information, which is supplied by a number of different data services. Travel agents use all the airline schedules. The defense industry is provided with information regarding who has received which government contracts and for how much. Even the broadcasting industry has a number of services aimed at it. Audience measurement figures are available on data banks, as is weather information and "morgues" (news stories from the past).

Other information available in various data banks includes the current status of any congressional legislation, hundreds of specialized newsletters that deal with such information as tobacco exports, organized religion, ski conditions, Japanese baseball scores, book and movie reviews, home decorating, and trivia.[54]

Obviously, some of this material is intended more for individuals than companies. Several companies, such as The Source and CompuServe, have established services aimed at individuals for home use; they also serve busi-

consumer information

nesses because those customers can pay more. Generally there is one rate for accessing information during normal working hours and a lower rate for accessing information at night.

Computer buffs have set up their own interactive networks, some of which are connected to formalized data banks and others of which are ad hoc, just to communicate with each other in much the same way that ham radio operators do. One of the interactive services that CompuServe provides is called Litforum. It is designed primarily for writers so they can show each other material such as scripts, poems, and stories.[55]

Data banks are still a growing, unstable business. They have achieved popularity, but not to the degree that some people predicted.[56]

One of the problems with various data banks is that each has its own system for entry and for obtaining information. This makes it difficult for someone to learn how to access all the information that might be needed. At first the only people who accessed information were trained librarians. This meant that the inducer had to work through an intermediary. To circumvent this, a few companies were set up that combined various data services into one

easy access

and made the access and retrieval the same for all the information. Dialog is one such company, consisting of over three hundred data banks.[57]

security

Another problem with data banks is security. Obviously, the companies that compile and then sell their information do not want people obtaining it for free. Other data base information, such as credit ratings, is confidential and should not be given out unless there is a bona fide need to know. Computer **hackers** have broken into various banks, so tighter security measures have been devised and have been at least partially successful.

inputting data

Another problem is what is referred to as **keyboarding.** In order for information to get into a data bank, someone has to actually type it. For newer information, this is not a great problem. Most books and magazines are now typeset using something that is computer based, so this material can be integrated into a data bank without retyping it. Therefore, many of the indexing and magazine article services only go back to the 1970s when this technology was adopted. Getting earlier information into data banks is a huge job. Only a fraction of the books in the Library of Congress are in data banks because of the huge task involved with totally retyping them.

Several technologies are developing that can solve this problem. One is **electronic scan** and the other is **voice-activated computers.** With electronic scan, a machine similar to a Xerox machine scans the page of a book or magazine and translates that information into the digital matter needed for the computer. This means that, instead of retyping material, it can simply be copied. With voice-activated computers, someone can read the information and the computer translates the voice into bits and bytes.

All of the data base actions discussed so far involve retrieving information. There are also services called **transactional services** or interactive video that are incorporated within data banks. These services allow the inducer to take some action, such as buying a ticket, making an airline reservation, or buying a shirt. The banking industry has involved itself with this service, setting up systems by which people can transfer money from one bank to another and pay bills without writing checks.

Another interactional service is **electronic mail.** Through this service, companies or individuals can send letters, advertisements, or other information to another person, company, or group. Each entity on an electronic mail network has an "address" to which others can send material. The person looks at this mail by going to a computer and typing in the code needed to retrieve the messages. This method has advantages over the present mail system because of its speed. It also saves trees because very little of it needs to be committed to paper.[58]

Information that does need to appear on paper can be printed out from data banks by connecting the computer to a printer. This, of course, raises questions of copyright, especially if articles or books are copied.

Video Games and New Advances

At the home level, computer technology is used for **video games** so that interaction can take place between the person playing the game and the TV set or computer screen. Video games have had their ups and downs. During the early 1980s arcades were abuzz with video games, and the home TV set was often used more for the games than for watching TV. Companies, especially **Atari,** had huge profits for several years. Then, as quickly as the video game craze came, it vanished, a victim of the fickle public. However, it was revived again during the late 1980s, on a somewhat less frenetic level, primarily by the Japanese company, **Nintendo.**[59]

Overall, most facets of the computer industry have had ups and downs during its short history. Even the giants, IBM and Apple, have had downturns because they have overestimated the public's appetite for computers in general, or for certain computer features.[60] Anyone involved with this business, including the consumers who buy the products, must keep up with the times.

interactive services

electronic mail

video games

new ideas

Almost daily there are new advances that make computers smaller, more powerful, more lightweight, and more flexible. The software to operate computers gains in both complexity and simplicity all at the same time. New terms, such as CD-ROM, DVI, and CD-I fill computer magazines, promising even more sophisticated uses of computer technology.[61]

Probably most important of all for people involved with personal and organizational communications are the myriad of ways in which the various media—videocassettes, video discs, telephones, and computers—are being combined in order to enhance the overall importance of the telecommunications industry.

Personal and Organizational Telecommunications Chronology

1876   Telephone invented by Alexander Graham Bell
1938   Vacuum tube computer developed at Iowa State
1952   UNIVAC predicts Eisenhower election
1956   Consent decree banned AT&T from entering nonregulated business
1958   IBM introduces computers that use transistors
1964   IBM introduces computers that use chips
1968   Educators and industry began using small nonbroadcast cameras and tape recorders
1970   Three-quarter-inch cassette tape available
1971   Groups formed that eventually became ITVA
1971   FCC authorizes ITFS
1974   FCC authorizes cellular phone concept
1975   Home computers advertised
1975   Sony Betamax introduced

1976 VHS introduced
1976 Universal and Disney sue Sony over copyright
1978 DiscoVision laser disc introduced
1979 Teleconferencing starts
1980s Many data banks offered
1980 Consent decree breaks up AT&T
1981 Selectavision capacitance disc introduced
1981 FCC invites applications for cellular phone systems
1981 More than 3 million consumer VCRs sold
1982 Pioneer buys laser disc system
1983 FCC gives some ITFS frequencies to MMDS
1983 Knight-Ridder sets up Viewtron
1984 Times-Mirror sets up Gateway
1984 Supreme Court says home taping does not violate copyright
1984 Consumer camcorders introduced
1984 RCA stops manufacturing Selectavision
1985 Sony introduces 8mm format
1985 Laser videodiscs join with CD technology
1986 Viewtron and Gateway go out of business
1988 Sony begins selling VHS
1990 Phone companies try to expand into electronic publishing

Conclusion

The various telecommunications forms discussed in this chapter—cameras and VCRs, videodiscs, telephones, cellular phones, videotext, computers, and data banks—are used both personally and organizationally.

Cameras and VCRs began being used by industry and individuals when they became small enough and inexpensive enough. Organizations use them to train, to enhance instruction, to aid internal communication, to demonstrate products, and to improve public relations. Programs have been shown through closed-circuit TV, ITFS, and teleconferencing.

Individuals use cameras and VCRs to tape programs off the air, to play pretaped movies or other material bought or rented from video stores, and to make home videos. A wide variety of formats have been developed for consumers over the years including Betamax, VHS, Super-VHS, Video-8, and Hi-Band 8.

Videodiscs were developed for consumers but are being used more extensively by businesses. Companies use them for data storage, training, kiosks for customers, and travel information. Consumers use them mainly to view movies, but neither the now-abandoned capacitance format or the current laser format have been nearly as successful as VCRs for this purpose.

The telephone has been used personally and professionally for more years than any of the other technologies discussed in this chapter. The phone business has changed from the early days when AT&T was a monopoly to the post-1980 consent decree days when a large variety of services and features exist, such as redial, FAX, toll-free numbers, call waiting, and dial-a-message services.

Cellular phones are also used for business and personal reasons. Salespeople and others who travel a great deal can conduct business from their cars. Individuals can use cellular phones just to keep in touch while they are driving.

Videotext was planned as a consumer service. Had it been successful, it might have moved into the commercial realm. But the Knight-Ridder and Times-Mirror systems failed. Videotext's aim was to give subscribers information similar to what is contained in a newspaper.

Computers that were tube, transistor, and simple silicon chip based were designed for the education, military, and corporate worlds. Only later in their history, when VLSI was developed, did computers come to the individual. In the business world they are used for processing words and for processing all types of data, much of it financial in nature. Individuals use personal computers for such purposes as preparing letters, keeping track of investments, and playing games.

Data banks deliver specialized information to professionals such as stockbrokers, lawyers, librarians, and broadcasters. Individuals can subscribe to a vast array of services that provide information on sports, news, books, and trivia. They can also use transactional services such as banking. Electronic mail is presently business-oriented but could become a way to reach homes also.

Personal and organizational telecommunications is in a state of transition with exciting possibilities for the future.

Thought Questions

1. What could be done to make videodiscs more successful at the consumer level?
2. Do you think there is any chance that videotext services will be revived? Why or why not?
3. What types of business actions do you think can be conducted well with teleconferences? What types cannot?
4. What do you think will be the status of data banks ten years from now?

Business Aspects of Telecommunications

Telecommunications is a business. Like any other business, it must maintain an awareness of income and outgo, and of profit and loss. Unlike many businesses, the product can be elusive, with programs disappearing from consumers the moment after they receive them. Some people say the product is not really programs at all, but rather commercials. Although some of the electronic media forms do not involve advertising, in most cases the sponsor's dollar is the ultimate source of revenue. Sponsors are very interested in how many people receive their messages. For that reason, audience measurement has arisen. Although most companies desire to be aware of who is using their products, the ratings methods used to accomplish this in radio and television are rather unique. The interactions involved with standard business practices, advertising, and audience research are ongoing throughout the telecommunications industry.

PART 3

Business Practices

Introduction

The major duties that need to be undertaken in most electronic media companies include general management, business, finance, legal, human resources, programming, news, engineering, sales and marketing, and public relations.

Just how these duties are undertaken varies greatly from one company to another. Size is one important factor. In a small radio station, these duties are combined so that seven or eight people fill all job responsibilities. For example, the general manager may hire employees, write the checks, sell ads, produce promotional spots, handle an on-air shift, and even sweep the floor. In a large network, the programming department may have hundreds of people with specific employees in charge of comedy, drama, movies, daytime soaps, and children's programming.

The type of media is another factor. The sales department at a commercial TV station is primarily concerned with selling to advertisers. Within a cable TV system, the primary selling is aimed toward customers. A public TV facility does not have an actual sales department but employs people who try to obtain grants.

In addition to the internal management functions of media companies, there are a number of external organizations such as consultants, brokers, agents, and unions that impact upon the media companies. Obtaining employment within any area of the electronic media is not easy, but it is certainly possible and can be very rewarding.

CHAPTER 8

I've been fired twice, canceled three times, won some prizes, owned my own company, and made more money in a single year than the President of the United States does. I have also stood behind the white line waiting for my unemployment check. Through television I met my wife, traveled from the Pacific to the Soviet Union, and worked with everyone from President John F. Kennedy and Bertrand Russell to Miss Nude America and a guy who played "Melancholy Baby" by beating his head. With it all, I never lost my fascination for television nor exhausted my frustration.

*Bob Shanks
when vice-president of ABC*

General Management

titles

Regardless of the size or scope of a telecommunications business, someone must be in charge. This "boss" does not always have the same title. At radio and TV stations and cable systems, the person is usually called a station manager or a general manager. At networks the person is usually called a president. Often the term **chief executive officer** (**CEO**) is used to designate the person at the top of the organization.

reporting procedures

Although the people engaged in general management usually have a great deal of autonomy in terms of making decisions, most of them have a higher authority overseeing what they do. General managers of cable TV systems usually report to presidents of **multiple system owners** (**MSOs**); for example, the general manager of a local TCI cable system reports to management within the parent TCI company. Likewise, radio and TV station managers often report to the president of a **group owner,** a company that owns a number of stations; for example, the general manager of KYW-TV in Philadelphia reports to Westinghouse Broadcasting because Westinghouse owns KYW and other stations. Network presidents report to company presidents; for example, the president of ABC reports to the president of Capcities. Company presidents report to Boards of Directors.

duties

The duties of chief executive officers vary, but generally these are the people who provide the leadership and direction for the unit they oversee. They set goals for the company directed at both the long-term and the short-term success of the organization. If a crisis occurs, CEOs stand at the front of the firing line. They are the ones who must decide what to do if the competition is trying to hire away a well-liked anchor, if sales are lower than predicted, or if lightning knocks out the transmission tower.

General managers also deal with the outside world. They talk to and handle complaints from people in the community, they join professional and community organizations, and they try to anticipate the competition's next move. Internally, one of their main jobs is to hire and motivate good people. They have to create and oversee a team of managers who will work together for the overall good of the company. If different departments collide, general managers must mediate. When times are bad, they must find means other than pay increases to keep employees involved and productive.

styles

Different general managers have different styles. Some like to become very involved in the details of the station or network. Others delegate work extensively and let the people below them have a great deal of responsibility. Some are flamboyant and make their presence known at every turn. Others are more reserved and businesslike. General managers are very influential in setting the overall atmosphere of the organization.[1]

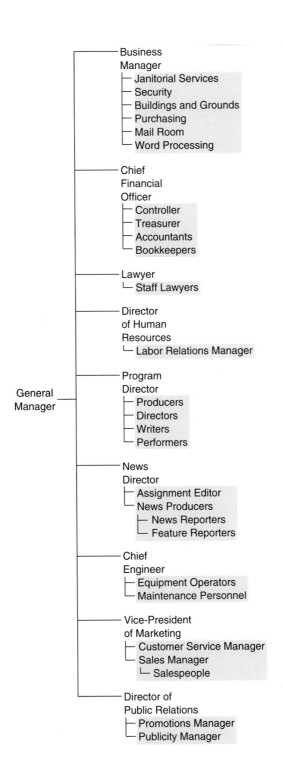

FIGURE 8.1
Theoretical electronic
media organization
chart.

Business

functions

The functions usually referred to as business functions are those that involve the company "housekeeping." Janitorial services are in the business department, as are security guards and the grounds and buildings maintenance personnel. The purchasing of supplies and equipment, such as paper and copy machines, is handled under business. The mail room and general word processing and computer inputting falls in this area. If drivers are needed to pick up celebrities at the airport, this chore, too, is often handled by business.

variations

Usually the business department is headed by a business manager who reports to the general manager or other CEO. Sometimes there is no business department, and the general manager or his or her administrative assistant handles the business functions. This is particularly likely in a small organization that rents its facilities from a landlord who takes care of repairs and security. Other times the business department encompasses functions other than housekeeping, such as finance, legal, and human resources.

Finance

status

The status of finance has increased greatly in recent years. Once dismissed as "those bean counters" by the creative element of broadcasting, finance people are now calling the shots. Some of this has to do with the fact that broadcasting is now a mature business. When most businesses are in their early stages, engineers are king because the product has to work before anything can be done with it. As a business enters its second stage, the sales department is at the top of the heap because the most important thing is selling the product. When the business matures, finance is very important because the nuances of making a profit become necessary for the long term. Broadcasting has progressed through these stages. Cable TV and other forms of media, even though they are newer, are also placing greater emphasis on financial statistics.

Finance has also increased in status because budgets have become tighter for all the media. A slip in the production schedule or a need for extra sets could be covered easily when profits were high, but now the pennies need to be accounted for more carefully.

job titles

A number of people are involved with keeping track of a company's finances. Large companies often have a **chief financial officer (CFO),** who is considered to be very high ranking within the company and is involved in many of the decisions that affect all aspects of the company. Other companies may have a vice-president (or director or manager) of finance. Usually this title is not as lofty as CFO, but the person holding it is still in the decision-making ring. The **controller** of a company is charged with overseeing the profit and loss and the expenditures. Some of this involves merely record keeping, but more and more the controller is brought into the decision loop when it appears that profits will be tight or nonexistent. The **treasurer** handles cash, making sure the company has enough money on hand to pay its bills. He or she makes sure that money owed to the company by others is paid so that the company can pay its bills. The treasurer also invests the company's money so that it will earn interest. Accountants establish and oversee ledgers that various department heads use to keep track of their expenses. For example, 6 may be designated as the number to be used for all expenses in the sales department. The number 6.1 might be entertainment expenses in sales, and the number 6.13 might be gasoline expenses for the sales department. Bookkeepers put the various numbers in the correct books, and although they are still often called bookkeepers, they are now really computer keepers because most companies' financial records are highly computerized.

Many duties of the finance department involve record keeping. It regularly issues a **balance sheet** that lists all assets and liabilities of a station or network at a specific point in time, and a **profit and loss statement** that shows how much the company received and spent over a period of time to show whether expenses are exceeding income. Many other forms are stored in a computer, including ones that keep track of cash receipts and others that list cash disbursements. record keeping

The finance department pays the bills and collects the money owed to the station. If someone (such as an advertiser) does not pay a bill, the finance department's odious job is to see that the money is paid. For large overdue debts, the finance department might hire a collection agency. bills and collections

Finance also handles payroll. This department does not decide how much employees will be paid, but it issues the checks and keeps the necessary records. When financial reports need to be made for the FCC, social security, or health insurance agencies, the finance department does so. other duties

Finance departments within public broadcasting are slightly different because the income does not come from advertising. Much of the income comes from grants, so the finance department must make sure that expenses are attributed to the proper grant. Also, public broadcasting, by nature, cannot show a profit. public broadcasting

The money for corporate television departments is usually allocated through the company involved. The manager of the corporate television department must keep track of the department expenses, but the overall financial control comes from the company's finance department.[2]

Legal

Only the largest of the electronic media companies have legal departments. These are headed and staffed by lawyers who handle any legal actions brought against the company or any legal actions the company wishes to take against another company. The lawyers also keep abreast of the new laws and regulations that affect the company so that they can advise management on changes that may be in the offing. In addition, they lobby in the interest of the company so that unfavorable legislation does not pass. duties

Smaller companies hire lawyers on an as-needed consultant basis. Sometimes they have a law firm on a **retainer**—they pay the firm a certain amount per year, and the firm then handles any problem that might arise. structure

Balance Sheet

As of _____

(Date)

Assets	This Year	Last Year
Current Assets		
Cash	$	$
Current investments		
Receivables, less allowance		
for doubtful accounts		
Program rights		
Prepaid expenses	_____	_____
Total current assets		
Property, plant & equipment at cost		
Less: Accrued depreciation		
Net property, plant & equipment		
Deferred charges and other assets		
Programming rights, noncurrent		
Intangibles	_____	_____
Total assets	$_____	$_____

Liabilities & Stockholders Equity		
Current Liabilities		
Accounts and notes payable		
Accrued expenses		
Income taxes payable	$_____	$_____
Total current liabilities		
Deferred revenue		
Long-term debt		
Other liabilities	_____	_____
Stockholders Equity:		
Capital stock		
Additional paid-in capital		
Retained earnings		
Treasury stock	_____	_____
Total Stockholders Equity	$_____	$_____
Total Liabilities & Stockholders Equity	$_____	$_____

II-1

(a)

Profit and Loss Statement

Current Month							Year To Date				
Actual			**Variances**				**Actual**			**Variances**	
This Year	Prior Year	Budget	Prior Year	Budget			This Year	Prior Year	Budget	Prior Year	Budget
					Gross Revenues:						
					Local time sales						
					National rep time sales						
					Network						
___	___	___	___	___	Other		___	___	___	___	___
___	___	___	___	___	Total Gross Revenue		___	___	___	___	___
					Direct Expenses:						
					Agency commissions						
					National rep commissions						
					Local sales commissions						
___	___	___	___	___	Music license fees		___	___	___	___	___
___	___	___	___	___	Total Direct Expenses		___	___	___	___	___
$___	$___	$___	$___	$___	Net Revenue		$___	$___	$___	$___	$___
					Operating Expenses:						
					Program						
					Outside production						
					News						
					Technical						
					Selling expenses						
					Research						
					Advertising & promotion						
					General & administrative						
___	___	___	___	___	Depreciation		___	___	___	___	___
___	___	___	___	___	Total Operating Expenses		___	___	___	___	___
					Income Before Federal Income Taxes						
___	___	___	___	___	Provision for Federal Income Taxes		___	___	___	___	___
$___	$___	$___	$___	$___	Net Income		$___	$___	$___	$___	$___

(b)

FIGURE 8.2

An example of what might be included on (*a*) a typical balance sheet and (*b*) a typical profit and loss statement.

Large group owners or MSOs often have a legal department that services all of the stations or cable systems that it owns. This is particularly handy during license or franchise renewal times when legal documents must be prepared.

Human Resources

Another common name for the human resources department is the employee relations department. These departments used to be called personnel departments, but this word became rather tainted because it was used for departments that did little more than handle the paperwork related to hiring and firing. As personnel departments expanded their activities, their titles were changed to more accurately reflect their duties. The head of the department is usually called a director or manager of human resources, although in large companies this person may have a vice-president title.

Human resource departments still handle procedures related to hiring and firing. They place ads to fill openings within the company, screen resumes, and recommend a number of candidates to the person who is trying to fill the vacancy. Occasionally people in human resources are actually responsible for making the decision as to who is hired, but more often the head of the department with the need will make the decision. If a person is to be fired, the human resources department makes sure all the proper procedures have been undertaken, such as giving the person several notices of poor performance before the actual firing.

Human resources also has responsibility for handling the company benefits package, including health insurance and retirement. This department makes sure the company abides by the equal employment guidelines in hiring, promotion, and layoffs. If there are unions within the company, human resources is often in charge of overseeing union-management negotiations. If there is no union, the department tries to make sure management sets up procedures in such a way that employees do not feel the need for a union. People in human resources are often called upon to mediate disputes between employees on an informal basis.

Like many of the other departments, human resources is much more likely to exist in large companies than in small ones. A small radio station may only hire someone every two years, so the general manager usually handles personnel matters. However, in a large, highly-unionized network, the human resources department may be subdivided with one division called labor relations, which deals solely with unions.[3]

Programming

The programming department is in charge of deciding what should be shown to the public and when it should be shown. Some programmers have very little choice; they simply pass through what the network or networks send them. A programmer for a local cable TV system, for example, has virtually no say as to what is shown on ESPN, MTV, or any of the other cable networks. This

Business Aspects of Telecommunications

POSITION ANNOUNCEMENT
WXYZ-TV

JOB TITLE: Electronic Graphics Assistant

LOCATION: This television position is based at WXYZ-TV, 860 Park Avenue, Fulforth, Wisconsin.

POSITION   The Electronic Graphics Assistant is responsible to the Electronic Graphics Manager for
DESCRIPTION: operating electronic design equipment and assisting in the training of other operators.

SPECIFIC DUTIES AND RESPONSIBILITIES

1.   Execute electronic graphics requests and set up file discs based on instructions
     from designers.

2.   Operate electronic design equipment including Graphics V, Chyron, Macintosh, and
     Laird. Operation will occur during production, post-production and live feed situations.

3.   Serve as principal back-up for electronic design coordinator.

4.   Perform other duties as assigned by the Electronic Graphics Manager.

SALARY: $15,885 plus fringe benefits and two weeks vacation.

QUALIFICATIONS: 1.   B.A. in communications, journalism, or related field required.

2.   Familiarity with basic principles of type and graphic design.

3.   Experience in working with computers or word processors required. Familiarity with
     television character generators preferred (Thompson Graphics V or Chyron).

APPLICATION: 1.   All of the following must be RECEIVED ON OR BEFORE August 16, 1990:

     a.   A completed WXYZ application form.

     b.   A detailed resume including names, current addresses, and telephone numbers of
          three (3) persons who can attest to professional qualifications as cited above.

     c.   A brief narrative describing how your experience relates to this position.

2.   Your application will be reviewed by a balanced screening committee.
     Interviews are tentatively scheduled for August 23.

3.   Contact the person indicated for application forms, inquiries, and submission of
     application materials:

          Marie Marshall, Secretary to the General Manager
          WXYZ-TV
          860 Park Avenue
          Fulforth, WI 53706
          (608) 253-2222

WXYZ provides equal opportunities in employment and programming, including Title IX requirements. Discrimination on the basis
of age, race, color, creed or religion, national origin, ancestry, sex, sexual orientation, veterans' status, arrest or non-job program
related conviction record or qualified disability,, including AIDS, is prohibited. Prospective employees may notify the Federal
Communications Commission, Washington, D.C. 20554 or other appropriate federal, state or local agency if they believe they have
been discriminated against.

An offer of employment is contingent upon establishment of identity and verification of employment eligibility as required by the
Immigration Reform and Control Act of 1986.

FIGURE 8.3
An example of what a job announcement from a TV station might look like.

PERFORMANCE APPRAISAL OF _____

Review Period: From _____ to _____ Date Reviewed _____

RATING SCALE:
    5 - Outstanding; exceeding nearly all others
    4 - Above average; above majority
    3 - Average; satisfactory performance for majority
    2 - Below average; needs improvement to meet standard
    1 - Marginal; unsatisfactory

FACTORS	APPRAISAL	DEFINITIONS
Knowledge		extent of technical and production knowledge and ability and how this knowledge and ability is applied relative to the job requirements
Quality of work		performance in meeting quality requirements of accuracy, appearance, thoroughness, and artistic level
Quantity of work		volume of work produced; speed and consistency
Dependability		extent of being able to be counted on to accomplish tasks on time and effectively; working above and beyond the call of duty when necessary
Attitude		general approach to work and co-workers; ability to work effectively with others in a positive manner and on a continuing basis
Attendance		lack of tardiness and absenteeism on the job
Human Relations		effectiveness in interaction with others in accomplishing job responsibilities; ability to establish productive working relationships and deal smoothly, effectively, and cooperatively with people
Organizational Skills		ability to organize thinking, plan and schedule what must be done, and set realistic completion dates; extent of follow through on tasks
Judgment		common sense; ability to profit from experience; ability to sort out facts to draw logical conclusions
Initiative		degree of "self starting" in satisfying objectives of the job; extent of seeking new responsibilities and challenge
Adaptability		ability to adjust to new or unforeseen situations or unusually heavy pressure; ability to learn new duties and adjust to new situations
Overall Appraisal		general ranking of abilities and accomplishments

FIGURE 8.4

An example of a form that might be used to evaluate employees.

person can, however, be influential in determining which of the cable channels should be carried on the system and which should not. Also, if a cable system has a local origination channel, the programmer can decide what that channel will cablecast.

The heads of programming—usually called program directors, program managers, or vice-presidents of programming—do not make major programming decisions all by themselves. The sales department, in particular, has input in terms of what programs will be the easiest to sell to advertisers.

The main duty of the programming department is to fill hours with programs by acquiring or producing them. If a great deal of programming needs to be acquired, the heads of programming have people to help them with different types of programming such as children's programming, drama, and soap operas. Material can be acquired from networks, syndicators, production companies, or individual producers. This process is detailed in chapter 13. acquiring

If programs are produced, they go through a fairly set procedure that is similar for all shows, be they public-access talk shows, corporate training tapes, or network movies of the week. During a **preproduction** period, the idea is created, a script is generated, and the show is budgeted. If need be, a cast is chosen, sets are built, props are obtained, shooting locations are selected, and a crew is scheduled. During **production,** actors, crew, and all the necessary supplies come together to tape or film the show. In most instances material is rehearsed, or at least talked through, before it is actually shot. In **postproduction,** the material that has been shot is edited, and often the sound is enhanced with music and sound effects. producing

Any electronic media entity that produces much of its own original material has producers and directors. Producers are people who organize productions, making sure that everything is at the right place at the right time, that all is finished on time, and that the production does not run over budget. Directors are in charge of the cast and crew during production and make most of the artistic decisions. Sometimes producers and directors have permanent jobs with a network or station, and sometimes they are hired on a **free-lance** basis, where they work only on a particular program. Likewise, writers and performers may be permanent or free-lance. A disc jockey at a radio station would have a permanent job, but the lead actress for a movie of the week would not. producers and directors

The programming department handles the product of the station or network. Because the success of most businesses is dependent on the product, the programming department is considered to be very important. In addition, it is associated with the glamor part of the business, so many people aspire to enter the field.[4]

News

Sometimes news is included within the programming department, but more often, because of its deadline-related complexity, it is a department unto itself. However, a small radio station that depends primarily on a wire service for its news stories would not have a separate news department. location

Sometimes the news department may oversee other forms of programming such as documentaries, sports programming, and editorials. In these cases, the programming department may be very small. In fact, it may consist of only one person who acquires programming; all of the in-house production is done through the news department. All-news radio stations and CNN revolve around the news department and do not have actual programming departments. But the most common pattern is that the news department handles news while the programming department takes care of most of the other forms of programming, such as entertainment and public affairs.

The news department is headed by a news director who has overall responsibility for the newscasts and determines the general approach that all the newscasts will take. Usually each particular newscast has a news producer who organizes the material and decides what stories will be aired. Reporters cover stories out in the field, and an assignment editor keeps track of the reporters and assigns breaking stories to the appropriate people. Anchorpeople, some of whom are also reporters, present the news. Many radio and TV stations and networks have special feature people who research and present material on such subjects as the weather, sports, consumer affairs, or the entertainment scene.

The news process involves obtaining news leads, gathering news, preparing copy, and presenting material. All of this is detailed in chapter 12.[5]

Engineering

The people who operate the equipment needed for producing material for both the programming and news departments are often in the engineering department. Sometimes, however, they are assigned to programming or news departments and, under other forms of organization, they are in a separate department known as production or technical support.

The engineering department includes not only those technicians who operate the equipment but also those who maintain and install it. The head of the engineering department is called the chief engineer. This person schedules the people and equipment needed by the programming and news department, oversees maintenance, and makes recommendations to the general manager regarding new equipment that is needed. In the cable TV business, the chief engineer oversees the installation of the cable in the community. This is a very complex job when the cable system is starting up or rebuilding its system.

If a station or network does a great deal of its own production, it needs a large number of equipment operators. These include technical directors (who operate the switcher), camera operators, audio technicians, character generator operators, videotape recorder operators, tape editors, and lighting crews. (For more information on the equipment that these people operate, see chapter 14.)

As mentioned previously, the engineering department is very important when an electronic media entity is starting up. Everything must operate technically before programming can be sent out to listeners or viewers. Because broadcasting is an older industry, engineering is not as important as it used to be. This is exacerbated by the fact that budgets are tight and new equipment is not purchased as frequently as in the past. In addition, the trend is to hire equipment operators on a free-lance rather than permanent basis. Freelancers are more expensive for each day that they work, but they are not paid for days that they are not needed, and they are not given benefits. In the newer industries, such as DBS, engineers are still very important because they are needed to design and test the equipment.

As with other departments, the number of people needed in the engineering department varies according to the size and functions of the company. Small radio stations often operate with just one engineer, making him or her the chief engineer. Large TV stations or networks have numerous people operating and maintaining equipment. One of the problems in large organizations is finding a good chief engineer. Engineers are comfortable working with equipment, but when they become chief engineers, almost all their time is taken up with scheduling and supervising. Sometimes this makes for unhappy chief engineers because their background and interests lie with schematics, levers, and knobs, but they must deal with paper and people.[6]

variations

Sales and Marketing

Sales and marketing are generally in the same department. Most commonly, sales involves the direct selling of advertising time to clients. **Marketing** involves everything else that will entice someone to buy time on a station or network, such as developing materials that give a station a positive image, hosting lunches for clients, and working with the programming department to develop shows that will have appeal to potential customers.

marketing

If sales and marketing are in the same department, that department is usually headed by a vice-president of marketing. The head of sales, usually called a sales manager, is under the vice-president. Small telecommunications firms usually do not have anything referred to as marketing, so the sales department is headed by a sales manager or sales director. This person hires, trains, and evaluates salespeople, sets sales policies, controls sales expenses, and communicates with other departments within the company.

titles and duties

Some media entities sell hookups or subscriptions directly to consumers. For example, cable TV systems, MMDS, computer data bases, and SMATV systems sell their services to individuals. Although this is sales, companies sometimes place this function in a department called customer service. This is a softer term intended to make the customers believe they are being serviced rather than sold a bill of goods. Customer service is also supposed to handle complaints and questions that people have about their cable service. Cable TV systems often have both types of sales. They sell hookups to customers, and they also sell advertising for their local origination channels or to insert within cable network shows. Sometimes both functions are placed in a sales department, but more often cable TV systems have both a sales department and a customer service department.

selling to consumers

Sometimes the sale of cable TV to customers is complicated because of the number of channels involved. Early cable systems charged an installation fee and a monthly fee for bringing in the local channels. With the advent of HBO, cable systems added a special additional fee for the commercial-free movie services. As cable programming proliferated, cable systems adopted **tiering.** Subscribers can select the tiers of programming they wish to receive. Different cable systems tier in different ways, but often they have a very inexpensive tier that includes local stations and cable access channels. Then they will have a more expensive tier that includes basic networks, such as ESPN, USA, and Discovery. (Some systems have several basic tiers that each include different networks.) Then they have other, more expensive tiers that include one or more pay services such as the Disney Channel, Showtime, or a regional sports channel. In addition to all of this, some systems have pay-per-view. For this, customers buy specific programs on a one-time only basis.

Public broadcasting does not have a sales department as such; however, public radio and TV stations often have development departments. These departments are responsible for obtaining money through grants and underwriting and often handle campaigns to obtain money from listeners or viewers.

The videocassette business operates the same way that most consumer product companies do—through a distribution network of wholesalers and retailers. Film companies and others distribute the cassettes either for sale or for rent through various video stores. In order to induce people to buy, the companies engage in advertising campaigns and special sales. Corporate video producers have a built-in customer, the institution for which they produced the tape. Therefore, they have no real identifiable sales function.

But the bulk of what is usually referred to as "sales" within the various electronic media involves the selling of advertising. Station or network salespeople (often euphemistically called account executives or sales representatives) approach companies or advertising agencies and attempt to sell them commercial time on the station or network. Details involving this process are discussed in chapter 9.

The people who sell advertising time must be well versed in the needs of the potential customers. In other words, they must know the customer's product well and must be able to detail how placing an ad on the station or network will aid the customer. Sales representatives must also know the competition (other stations, newspapers, magazines) so that they can tell customers why theirs is the best buy. Of course, to do this, they must also know the programming and personalities of their own station well. Probably, even more important, they must know who comprises their audience. The selling of advertising time really involves the selling of consumer eyes and ears to people who want the owners of those eyes and ears to buy their product.

For this reason, audience research activities are usually located in the sales department. Ratings are gathered by outside companies such as Nielsen and Arbitron (see chapter 10), but people at a station or network need to know how to read the ratings and interpret them so that the sales force can use the figures that most enhance the station.

```
              OMNIFRONT CABLE TV

                   RATE CARD
                  (Monthly fees)


Local TV Stations Only · · · · · · · · · · · · · · · · · · · · · · · · · ·  $5.40
Basic Cable (Local Stations and 20 Cable Services) · · · · · · · · · · · · ·  18.90
Premium Channels
    HBO · · · · · · · · · · · · · · · · · · · · · · · · · · · · · · · · ·  13.95
    Showtime · · · · · · · · · · · · · · · · · · · · · · · · · · · · · · ·  13.95
    The Movie Channel · · · · · · · · · · · · · · · · · · · · · · · · · · ·  13.95
    Cinemax · · · · · · · · · · · · · · · · · · · · · · · · · · · · · · · ·  13.95
    Disney Channel · · · · · · · · · · · · · · · · · · · · · · · · · · · · ·  10.95
    Sports Center · · · · · · · · · · · · · · · · · · · · · · · · · · · · ·  12.95
Premium Channel Discounts
    HBO and Cinemax (save $5.00) · · · · · · · · · · · · · · · · · · · · · ·  22.90
    Showtime and The Movie Channel (save $6.00) · · · · · · · · · · · · · ·  21.90
    Disney Channel as a 2nd pay service (save $2.00)  · · · · · · · · · · ·   9.95
New Installation · · · · · · · · · · · · · · · · · · · · · · · · · · · · · ·  50.00
New Installation if two Premium Channels are Purchased  · · · · · · · · · ·  30.00
Reconnect Current Line · · · · · · · · · · · · · · · · · · · · · · · · · · ·  20.00
Additional Outlets · · · · · · · · · · · · · · · · · · · · · · · · · · · · ·  15.00
Converter Box · · · · · · · · · · · · · · · · · · · · · · · · · · · · · · ·   5.00


City Taxes Not Included
All Rates Subject to Change
```

FIGURE 8.5
An example of a cable
TV rate card.

Another responsibility frequently placed in a station's sales department
is **traffic.** This responsibility includes listing all the programs and commercials
to be aired each day in a station log. Traffic is usually part of the sales department because the most important job involved with the log is scheduling
the various commercials. As time sales are made, someone, with the aid of a
computer, must make sure that the ads are actually scheduled to air under
the conditions stipulated in the sales contract.[7]

traffic

```
                                      DAILY LOG


Station:    WAAA      Port, Connecticut     12/25/92     Page 2 of 15


Time           Programs/            Length      Source      Type     Actual       Announcers
Scheduled      Announcements                                         Time         Initials

8:00:00        News of the hour     5 min.      Net         N
               Cancer Fund          30 sec.     LS          PS
               Mason Ford           30 sec.     Rec         C
8:05:00        Steve Stevens Show   23 min.     LS          E
               Pop-a-Hop            30 sec.     Rec         C
               Arbus Tires          30 sec.     Rec/LS      C
               C-C Cola             15 sec.     Rec         C
               Hank's Hardware      30 sec.     LS          C
8:28:00        Station Break        2 min.      LS          SI
               Crisis Helpline      30 sec.     LS          PS
               Richers              30 sec.     Rec         C
8:30:30        News and Weather     2 min.      LS          N
               Sid's Market         30 sec.     LS          C
8:32:00        Steve Stevens Show   13 min.     LS          E
               Joe's Garage         30 sec.     Rec         C
               Coming Events        1 min.      LS          PS
8:34:00        Sports Special       30 min.     LR          S

Remarks:

Source Abbreviations:   LS-Local Studio    LR-Local Remote    Net-Network    Rec-Recorded

Type Abbreviations:   PS-Public Service Announcement   C-Commercial   PR-Station Promo
                      SI-Station Identification   E-Entertainment   N-News   S-Sports   R-Religious
                      PA-Public Affairs
```

FIGURE 8.6
An example of a radio
station log.

Public Relations

public relations

Public relations is sometimes included with sales and marketing because its purpose is to build general goodwill that can help to enhance sales. It is similar to marketing except that it is usually directed more toward the audience than toward the advertiser. For this reason, some public relations departments are called **community relations.** Typical public relations activities include giving station tours, answering viewers' letters, and engaging in public service activities such as conducting campaigns to stop pollution.

Public relations often encompasses two areas known as promotion and publicity. **Promotion** is something a station pays for in order to enhance its image. **Publicity** is free. For example, an advertisement in a local newspaper urging people to tune in to a particular disc jockey would be promotion because a station would need to pay for the ad. A review of the disc jockey's show in the same local newspaper would be publicity because the station would not pay for it. Publicity is preferred over promotion because it is free, but the problem with publicity is that it may be negative. A positive review of the disc jockey's show is helpful, but a negative one is usually worse than no review at all.

promotion and publicity

Public relations departments are now often separate from sales because they have taken on increased importance. As the number of channel choices grows, promotion is much more important than it used to be. When the three networks dominated, they could do most of their promotion during late August right before they started their new season of shows. Networks used their own airwaves to promote programs and supplemented this with ads in *TV Guide* and other magazines and newspapers. But now programs change frequently, and people have so many listening and viewing choices that they must constantly be reminded to tune in a particular program or station. Promotion has become such a consuming business that some stations have promotion departments, separate from public relations.

importance

The person heading public relations (or community relations or promotion) is usually called a director or manager. One of the biggest responsibilities of this person is to set goals. Public relations directors have to decide what audience will be attracted to a particular format. For example, if a radio station is trying to change its image so that it attracts college students, the promotion and publicity efforts should not be directed toward corporate executives. Sometimes public relations efforts are geared toward increasing the size of the audience; other times efforts are geared toward retaining audience members.

goals

Usually it is the job of someone within the public relations area to write and produce the **promotional spots (promos)** that appear on a station to advertise its own programs. To do this, the person might screen the program, select part of it to include in the promo, and write catchy copy.

promos

The public relations area also designs advertising for other media such as newspapers, billboards, and the sides of buses. Contests that will attract audience members usually stem from public relations, and it often helps nonprofit organizations, such as the Red Cross or the American Heart Association, produce **public service announcements (PSAs)** aired by the station for free.

other duties

Some of what this area does is directed toward advertisers rather than audience members. For example, public relations people will often prepare slide presentations or written packets that the salespeople can use in their presentations to clients.

Overall this is an area that is growing in importance, although it is still not considered as important as many of the other functions of electronic media such as programming or sales.[8]

FIGURE 8.7

A typical promotional
ad, this one concerning
a PBS program.
*(Courtesy of Public
Broadcasting Service.)*

Allied Organizations

In addition to people who work directly for commercial broadcasting stations and networks, public broadcasting stations and networks, cable TV systems and networks, and other electronic media, there are many other people who spend time engaged in telecommunications work: television program producers and distributors; radio program producers and distributors; commercial and jingle producers; advertising agencies; station representatives; audience measurement companies; brokers; management consultant firms; talent agencies; employment services; communication attorneys; equipment manufacturers; consulting engineers; music licensing groups; unions; news services; and radio and television departments of colleges and universities.[9]

Of course, the activities of these people often extend beyond telecommunications. For example, advertising agencies are concerned with print media as well as broadcast media; their job is to find an advertising mix for their

ad agencies

customers, not just to arrange commercials for radio or television. Nevertheless, it is interesting to note the extent to which telecommunications, itself a service, is served by other organizations.

Many companies produce radio or TV programs or commercials. The number of jobs available in this area, however, is somewhat deceptive. A few very successful production companies—Paramount, Universal, and Columbia Pictures Television—supply large numbers of programs to the TV networks. However, most production companies are small groups who hope to hit on something that will bring financial reward. Many of them never make it, although they may reappear some years later with a new name and the same people involved. For the most part, independent production companies cannot afford to own their own facilities, so they rent regular station or network facilities and personnel in order to produce their programs. Therefore, the camera operator who works on the channel 5 noon news may at one o'clock begin working on an entertainment show to be aired on CBS. Independent production, therefore, does not necessarily create a large number of jobs for new people.

production companies

Most of the allied organizations are dealt with in other sections of this book, but a few bear describing in slightly more detail. **Brokers** aid in the buying and selling of radio and television stations and cable systems in much the same way that real estate agents do for homes. Individuals or organizations who wish to sell a facility may list it with a broker, who then tries to find a suitable buyer under the guarantee of a commission on the station sale price. Of course, stations can be sold without the aid of a broker.

brokers

Consulting engineers and management consultants are usually hired by stations on a short-term basis. A new cable system may hire a consulting engineer to determine the best location for a headend. A radio station that has been given permission to increase its power may hire a consultant so that all the technical procedures are properly accomplished. A station having personnel problems may hire a management consultant to suggest a new organization.

consultants

Agents

A word should be said about agents; they are generally controversial in a "needed by all, resented by all" role. Their function is to find work for creative people in return for a percentage—usually 10 percent—of the amount earned. The largest talent agency is **William Morris,** founded in 1898 to represent talent in the legitimate theater and vaudeville. It has grown steadily, adding representation for motion pictures, radio, music, and television. In its earlier days it was all but eclipsed by **MCA,** which was founded in 1924 as the Music Corporation of America to represent dance bands and orchestras. MCA grew rapidly, representing not only musicians, but actors, writers, producers, directors, composers, and others in the creative fields. It also began packaging shows with its own talent (for which it collected fees for the entire shows) and it acquired **Universal Studios** as its production house. In 1962 this situation

William Morris

came to the attention of the Justice Department, which invoked the **antitrust act** on the grounds that one company should not represent talent on one hand and produce programs with that talent on the other. Because MCA made more money collecting on its shows than on its individual stars, it kept the Universal production unit and gave up the talent agency business, thus leaving William Morris as the largest agency.[10]

It is virtually impossible for an actor or actress or a free-lance writer, director, producer, or other creative person to become successful without an agent because networks and most production companies deal only through agents, not directly with talent.

advantages

In some ways this can be advantageous to the person. An agent can describe a client in glowing terms that the client could not say about herself or himself. An agent is constantly out in the field representing people and possesses a wealth of information about what jobs will be available. The agent can also negotiate salary and contract, leaving the talent "above it all." Many agents can give sound professional advice to help a person improve. On the

disadvantages

other hand, agents who are so inclined can abuse their power. Critics of agents sometimes contend that the agent receives 10 percent of everything the talent earns and sometimes does nothing to deserve it. Because an agent represents a large number of people, he or she will not always have one particular client's interest at heart and sometimes will even sell out a client or make an undesirable deal for the client in order to continue doing business with a particular production company. Sometimes an agent will get a client involved in a disastrous project simply to obtain the 10 percent commission.

Networks and production companies also have ambivalent feelings toward agents. They would rather deal with agents than with hordes of aspiring actors, but once an actor becomes known and desired by the production houses, they would rather the agent disappeared so that they could deal directly with the person.

Unions

union functions

Another broadcasting phenomenon very important to the production scene is unions. Like agents, unions are much more important in the large cities than in small towns. Unions operate only where they are voted in by the employees—usually where there is a concentration of activity. At present about 33 percent of the TV stations, both commercial and public, and 8 percent of the radio stations deal with one or more unions. Of course, the major networks are unionized. The unions are involved with cable satellite programming, but local cable systems rarely deal with unions. Unions negotiate wages and working conditions. They also prosecute members of management who violate provisions of the union contract and union members who work for less than union scale or otherwise violate contract agreements.

AFTRA

One of the most prominent of the broadcasting unions is the American Federation of Television and Radio Artists (**AFTRA**), a union for performers in live and videotaped programs. AFTRA was started in 1936 as a union of radio artists in New York, Chicago, and Los Angeles and was called the American Federation of Radio Artists (AFRA). Today its jurisdiction extends to all fifty states and covers music and slide-tape recordings, as well as live and taped radio and TV programs and commercials.

Another performance union, the Screen Actors Guild (**SAG**), was organized in 1933 by actors discontent with the poor pay of motion picture studios. SAG and AFTRA have had several jurisdictional disputes over the years, the most bitter occurring when videotape was introduced. SAG felt it should have the jurisdiction because the material was being recorded to be kept much as film is kept. AFTRA felt it should have control because taping was accomplished in a TV facility. AFTRA won the battle, but at present SAG has jurisdiction over anything produced at one of the traditional film studios, even though the medium used is videotape.

SAG

Another acting union is the Screen Extras Guild (**SEG**), which tries to ensure that people with walk-on parts receive adequate pay. At various times plans to merge SEG, SAG, and/or AFTRA have been proposed, but none of these plans have made it through the unions' political process.

SEG

Historically, acting unions have had an extremely high unemployment rate. It is estimated that 80 percent of the members earn less than $2,000 a year from acting. Most of these people find other types of employment elsewhere, many in menial jobs that will allow them the freedom to look for acting work.

If a station or network is a signatory to a performers' union, it must pay at least the minimum wage negotiated by the union to all performers. Complicated formulas specify that a principal performer is to be paid more than someone who has less than five lines, who in turn is paid more than an extra. And, of course, the few stars of the business can demand far more than the minimum wage. The union also negotiates such matters as meal and rest periods, credits, wardrobe, use of stand-ins, travel requirements, dressing room facilities, and **residuals.** This last is a thorny issue involving the formula for paying performers for shows or commercials that are rerun. Performers feel the residual percentage should be high because reruns eliminate jobs. Networks and stations want to keep the residuals low so that rerun costs are low.

Another performance union is the American Federation of Musicians (**AFM**), which has jurisdiction over musicians who perform live or taped on radio or TV. AFM is not as important to broadcasting now as it was when radio networks had their own orchestras. During the late 1930s and early 1940s AFM forced stations to employ musicians even though they had no work for them by making the stations sign contracts stating that they would hire a certain quota of musicians based on the station's annual revenue. These contracts were declared illegal in 1940 and were not renewed, but the AFM threatened networks and record companies with strikes if they supplied music to stations that did not employ musicians, so many of the quotas continued. In 1946 Congress specifically outlawed the use of threats to require broadcasting stations to employ individuals in excess of the number needed to do the job.

AFM

The first established union for broadcasting engineers was the International Brotherhood of Electrical Workers (**IBEW**), which had originally been formed during the late 1880s by telephone linemen. This union successfully struck a St. Louis radio station in 1926 and later obtained network contracts. In 1953 NBC technicians formed their own union, which became known as the National Association of Broadcast Employees and Technicians (**NABET**). Now NABET and IBEW compete, sometimes vigorously, to convince employees of stations to vote them into power.

A third technical union, International Alliance of Theatrical Stage Employees and Moving Picture Machine Operators of the United States and Canada (**IATSE**), moved into television from motion pictures. In some stations this union only has jurisdiction over stagehands, and in other stations it includes stagehands and some equipment operators.

For all of these unions, members must pay an initiation fee and then regular monthly dues, usually an amount related to income earned. One of the points of frustration to people trying to obtain a first job in broadcasting is that one must have a job to join a union but must join the union in order to get a job. Breaking into that vicious circle is often difficult.

There are several broadcasting guilds that do not operate in exactly the same way as unions, but they are involved with setting pay rates and negotiating. Three of the most important are Directors Guild of America (**DGA**), Writers Guild of America (**WGA**), and Producers Guild of America (**PGA**).

In addition to the major telecommunications unions and guilds—AFTRA, SAG, AFM, IBEW, NABET, IATSE, DGA, WGA, and PGA—there are many smaller unions for groups such as costumers, model makers, set designers, makeup artists, and carpenters.[11]

Job Preparation

Gaining employment in telecommunications is not easy—there are many more people who would like to be so employed than there are jobs. A recent survey conducted by the Broadcast Education Association showed that more than 300 colleges have programs leading to degrees in radio and television. These schools graduate approximately 12,000 students a year, ostensibly into an industry that directly employs only slightly more than 240,000 people.[12]

Obviously, however, the persevering can be employed in broadcasting. Although obtaining a college education is not essential for most of the entry-level jobs, it is generally a wise course to take. Stations, networks, and cable systems are interested in hiring people with promotion potential and generally feel that a college education makes people more promotable.

Enrolling in one of the 300 colleges with programs leading to a broadcasting degree is advisable, but equally important is having broad knowledge in other fields. Someone wishing to enter the news field would be almost useless if he or she were an outstanding videotape recorder operator but knew nothing about national or international affairs. Political science, history, writing, and journalism courses should be a must for reporters. Likewise, accountants and salespeople should have proper business courses; engineers should know electronics; writers should emphasize composition and literature; and directors should be knowledgeable about drama, music, and psychology.

A student, while in college, should make every attempt to obtain experience in the industry through part-time jobs, or **internships.** Many telecommunications openings have an "experience required" tag attached to them, and people who have at least made the attempt to rub elbows with the industry can sometimes get a foot in the door for these jobs.[13]

Students would be well advised to join telecommunications organizations in order to meet people in the field. The old saying "it's not what you know but who you know" is often what enables a person to obtain employment. Many of these organizations have special reduced rates for students.

Students should also begin reading the trade journals of the industry. *Broadcasting* magazine, a weekly published in Washington, D.C., is the "bible" of the industry and covers the gamut of telecommunications industry news. Once a year the same organization that publishes *Broadcasting* publishes *Broadcasting Yearbook.* This supplement includes names, addresses, and facts about all radio and TV stations in the country and all cable systems, advertising agencies, networks, program suppliers, equipment distributors, and other telecommunications-related groups. Other trade publications of value include *Variety, Hollywood Reporter,* and the myriad of specialized journals put out by media organizations.

The first job is the hardest to obtain and usually involves a great deal of letter writing and pavement pounding. Many of the trades contain want ad sections that provide excellent leads. Letters of introduction and resumes can be sent to stations, networks, cable systems, and allied organizations. *Broadcasting Yearbook* can serve as an excellent source for names and addresses for these potential employment sources. Phone calls or visits to as many facilities as time and money will permit can also lead to employment.

The nature of the first job is not nearly so important as getting in the door. Most facilities promote from within, making it much easier to obtain the desired job from within than without. It is also easier to move from one facility to another if some experience has been gained, regardless of what the experience was. Patience tempered with aggressiveness and geographic flexibility are traits that are likely to lead to advancement.

Career Compensation

On the average, broadcasting is not an especially high-paying industry. True, there are the superstars who make millions, but they are few and far between. Others earn a decent but not sensational living. There is no standard pattern on fringe benefits, but most employers provide medical insurance and life insurance, as well as paid sick leave and vacations for full-time employees.[14]

What undoubtedly attracts most people to the telecommunications industry is the glamour, excitement, and power of the profession. All of these are present (albeit to a lesser degree than most people think), and they do

experience

organizations

trades

first job

pay

glamour

make for a richer, fuller, more rewarding life than many people experience in other day-to-day occupations.

insecurity

One by-product of this glamour, excitement, and power, however, is extreme insecurity. Except for those in a few highly-unionized jobs, no one is shielded from the pink slip. Last year's superstar can be this year's forgotten person—and, worse yet, have no means of support. Many people who "make it" in the business tend to spend like millionaires in order to maintain an image. Then, when the image is only a fad of the past, they have no nest egg to comfort them. The industry encompasses many free-lancers who are constantly in the state of being fired and hired and whose outrageously high salaries for one week's work are balanced by two years of unemployment. It is a hard field to break into and a hard field to stay in, but the rewards are worth it to those who endure.[15]

Conclusion

The various business practices needed for most telecommunications organizations can fall under general management, business, finance, legal, human resources, programming, news, engineering, sales and marketing, and public relations. The ways that these are organized depends on the size and type of the electronic media company.

The general manager's duties include giving leadership and direction, handling crisis, dealing with the outside world, and hiring and motivating good people. The business area handles housekeeping chores such as janitorial services, security, and the mail room. The finance department oversees profit and loss and expenditures, handles cash, and establishes and oversees ledgers. It prepares the balance sheet and the profit and loss statement and makes sure bills are paid and money is collected. In addition, finance handles payroll and various reports. In public broadcasting the finance department must make sure expenditures are aligned to the right grant. The legal department handles suits, keeps abreast of laws and regulations, and lobbies for bills favorable to broadcasters. Human resources handles procedures related to hiring and firing, arranges benefits, and makes sure that the company abides by equal employment policies. It also deals with the various guilds and unions—AFTRA, SAG, SEG, AFM, IBEW, NABET, IATSE, PGA, DGA, and WGA.

Those in programming are responsible for filling airtime by obtaining and producing programs. These programs can be obtained from networks, syndicators, and others. When they are produced, programs go through pre-production, production, and postproduction cycles. Programmers often work with production companies and with agents. The news department obtains leads on stories, gathers news, prepares copy, and presents the news. Sometimes it handles other forms of programming such as sports and documentaries. People who operate equipment are most often in the engineering department. This department also undertakes the maintenance and installation of equipment and does the cabling for cable TV systems.

The main job of the sales and marketing area is to sell commercial time by knowing the customer, competition, and station. Often this department works with advertising agencies. Some sales departments sell to customers, such as those who sell the various tiers of programming to cable TV households. Public broadcasters do not sell commercials but try to obtain grants and underwriting. The videocassette business sells to stores. Both research and traffic functions are usually located in the sales department. Public relations involves such duties as giving station tours, answering letters, holding contests, and providing public service. It often includes both promotion and publicity. These departments also produce promotional spots and PSAs.

The head of an organization is the chief executive officer, often called the general manager, the station manager, or the president. The person who heads the business department is usually called the business manager. The chief financial officer often heads the finance department, which also contains a controller, treasurer, accountants, and bookkeepers. The legal department is headed and staffed by lawyers. The director of human resources may have a labor relations manager reporting to him or her. The program director often oversees producers, directors, writers, and performers. The news director is in charge of news producers, news and feature reporters, and assignment editors. Chief engineers have equipment operators and maintenance personnel working for them. If a company has a vice-president of marketing, the sales manager and customer service manager report to him or her. The department also consists of salespeople. A director of public relations may have promotion managers and publicity managers in the department.

No two organizations are exactly alike, but most include these various functions in one way or another.

Thought Questions

1. Explain why you think all radio and TV stations should or should not be unionized.
2. Do you feel that, overall, an agent is an asset or deterrent to a performer? Why?
3. Should the telecommunications departments in colleges and universities cut back on the number of students they admit so that fewer people will be prepared to enter the job market? Why or why not?

Advertising

Introduction

All telecommunication entities are dependent upon money in order to operate. Sometimes a small amount of this money can be earned if the station, network, or corporation is willing to rent out its facilities for others to use. Sometimes these entities can also sell their programs at home or abroad. In other situations, such as that of pay-cable networks or the video disc industry, money comes from charging the customer. Corporate video production is funded through company coffers, and public broadcasting receives its money from the government, businesses, and endowments.

But for most electronic media, advertising is the primary source of income. This has been true since the very early days of radio, and although advertising practices have changed over the years, they have not diminished in importance. Simply stated, radio and television stations and networks obtain money from advertisers and use it to produce and transmit programs. The intent is for programs to be watched by as large an audience as possible so that the ads, in turn, reach as large an audience as possible.[1]

CHAPTER
9

The root of all television is money.

Anonymous

Rate Cards

Basic to much of advertising is the rate card, a chartlike listing of the prices that a station charges for different types of ads. One of the many variables

number of commercials

that affects the prices listed on the rate card is the number of commercials the advertiser wishes to air. Stations rarely accept only one ad at a time—it would be too expensive in relation to the time taken by the salesperson to sell the ad. What is more, a buyer would not realize satisfactory results if his or her product was mentioned only once. Usually a station requires that an ad be aired at least twelve times and will try to induce the advertiser to buy even more air-time by offering frequency discounts, which are lower prices per ad as the number of ads increases. For example, a company that places an ad twelve times on a radio station might be required to pay $100 per ad, or a total of $1,200. If it placed the ad twenty-four times, it would pay $80 per ad, or a total of $1,920. And if its ad ran thirty-six times, each ad would be reduced to $70, for a total of $2,520.

length of ad

Another variable is the length of the ad—usually fifteen, thirty, or sixty seconds. Obviously, a one-minute commercial will cost more than a thirty-second commercial, but usually not twice as much. For example, twelve one-minute ads might cost $100 each, and twelve thirty-second ads might cost $75 each. This is because station costs, such as handling, are more expensive for two separately produced thirty-second commercials than for one sixty-second commercial.

type of facility

Another major cost variable is the type of facility for which the ad is purchased. A thirty-second ad that costs $50 on a cable TV local origination channel might cost $100 on a radio station; $1,500 on a radio network; $2,000 on a TV station; $25,000 on a cable TV network; and $100,000 on a broadcast TV network. The underlying basis of all rate cards is the number of station listeners or viewers. In all probability, a 3,000-watt FM station would not charge as high a rate as a 50,000-watt AM station, and a small-town TV station would have a much lower overall rate card than a similar station in a metropolitan area.

national or local

Costs to national advertisers may be higher than costs to local advertisers. The rationale here is that the national advertiser can profit from reaching all the people in a station's coverage area while a local advertiser might not. For example, a national automobile manufacturer has the potential for selling cars to people in an entire city that a station covers, but the people in the southern part of the city are not likely to respond to an ad for a car dealer in the northern part of the city, especially if there is a similar dealer in the southern part of the city. For this reason, part of the station's coverage area is virtually useless to the local car dealer advertiser; therefore, the station charges the local advertiser a lower rate than it charges the national advertiser.

time of day

Audiences are of varying sizes at different times of the day, so this factor also becomes a variable on rate cards. A radio station or network would probably have the greatest number of listeners from 7:00 to 9:00 A.M. and from 4:00 to 6:00 P.M., when people are in their cars. This is referred to as Class AA time and would cost the most money. The hours between 9:00 A.M. and

CLASS "AAA"				CLASS "AA"		
3 PM — 12 MIDNIGHT, Monday - Friday				6 AM — 10 AM, Monday - Sunday		
10 AM — 12 MIDNIGHT, Saturday - Sunday						
	60's	30's			60's	30's
1x	$90	$76		1x	$70	$60
6x	$80	$68		6x	$65	$55
12x	$70	$60		12x	$60	$50

CLASS "A"			CLASS "B"	
10 AM — 3 PM, Monday - Friday			12 Mid — 2 AM, Monday - Sunday	
	60's	30's	Flat Rate:	$25
1x	$60	$50	CLASS "C"	
6x	$55	$45	2 AM — 6 AM, Monday - Sunday	
12x	$50	$40	Flat Rate:	$15

FIGURE 9.1

A small station rate card.

4:00 P.M. and 6:00 to 10:00 P.M. would have a lower fee—these hours are referred to as Class A time. The lowest rate of all is Class B time, which would be 10:00 P.M. to 7:00 A.M., when many people are sleeping. Television stations are a different story, with most viewers congregating between 7:00 and 11:00 P.M., the Class AA time. Classes A and B vary from station to station, depending on programming. Some stations fine-tune their times to AAA, AA, A, B, and C degrees.

programs

Television stations often sell ads on particular programs rather than for particular times. An ad bought for 9:00 P.M. Tuesday during "The Tuesday Night Movie" might cost more than an ad purchased at 9:00 P.M. Wednesday during "This Week's Report." In this case, the rate card lists programs rather than times. Television networks usually do not employ a rate card at all but sell ads on the basis of what the market will bear.

other factors

Other variables taken into account when rate cards are established include the number of months or years an advertiser utilizes a particular station and whether an advertiser is willing to be flexible concerning the advertising times or whether a fixed position is wanted to assure that an ad will be run at a particular time. Sample rate cards are reproduced in figures 9.1 and 9.2. These show some of the variances, such as number of commercials, length of commercials, and time of day.[2]

Financial Arrangements

Theoretically, a station time sales representative approaches a potential advertiser, convinces him or her to buy a group of ads at the price established on the rate card, and then sees that a bill is sent to the advertiser for the correct

FREQUENCY ANNOUNCEMENTS

Announcement rates are based on the number used during an established twelve-month period, and become effective from the beginning of service on firm contracts or as earned.

60-Second Announcements

(150 Words, Live)

Times	AAA	*AAA Combo	AA	A	B	**C
1	$320	$250	$205	$131	$100	$44
15	309	247	194	124	95	42
50	296	237	179	116	89	39
150	275	220	166	107	84	36
300	261	209	159	100	79	34
500	253	202	154	97	77	33
750	246	197	151	95	75	32
1000	240	192	149	93	74	31

30-Second Announcements

(75 Words, Live)

		*				**
1	$256	$205	$164	$105	$80	$35
15	248	198	155	99	76	34
50	238	190	143	92	71	31
150	220	176	132	86	67	29
300	209	167	127	80	63	28
500	203	162	124	78	61	27
750	195	156	121	76	60	26
1000	193	154	120	75	59	25

10-Second Announcements

(25 Words, Live)

		*				**
1	$160	$130	$103	$66	$50	$22
15	155	124	98	62	47	21
50	148	118	89	58	45	20
150	138	110	83	54	42	19
300	130	104	80	50	40	18
500	126	101	78	48	39	17
750	124	99	76	47	38	16
1000	121	97	75	46	37	15

*Combo: Rates apply only to the number of AAA announcements that are ordered and broadcast in combination with an equal or greater number of announcements of the same or greater length within other time classifications except Class C.

Frequency Announcements Rate based on total number of announcements used during an established twelve-month period. Rate applies as earned.

Weekly Package Plan Rate based on a consecutive seven-day period.

WEEKLY PACKAGE PLAN

Weekly Package Plan rates apply only to the number of 60-second, 30-second and 10-second announcements broadcast for one product within a consecutive seven-day period on run-of-station schedules. Such announcements are subject to immediate preemption for frequency announcements without notice. Further, without liability to the Station, Weekly Package Plan announcements are subject to omission by Station without charge to Advertiser, or to rescheduling by the Station in a time period considered equivalent by the Station.

Frequency announcements may be combined with Weekly Package Plan announcements to earn lower plan rates, but Weekly Package Plan announcements may not be combined with frequency announcements to earn lower frequency rates.

60-Second Announcements

(150 Words, Live)

Weekly Plan	AAA	*AAA Combo	AA	A	B	**C
6	$291	$233	$176	$117	$94	$40
12	284	227	171	112	89	38
18	275	220	166	107	84	36
24	268	214	161	102	79	34

30-Second Announcements

(75 Words, Live)

		*				**
6	$233	$186	$142	$94	$75	$32
12	226	181	137	90	71	30
18	220	176	132	86	67	29
24	214	171	128	82	63	27

10-Second Announcements

(25 Words, Live)

		*				**
6	$146	$117	$89	$60	$48	$20
12	143	114	86	57	45	19
18	138	110	83	54	42	18
24	135	108	81	51	40	17

TIME SIGNALS

(Flat and not combinable with other announcements)
(12 Words of Commercial Copy or Six Seconds Transcribed)

AAA	*AAA COMBO	AA	A	B	**C
$104	$83	$62	$41	$30	$13

**C May be counted toward Frequency Announcements or Weekly Package Plan, but may not count toward Combo.

amount. However, in the real world of economics this theoretical case is not always the practiced one.

For example, many radio stations, particularly smaller ones, engage in what is called **barter** or **trade-out**—they trade an advertiser airtime for some service the station needs. Perhaps the station owns a car that needs occasional service; in order to receive this service free from Joe's Garage, the station will broadcast twelve ads a week for Joe's Garage with no money changing hands. Similar barter arrangements are often negotiated with restaurants, stationery suppliers, gas stations, audio equipment stores, and the like.

One of the most common methods of selling commercial time is called **run-of-schedule (ROS).** The salesperson and advertiser decide on how many ads should be run and sometimes make up a package that consists of ads in different times—A, B, and C. The station is the one that decides on the specific times the ads should run based on the availability of commercial time. For this privilege, the station gives the advertiser a discount so that he/she does not

Business Aspects of Telecommunications

FIGURE 9.2

A larger station rate card.

barter

run-of-schedule

pay what the rate card designates for the times selected. In some instances, an advertiser who buys ROS will get an even better deal because the station will run an ad that was supposed to be in B time in A time if an A time spot is available. The station does not charge the advertiser extra for this.

Somewhat the opposite of ROS is a **fixed buy,** wherein the advertiser very specifically states the exact time that each ad should run. For this, the advertiser pays the premium price. Of course, this often involves some negotiating. If two advertisers want 9:00 A.M. Monday, someone must compromise.

Co-op advertising involves shared costs, usually by a national and a local advertiser. A commercial for an "Instant Pleasure" camera may end with, "To purchase this sensational camera, make a short trip to Lou's All-You-Need Photo Stop at 160 Main Street." The cost of this ad would be divided on an agreed-upon formula between Instant Pleasure and Lou's.

Some stations offer **rate protection** to their regular customers so that even if the station's rate card costs increase, these good customers can still buy ads at the old prices.

Although it is poor business practice, some salespeople, in order to obtain business, simply cut the prices on the rate card and offer ads at lower prices than those indicated.

Financial arrangements vary with the times. When times are good and stations can sell more ads than they have airtime for, they stick rather firmly to the rate card. When times are bad, stations offer a great number of incentives to bait advertisers into buying.

Advertising Practices

Regardless of the financial arrangements, there are several different ways that advertising is bought. For example, it can be divided into **network buying, national buying, regional buying,** and **local buying.** When an advertiser buys with a network—either broadcasting or cable—its ads go everywhere that the network program does. In the case of a commercial radio or TV network, such as CBS or ABC, the ads play on all the stations affiliated with the network. For a cable network, the ads appear on all the cable systems that carry that network. Network ads are heard or seen on all stations or systems at the same time, just as network programs are. Most network ads are for products that are needed by many people throughout the country, such as automobiles, toothpaste, or cereal.

National buying is also for products with wide appeal, but it involves buying time on individual stations. The advertiser places the ad on many stations, usually enough to cover the whole country, but the ads do not play at the same time on all stations. One reason an advertiser may take this route is that its product may desire a certain demographic audience. Maybe an automobile manufacturer wants to promote a car that it feels will particularly appeal to women with small children. It can select stations throughout the country that target that audience. But ads for products intended for everyone also often use national buying.

fixed buy

co-op

rate protection

rate cutting

network buying

national buying

Regional buying covers a particular area of the country and is most often used for products that do not have a national appeal but can be useful in a number of places. A tractor manufacturer, for example, might want to advertise throughout the rural areas of the Midwest. Regional buys can be made by purchasing time on a number of different individual stations in the region, or they can be made by purchasing time on a regional network. The latter are most common with radio and with cable TV sports programming.

Local buying involves placing commercials in a limited geographic area. It is used primarily by merchants who only service a local area—used car dealers, grocery stores, furniture stores. A local advertiser often places the same ad on many different local radio and TV stations and cable local origination channels. In this way, the ad can reach a maximum number of people in a particular area.[3]

Another way to discuss advertising buying is to divide it into **program buying** and **spot buying.** In the early days of radio and television most advertisers bought programs, paying all the costs to produce and air programs such as "Lux Radio Theater," "Colgate Comedy Hour," and "Kraft Television Theater." The advertiser had the advantage of constant identification with the program and its stars. In fact, commercials were often integrated into the program. The announcer would drop by Fibber McGee and Molly's house and extol the virtues of Johnson's Wax, or singers would serenade Jack Benny with a ditty about Lucky Strikes. Program buying is rarely done anymore for several reasons. One reason is that costs of television production have soared to the extent that not even the largest companies can afford the total underwriting of a program week after week. Also, after the quiz scandals of the late 1950s, the networks became leery of such overriding program control by advertisers and began to take greater control of content. Only occasional specials are now totally paid for by one advertiser.

What is common nowadays is spot buying. The advertiser bears none of the cost for or identification with a particular program, but rather buys airtime for certain times of the day. Hence, a disc jockey whose show runs from 6:00 to 9:00 A.M. may present ads from several dozen advertisers, all falling under the umbrella of spot advertising. In radio, spots can be aired at almost any time because program segments are short—three or four minutes for a musical selection or two or three minutes for a group of news items. In television, spots are aired at more definite times because the programming has natural breaks where ads are inserted. Spots can be network, national, regional, or local. In fact, the same commercial spot that is shown on MTV (a network) might be shown on a music video show produced by a local station or a music video program that is syndicated and shown on a number of TV stations throughout the country.[4]

Another way to categorize commercials is by their length. During the early radio and TV days, most commercials were 60 seconds long. But as commercial costs rose and the number of companies wishing to advertise on radio and TV also grew, 30 second commercials became more common. They cost the advertiser less, and more products could be advertised within a given period of time. In the mid-1980s, a new commercial form known as the **split-30** emerged. This is two 15 second commercials, but the commercials are for related products manufactured by the same company. Gradually the split-30 phenomenon evolved into individual commercials that are 15 seconds long. Although 15 second commercials are less than 10 percent of all commercials, they are increasing in number.[5]

The amount of time devoted to commercials on television is also increasing. The commercial networks that, in the early 1980s, were airing no more than nine and a half minutes of ads during prime time now show about eleven minutes of ads during that time. Other times of the day, such as afternoon, have more than sixteen minutes of each hour devoted to commercials.[6]

amount of commercial time

Sales Staffs

Radio and TV time selling—like insurance, shoes, and stock—needs a salesperson to interact between the product and the customer. Exactly how this function is executed varies greatly from station to station.

The advertising practices of a small radio station are not difficult to follow because most of the advertising comes from the local area. Usually the disc jockeys are also salespeople, spending four hours a day announcing and playing records and four hours going to the local hardware stores, grocery stores, and car dealers selling ads. Often the general manager of a small station doubles as a salesperson, too.

local salespeople

Within larger radio and TV stations the selling process is more indirect. TV talent and big-name disc jockeys usually do not sell time. Instead, members of a sales force contact local merchants and are generally paid on a salary plus commission or commission-against-draw basis. These salespeople are expected to service ads as well as sell them so that advertisers will repeat their business. Servicing ads includes making sure the ads are run at the appropriate times and facilitating any copy changes that may be needed. For example, a furniture store may be having a sale scheduled to end on April 30, and the ad copy that a disc jockey reads over the air states this. If the store decides to extend the sale to May 15, the salesperson should make sure the copy is corrected.

TV stations and larger radio stations are interested in securing ads from companies that distribute goods nationally—companies such as Procter & Gamble, General Mills, and General Motors. Since it would be very expensive for each station to send salespeople to the headquarters of these companies, stations generally hire **station representatives** to obtain these ads. Station representatives are national (or regional) sales organizations that operate as an extension of the stations with which they deal. They are middlemen between the company that wants to buy ads and the radio or TV stations that want to sell ads. Usually these reps handle national spot sales for dozens of stations around the country, some representing just radio stations, some just TV, and some a combination of the two. Often they try to line up stations that are similar (for example, a number of country music radio stations), so that they

station reps

can sell ads to companies wishing to attract that demographic group. For obvious competitive reasons, a station rep will not service two stations in the same listening area.

The network-owned and -operated stations have their own corporate reps, as do some large independent chains such as Westinghouse and Gannett, but most of the stations of the country utilize station representative companies such as Petry, Katz, or Blair.

A station rep earns income from commission on the sales obtained for the station. The percentage of commission varies because it is negotiated by each station and rep, but it generally falls between 5 and 15 percent. TV stations generally pay a lower commission because they sell millions of dollars worth of ads each year, while radio stations pay a higher percentage because the price per ad is lower. Needless to say, good communication must be maintained between the station and rep so that the rep can work in the station's best interest. In addition to selling ads, some station representatives help their stations with programming decisions, audience research, and station promotion. They do this, not out of the generosity of their hearts, but because they can sell time easier for successful stations with high ratings.[7]

Radio networks and cable networks maintain sales forces to obtain national ads for their programs. Advertising buying in both these areas has risen in recent times.

But the largest portion of advertising dollars go to the commercial networks. These networks maintain large sales forces that are devoted to selling and servicing major national advertisers. The competition among these sales forces is keen. Each network wants to ensure that all its programs sell—and sell at the highest rate possible. Network commercial time is a fairly limited commodity. Only a certain number of spots are available, and these cannot be expanded as easily as a newspaper can add pages when ad demand is heavy. A network has the same amount of time available during the holidays, a heavy season, as it does the rest of the year, so the sales force must be careful to sell in such a way that the overall needs are met.

Because the buyers and sellers of TV network time are generally a small, close-knit fraternity, a network salesperson should emphasize service. One way to do this is to make sure the client buys enough variety in the network schedule that if some of the programs fail, the client still has time in successful portions of the schedule. Generally network management is also deeply involved with sales and makes sure that a knowledgeable team presents programs and program ideas to potential sponsors.

For many years, TV network advertising sold out rather easily at high prices. However, as other forms of TV have eroded the audience, network sales have gone soft, and networks have found they cannot raise prices and still attract advertisers.[8]

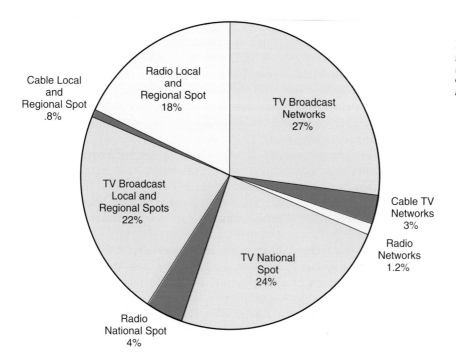

FIGURE 9.3
Distribution of
advertising dollars.
(Source: Statistical Abstracts
of the United States, 1990,
p. 557.)

Advertising Agencies

When salespeople or station representatives are vying for ads from major advertisers, they generally do not deal only with the company, but rather work through the company's **advertising agency.** An ad agency handles overall advertising strategies for a number of companies by advising them about newspapers, magazines, direct mail, and other forms of advertising, as well as radio, TV, and cable. Ad agencies came into existence more than a hundred years ago to act as middlemen between companies who needed to advertise their product in markets where they were distributed and advertising outlets that wanted to obtain as many ads as possible.

functions

Most major advertising agencies are generally termed **full-service agencies.** They establish advertising objectives for their clients and determine to whom the product should be sold and the best way to reach these consumers. They conduct research to analyze the audience or test the advertising concept and then attempt to obtain the best buy for the money available. Full-service agencies design the advertisement or oversee the commercial and determine when a company should initiate a campaign and how long it should continue. After ads have run, advertising agencies handle postcampaign evaluations. In all these processes they work closely with marketing and sales executives within the companies they represent.

For its efforts, the advertising agency generally receives 15 percent of the billings. In other words, a billing of $1,000 to its client for commercial time on a TV station means that $850 goes to the station and $150 is kept by the ad agency to cover its expenses. Because dollar amounts are fairly constant from agency to agency, the main "product" an ad agency has to sell to its customers is service. Advertisers sometimes change agencies simply because one agency has run out of creative ideas for plugging its product, and a new agency can initiate fresh ideas.

Because agencies generally handle all media, the broadcast salesperson who approaches an ad agency must convince the people handling accounts that broadcasting is the best arrangement for their clients and that the particular network or station he or she represents offers an exceptionally good deal. Cable TV representatives must, of course, tout their product over that of radio, TV, or other media.

Some of the largest ad agencies are Saatchi and Saatchi; J. Walter Thompson; Leo Burnett; Foote, Cone and Belding; and McCann-Erickson.[9] Ad agencies, like many other companies, have been purchased and consolidated. Most of these agencies are divided into departments that include account executives, media specialists, TV program buyers, copywriters, art directors, and marketing research specialists. Therefore, a team works on the ad campaign of any one company. Not all agencies are large, however. In fact, some are one-person operations with only a few clients who like the overall attention they receive from the person directly responsible.

Not all major advertisers employ an ad agency. Some maintain their own **in-house** services, which amount to an ad agency within the company. Other companies will not hire a full-service agency but will hire what are sometimes referred to as **modular agencies.** These agencies handle only the specific things a company asks for—perhaps the designing of an ad or the postadvertising campaign research. Usually this type of work is paid for by a negotiated fee rather than a 15 percent commission. There are also media buying services that concentrate only on buying radio and television spots.

Although there are many variations for advertising, the most common procedure is for a company to hire an agency to make the basic advertising decisions and to see that they are carried out. As far as broadcasting is concerned, these ad agencies deal with networks, station representatives, and occasionally individual stations to ensure that the ads of their clients receive the best treatment. Throughout the process, both company officials and broadcast executives are informed of the successes and failures of the advertising campaign.

Commercial Production

A company buys time on a station or network in order to air a commercial extolling the virtues of itself as a company or of one or more of its products. Production of commercials, then, is a very important facet of the whole advertising scene.

Companies rarely produce their own commercials because they simply are not endowed with the equipment or know-how to do so. They are in the bread-making, car-selling, or widget-manufacturing business, not the advertising business.

radio productions

Small businesses that place ads on small radio stations generally have the stations produce the ads. The people who sell the time to the merchant will write and produce the commercial at no extra cost. If they are good salespeople, they will spend time talking to the merchant to determine what he or she would like to emphasize in the ad and they will check the ad with the merchant before it is aired. Production is usually quite simple, frequently just written copy that the disc jockey reads over the air. Often the commercial is prerecorded with music or sound effects added, and occasionally the merchant will talk on the commercial. At any rate, the whole process is kept at a simple, inexpensive level.

Large companies that advertise on radio usually have much more elaborately produced ads that include jingles and top-rated talent. The production costs are generally in addition to the cost of buying time.

TV productions

Television commercials are still more elaborate. Usually they are slick, costly productions that the advertiser pays for in addition to the cost of commercial airtime. Local stations sometimes produce commercials for local advertisers and charge them production costs—the cost of equipment, supplies, and personnel needed to make the commercial. Much of local TV advertising and most network commercial production is handled through advertising agencies.

As part of its service to the client, the advertising agency will decide the basic content of the commercial, perhaps by trying to come up with a catchy slogan or jingle. Usually a **storyboard** is made—a series of drawings indicating each step of the commercial.

The advertising agency then puts the commercial up for bid. Various independent production companies state the price at which they are willing to produce the commercial and give ideas as to how they plan to undertake production. The ad agency then selects one of these companies and turns the commercial production over to it.

Some production companies maintain their own studios and equipment, while others rent studio facilities from stations, networks, or companies that maintain the facilities. It is possible that a station may lease its facilities for a commercial that winds up appearing on a different station.

The time, energy, pain, and money that is invested into producing a commercial often rivals what goes into TV programs themselves. One commercial often takes days to complete and involves fifteen or twenty people and costs well over $50,000. Take after take is filmed or taped so that everything will turn out perfectly. Once the commercial is "in the can," it may air hundreds of times over hundreds of stations—hence, the obsession with perfection.

approaches

Commercials can take many different approaches. Some attempt to be minidramas that show how the advertised product can solve a problem. Others place the accent on humor. In an attempt to establish credibility, the ad agency

OPEN ON 7-YEAR-OLD GIRL WEARING
A POTATO RING ON EACH FINGER.
DEMURELY, SHE TAKES ONE OF THE
RINGS OFF HER FINGER ...

MUSIC THROUGHOUT.

SINGERS: "Give a ring..."

PUTS IT IN HER MOUTH, AND
CRUNCHES.

SFX: LOUD CRUNCH.

SINGERS: "Give a crunch.
          Give Crunchi-O's."

CUT TO GORILLA FACE MASK.

SINGERS: "Give a ring..."

MASK IS REMOVED REVEALING
BEAUTIFUL WOMAN. SHE WINKS AT
CAMERA AS SHE BITES INTO
CRUNCHI-O.

SFX: LOUD CRUNCH.

SINGERS: "Give a crunch.
          Give Crunchi-O's."

CUT TO JEWELER LOOKING THROUGH
EYEPIECE AT POTATO RING.

SINGERS: "Give a ring..."

HE DECIDES RING IS GOOD ENOUGH
TO EAT. AND EATS IT.

SFX: LOUD CRUNCH.

SINGERS: "Give a crunch.
          Give Crunchi-O's."

DISSOLVE TO CRUNCHI-O CANNISTER.

MUSIC UNDER.

ANNCR. VO: "Introducing Crunchi-O's.
            The new potato snack from
            Nalley's...."

DISSOLVE TO CU OF PRODUCT.

ANNCR VO: "made in the shape of
           golden bite-size rings ...
           for a crunchy, crisp
           potato taste that's
           really different."

CUT TO TEENAGER REACHING
INTO CANNISTER FOR CRUNCHI-O.

SINGERS: "give a ring..."

HE TOSSES CRUNCHI-O OVERHEAD AND
WAITS TO CATCH IT ON THE WAY DOWN.

SINGERS: "Give a crunch..."

SUDDENLY, HUNDREDS OF CRUNCHI-O'S
POUR DOWN, FORCING TEENAGER TO
COVER UP.

SINGERS: "Give Crunchi-O's."

CUT TO MEDIUM CU OF CRUNCHI-O
CANNISTER.

MUSIC UNDER.

SUPER: AMERICA'S ONLY
       POTATO RING.

ANNCR VO: "Crunchi-O's.
          America's only
          potato ring.
          New from Nalley's."

FIGURE 9.4

A commercial
storyboard. *(Courtesy of
Della Femina, Travisano)*

might decide to have a well-known celebrity pitch the product with a testimonial of its virtues. Some ads demonstrate products in use. Others involve interviews with satisfied customers. Some ads use no people whatsoever but employ animation, special effects, or unusual design. Most ads try to appeal to some basic human instinct such as security, sex, love, curiosity, or ego inflation.[10]

Commercials have enabled television and radio to create their programs, but the "program content" of commercials has also had an effect on the vernacular and life-style of American society. Remember "Where's the beef?"; "Leave the driving to us"; "Look ma, no cavities"; "Greasy kid stuff"; "Double your pleasure, double your fun"; "Ring around the collar"; "Try it, you'll like it"; "I can't believe I ate the whole thing"; "Up, up and away"; "Flip your Bic"; "It's the real thing"; "I'd like to buy the world a Coke"; "Reach out, reach out and touch someone"; and "You deserve a break today"?[11]

past commercials

PSAs and Promos

Advertising time often includes segments that look like commercials but really aren't. Often these are **public service announcements (PSAs),** which are for nonprofit organizations or causes. Stations do not charge to air these, so they are often put at undesirable times. However, effectively produced PSAs are noticed and often lead to increases in desirable actions, such as more women going in for breast cancer check-ups, people contributing money to rape crisis centers, or people calling drug abuse hot lines. Stations sometimes help organizations produce PSAs but many of them are produced for a fee, or free, by the same production companies that produce commercials.

PSAs

Another type of ad that is really not an ad is a **promotional spot (promo).** This form of advertising promotes programming that will be aired on the station or network. Usually these contain short scenes (often of a teasing nature) of part of the program, along with information about when it will be airing. Other promos consist of a slide or graphic depicting the show. Promos are often put on the air at times that have not been sold for commercials. However, people at the station or network who are in charge of promotion usually feel these spots should be aired at select times when many people are viewing or listening so that they are made aware of future programming. These are the same times that advertisers prefer, so the promotions people and the salespeople often have an ongoing feud about the placement of the promotional spots. Salespeople say the time should be given to paying clients. Promotion people say that later sales will lag if the audience is not made aware of future programs.

promos

Advertising to Children

Although the whole field of advertising is filled with controversy, commercials aimed at children have yielded particularly strong controversies—one being the distinguishability of advertisements and program content. On early radio

FIGURE 9.5
(*a*) The Bing Crosby family doing a spot for Minute Maid orange juice. (*b*) Johny Philip Morris, who was part of the "Call for Philip Morris" cigarette ads on radio. (*c*) Betty Furness advertising for Westinghouse during the 1952 nominating conventions. *(a. courtesy of the Coca-Cola Company; b. courtesy of KFI, Los Angeles; c. courtesy of Betty Furness)*

(a)

(b)

(c)

and early TV, the hosts and stars of children's programs presented many of ads and programs the ads. For example, Captain Kangaroo, after doing a stint about growing vegetables, would walk over to a set featuring boxes of cereal and extol the virtues of the sponsor's product.

This practice came under fire when research showed that children have a great deal of difficulty distinguishing between program content and commercial message. Parents, particularly a group of Boston parents who formed an organization called **Action for Children's Television (ACT),** also complained about the ads. They pointed out that because of the ads, children were constantly asking their parents to buy them things. If a favorite cartoon character said Crackle Bars should be eaten every day for breakfast, young tots would become believers and badger parents until Crackle Bars were served for breakfast.

As a result of the research and the complaints, the FCC, in 1974, issued guidelines stating that programs' hosts and stars should not sell products and that the display of brand names and products should be confined to commercial segments. In other words, a box of Kellogg's Corn Flakes should not appear within the program itself. These guidelines also stated that special measures should be taken to provide auditory and/or visual separation between program material and commercials. This led to **islands** at the end of most children's commercials. These were still pictures with no audio followed by a definite fade to black. This was to help children understand that the commercial had ended and the program was about to begin again.

Starting about 1983, when the FCC was in a deregulatory mood, these guidelines were no longer touted or enforced, but most commercial producers abided by them anyway.

The amount of time devoted to commercials within children's programming has also varied in controversial ways over the years. For several decades, time for ads the National Association of Broadcasters had a self-regulatory code that stated the maximum number of minutes per hour that should be devoted to commercials within children's programs. During the 1960s and early 1970s, this number was sixteen. Then in 1973, the code was amended (largely because of the insistence of ACT) to reduce the number to twelve. Shortly after that it was further reduced to nine and a half minutes per hour. Although this code was not a law (it was just an advisory standard decided by people within the television industry), most stations abided by it.

However, the code was outlawed by the Justice Department in 1982, and the FCC, in its deregulatory mood, said there should be no limits on the amount of advertising on any programming. This led to increases in the number of ads in children's programs and led to a short-lived phenomenon known as **toy-based programming.** For these programs, a particular toy was the "star" of the show. Often the toy was created before the show with the idea that the program would sell the toy. This did not sit well with ACT and other parents who felt their children were being exposed to commercials that were thirty minutes long.

But in 1990, restrictions were once again placed on commercials during children's programs, this time by Congress. It passed a bill limiting commercials to twelve minutes per hour during weekdays and ten and a half minutes on weekends. This essentially outlawed toy-based programs, although they had already somewhat fizzled as a programming form because children were not watching them to any great degree.

The production techniques of children's ads have also been questioned, especially those that make the ads appear to be deceptive. Camera lenses that make a toy car look like it travels much faster than it does, sound effects that add to the excitement of the board game but are not part of the game, camera angles that make a toy animal look four feet tall when in fact it is four inches

tall—all came under criticism. Their use also followed the same regulatory-deregulatory mood of the FCC that affected other aspects of advertising to children. During the 1960s, this type of exaggeration and fluffery was common. During the 1970s children's ads were more low-key, and during the 1980s, flamboyance began to creep in again.[12]

Other Commercial Controversies

Not only children's ads have been accused of being deceptive. Ads aimed at adults also can prove to be less than what they seem to be. One classic case involved a Rapid Shave commercial that touted that the product could soften even sandpaper. Although it looked like the razor was shaving sandpaper, what was being shaved was a sheet of plastic covered with sand. The Federal Trade Commission made Rapid Shave stop airing the commercial because it was deceptive. In another instance, Campbell's soup put marbles at the bottom of a bowl so that the vegetables would rise to the top and make for a richer looking soup. Again the FTC intervened—no one puts marbles in soup. Even when demonstrations are valid, they may not be applicable. For example, a watch put through a hot and cold temperature test may operate perfectly at 110 degrees and -20 degrees, but not be satisfactory at normal temperatures.

Testimonials have had their share of criticism because the well-known stars or sports figures used for the commercials have no way of knowing whether what they are reading from the cue cards is correct or not. There is a ruling that stars actually have to use the product they are advertising (this arose after Joe Namath advertised panty hose), but they cannot be expected to know all of its assets and liabilities. Sometimes the implications given by or about the stars are misleading—"John Q. Superstar runs ten miles a day and eats Crunchies for breakfast." The implication, of course, is that the Crunchies enable him to run ten miles—in all probability a false premise.

Another area of controversy concerning commercials involves the types of products that are advertised. There are critics who claim that advertising nonprescription medicines encourages people to use them excessively. Likewise, beer, wine, and personal sanitary products are considered by some groups to be inappropriate TV fare.[13] A debate raged over whether or not condoms should be advertised. Stations and networks aired steamy love scenes within the programs without mention of any type of sexual protection but would not accept ads for condoms. Eventually many stations did accept the ads, primarily because of the threat of AIDS.[14]

Program-length ads, often referred to as **informercials,** are also criticized. Toy-based programming was a form of a program-length ad, but at least there was a story involved. Many of the informercials aimed at adults are half-hour interviews and anecdotes focused totally on promoting a certain diet, baldness treatment, or a pair of sunglasses. They are filled with plugs to "call now," "write for more information" or accept some free inducement in exchange for trying the product. Unlike toy-based programs, which were considered children's programming, informercials are considered to be commercials. As a result, the companies making the programs pay stations in order to air the programs. As stations' profits sink, many more are airing these program-length commercials because they bring in needed income.[15]

informercials

Pros and Cons of Advertising

Despite the fact that advertising achieves its purpose—that is, it sells products and it provides the money for program production—its very existence is highly controversial.

The overall structure of the broadcasting-advertising relationship is frequently questioned. Many countries have depended on government tax money to operate their broadcasting systems; commercials, if they existed at all, provided only supplementary income. Of course, this leads to greater government control because as the saying goes, he who pays the piper picks the tune. There are those who claim the piper analogy holds for American TV, too; however, the heads of large companies and advertising agencies control program content instead of the government. Critics point to specific instances of advertiser censorship and to the overall censorship that occurs simply because advertisers refuse to sponsor certain types of programs. Advertisers retort that they really keep hands off programming content. Now that most advertisers buy spots and do not sponsor entire programs, they contend that they do not even have an opportunity to influence the program content.

overall structure

For media that charge the customer, the question of the role of advertising becomes acute. Many people originally subscribed to cable TV so that they would not have to watch commercials. As cable increases its advertising orientation, the consumer is winding up paying for the programming and also seeing commercials. Even videocassettes rented at the local video store sometimes contain commercials.

paying fees

Advertisers feel, however, that they are getting less for their money, partly due to the services that charge the customer. With the vast array of program services available now, many people switch channels whenever the commercials come on. This decreases the effectiveness of commercials. When people videotape programs, they often fast forward through the commercials when they replay the programs. Again, commercials are losing audience. Advertisers feel they should not be charged high rates when, in fact, people are not exposed to their messages.

not viewing commercials

clutter

The public often criticizes the way commercials are aired. The most frequent complaint is that there are simply too many commercials, a situation often referred to as **clutter.** As stations and networks see their dollars shrinking, they are expanding the number of minutes devoted to ads, creating more clutter.

interruptions

The frequency with which commercials interrupt programs is another gripe often voiced. In some countries commercials are grouped only between programs or all at one specific time of day. Of course, rates cannot be as high for clustered commercials because they lose impact under that sort of airing situation. For this reason, advertisers are generally opposed to the clustering that now occurs.

length

Ads appear to be more numerous than they actually are because most commercials are much shorter than they used to be. Where one minute used to contain one advertisement, it now contains as many as four.

frequency and loudness

The frequency with which certain commercials are aired also grates on some people's nerves, and they become irritated with the repetitious sales pitches. The loudness of some commercials is also criticized. Commercials, especially on radio, seem to blare out in comparison to other programming. Advertisers and stations retort that they air particular commercials only as frequently as they appear to be effective and that there definitely is not an intentional volume increase for commercials.

effect on society's structure

There are those who see advertising as having an overall negative effect on the structure of our society. It makes our society dominated by style, fashion, and "keeping up with the Joneses," while at the same time it retards savings and thrift. It fosters materialistic attitudes that stress inconsequential values and leads to waste of resources and pollution of the environment. Advertising also fosters monopoly because the big companies that can afford to advertise can convince people that only their brands have merit. The counter argument to all this is that advertising drives the economy. If people were not enticed into buying things, the economy of the country (and world) would be stagnant at best.

A great deal of broadcast advertising criticism is aimed at the content of commercials, accusing them of being misleading, insulting, abrasive, and uninformative. However, it has been proven that ads that instill negative feelings in people often lead to sales of the product because people remember the brand name.

negative feelings

time constraints

Because fifteen seconds, thirty seconds, or even a minute is not enough time to explain the assets of a product in an intelligent manner, the intent of commercials is to try to gain the audience members' attention so that they will remember the name without really knowing much about the product. Of course, just as a writing school will discuss its successful graduates but never its unsuccessful ones, the liabilities of a product or service are never discussed.

wording

Words such as "greatest," "best," and "most sensational" may be good copy, but they have very little concrete meaning. Worse yet are terms such as "scientifically tested" and "medically proven," which are not followed by any description of what constitutes the test or proof. Sometimes advertisers claim that "our product is guaranteed to give 50 percent more satisfaction" or "ours is whiter and brighter"—50 percent more than what? whiter and brighter than what? Statistics can lie too: "Three out of four doctors" might represent a total sample of four doctors. Secret ingredients such as "DXK" and "KQ108" are thrown about without indication of what these ingredients contain. Advertisers retort that some hyperbole is needed in commercials; otherwise, no one will pay attention to them.

Commercials have also been criticized for fostering **stereotypes** of women and minorities—the dumb homemaker who can't get her wash clean without help from a detergent genie, or the lazy Mexican who doesn't want to do his work. Commercial producers claim that they do a better job of destroying stereotypes than programs do because commercials present women and blacks in responsible positions.

Although advertising is severely criticized from many quarters, it has its positive elements. Primarily, it supports program costs so that the public can receive broadcast radio and television without paying for it directly, and it informs people about products available to them.

stereotyping

Conclusion

Commercial practices of each telecommunications entity are different, but generalizations about what might happen at a typical small commercial radio station, large commercial TV station, commercial network, and basic cable TV network fairly well summarize the various facets of advertising.

Most small radio stations are totally dependent on advertising for income. Their rates are much lower than those of TV stations, with commuter hours being the highest-rated times. Small radio stations give quantity discounts and have particularly good deals for local advertisers who are likely to buy repeated spots. Sometimes they offer time well below the published rate card price, and often they barter by accepting goods or services instead of money. Members of the sales force usually have other jobs within the station, such as disc jockey or general manager, and most of the selling is to the local community in the form of local spots. The radio station staff builds lasting relationships with the advertisers and usually writes and prepares the commercials. They usually air a fair number of PSAs and some promos. Viewers are likely to complain if the station's ads are too frequent or too loud.

A large TV station receives most of its income from advertising but may also rent out its facilities or sell some of its programs. The most expensive time on its rate card is evening prime time, and it is interested in selling mostly thirty-second ads. The station may try to push run-of-schedule buying as a means of filling its commercial time. Most of the buying is on a local or national spot basis, although occasionally an advertiser will pick up the tab for an entire program. A sales staff covers the local territory, but a large TV station also hires a station representative to obtain national or regional ads. Occasionally the station sales manager deals directly with an advertising agency. Most of the commercials are delivered preproduced, but if the station does tape a commercial for an advertiser, it charges extra for that service. Stations air PSAs and promos as time permits. Of all forms of media, they are most likely to program informercials. Viewers are likely to become irritated by commercial interruptions and may complain to stations about misleading or vague commercials, or commercials that foster stereotypes.

Commercial networks, too, are dependent upon advertising for income. They do not publish formal rate cards, but work closely with advertising agencies to supply programming (for the ad agencies' customers) that will sell the most ads at the highest rates possible. The commercials that appear on network TV are meticulously and expensively produced, usually by independent production companies, not the network itself. The heat of advertising criticism is directed toward networks because they are so visible. Criticism includes the questioning of the very existence of an advertising-based entertainment structure. Other criticisms are directed at the advertising of personal products, the negative effects of advertising-induced materialism on society, the validity of testimonials and demonstrations, and the various issues involved with the inclusion of commercials within children's television programs.

Basic cable TV networks are becoming more and more reliant on advertising, although they also receive money from subscribers. Usually they do not have a set rate card but sell at what the market will bear. They deal with advertising agencies and go for national commercials, although some of the regional sports networks cablecast regional spots. Cable networks receive some of the same complaints about ads that commercial networks do, but the main complaint against them is that subscribers must pay for the service and also watch commercials.

Thought Questions

1. Explain why you think broadcasters should or should not be made to hold strictly to their rate cards in selling all advertising time.
2. Should ads be allowed on radio and TV regardless of product? Support your answer.
3. Should ads be banned on children's TV programs? Why or why not?

Audience Measurement

Introduction

Few aspects of telecommunications are criticized as vehemently as ratings, and yet they are all but worshiped within the executive realms of broadcasting. Ratings have arisen because of the desire of various telecommunications groups to know how many people are watching or listening to programs. Advertisers invest a great deal of money in commercials and commercial time and want to know that their messages are reaching an audience. Stations and networks generally charge advertisers at a rate based on the number of viewers or listeners. Both media forms want to know public reaction to their programs so that they can anticipate changes in trends or know when a program has passed its prime.

When companies advertise in print media, such as newspapers and magazines, they are also interested in knowing the audience size. The advantage of the print media is that it's possible to count the number of copies sold; however, the number of people watching TV or listening to the radio in the privacy of their homes or cars is much harder to count. As a result, audience measurement systems have been developed.

CHAPTER 10

There are only two rules in broadcasting: keep the rating as high as possible and don't get in any trouble.

anonymous television executive

Early Rating Systems

fan mail

Systems for determining audience size began very early in radio history, the first "rating method" being fan mail. Research showed that one person in seventeen who enjoyed a program wrote to make his or her feelings known. This "system" was effective enough while the novelty of radio lasted, but it was never representative of the entire audience.

free prizes

When radio became more commonplace, stations would offer a free inducement or prize to those sending in letters or postcards. This "system" was not considered an accurate measurement of the total number in the audience, but it could give comparative percentages of listeners in different localities. For example, if two stations in two different cities made the same offer, the number of replies to each station would tell which station had the larger audience.

Crossleys

Early in the 1930s, advertising interests joined together to support ratings known as the **Crossleys,** started by Archibald Crossley in 1929. He used random numbers from telephone directories and called people in about thirty cities to ask them what radio programs they had listened to the day before his call. This came to be known as the **recall method**—people recalled what they had heard at a previous time. Advertisers, seeing the value of this for determining how many people had heard commercials, formed a nonprofit organization called the Cooperative Analysis of Broadcasting and hired Crossley. These Crossley ratings existed for more than fifteen years, but were discontinued in 1946 because some for-profit commercial companies offered similar services that were considered to be better.

Hooper ratings

The most significant of the commercial company ratings were the **Hooper** ratings, started by C. E. Hooper during the mid-1930s. These were similar to the Crossley ratings except that respondents were asked what programs they were listening to at the time the call was made—a methodology known as **coincidental telephone technique.** This was considered to be more accurate than the recall method because it did not depend on people's memories. The Hooper ratings fell on hard times, mainly because radio fell on hard times due to the introduction of television. In 1950, most of Hooper was purchased by the A. C. Nielsen Company. Part of it still exists, providing customized services for radio.

The Pulse

Another radio rating service, **The Pulse,** Inc., that began in 1941 utilized face-to-face interviewing. Interviewees selected by random sampling were asked to name the radio stations they had listened to over the past twenty-four hours, the past week, and the past five midweek days. If they could not remember the stations they had heard, they were shown a roster containing station call letters, frequencies, and identifying slogans. This was generally referred to as the **roster-recall method.** The Pulse was a dominant radio rating service for many years but went out of business in 1978.

BMB

In 1946 the **Broadcast Measurement Bureau (BMB)** was started and supported by the stations themselves. Postcard questionnaires were used to obtain county-by-county information about the audiences of each station. However, the return rate on the postcards was not very high, and BMB published only two reports before going out of business for lack of station support.[1]

FIGURE 10.1

A 1936 audimeter utilizing punch tape. *(Courtesy of A. C. Nielsen Company)*

Nielsen

The name most readily associated with ratings today is **Nielsen.** This company was established in 1923 by A. C. Nielsen, Sr., primarily to conduct market research for industrial and consumer goods companies. Its primary research during the 1930s involved drugstores. Nielsen had a sample of drugstores save their invoices; the company then analyzed these and sold the information to drug manufacturers so they could predict national sales.

In 1936 Nielsen acquired a device called an **audimeter** from two M.I.T. professors. This device provided a link between a radio and a moving roll of punched paper tape in such a way that a record could be made of the station that was tuned in on the radio. Nielsen perfected this device and in 1942 launched a National Radio Index—a report that indicated how many people listened to various programs. For this report, Nielsen connected the audimeter to radios in one thousand homes. Specially trained technicians visited these homes at least once a month to take off the old punched tape and put on new— a very involved process. Nielsen analyzed the information on the tapes and sold it to networks, stations, advertisers, and others interested in knowing how many people were listening to radio programs.

By 1949 the audimeter had been perfected to the point that ordinary people could remove the tape and mail it in to Nielsen. The tape was eventually changed to film, which made the operation even easier.

In 1950 Nielsen began attaching its audimeter to television sets and preparing reports about the television audience as well as the radio audience. Then in 1964, due to economic considerations and the changing nature of radio, Nielsen dropped its radio research and concentrated specifically on television.

The sophistication of the audimeter increased, and by 1970 most of them were connected directly to phone lines that led to a central computer in Florida. These were called **Storage Instantaneous Audimeters (SIAs).** With them, no

radio audimeters

television audimeters

SIAs

FIGURE 10.2

A Nielsen Storage
Instantaneous
Audimeter (SIA).
*(Courtesy of A. C. Nielsen
Company)*

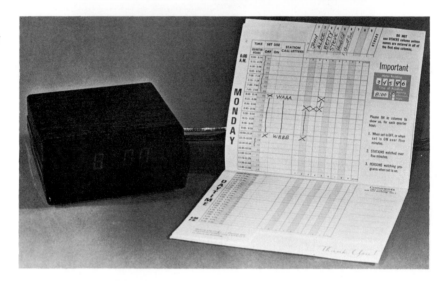

FIGURE 10.3

A Nielsen recordimeter
and diary. *(Courtesy of
A. C. Nielsen Company)*

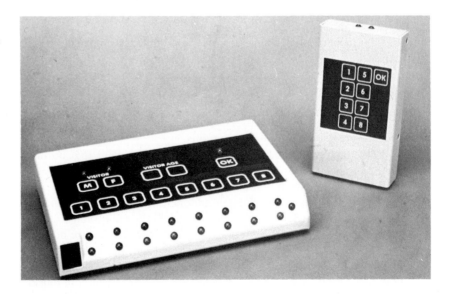

FIGURE 10.4

A Nielsen peoplemeter.
*(Courtesy of A. C. Nielsen
Company)*

Business Aspects of Telecommunications

one had to mail anything to Nielsen. The computer dialed all SIAs several times an hour and gathered information from them.

The SIAs could only indicate whether or not a TV set was on and what channel it was tuned to. It could not tell whether or not anyone was actually watching the set. To solve this problem, Nielsen had a different sample of people keep **diaries.** In these diaries, people listed what programs were watched by which members of the family. These people had a different kind of device, called a **recordimeter,** attached to their sets. It was used as a check to make sure the entries in the diary were not overly different from when the TV set was actually on. The recordimeter operated somewhat like a mileage counter on a speedometer in that digits turned over every six minutes that the set was in use. People keeping the diary had to indicate the "speedometer" reading at the beginning of each day. They also indicated **demographic** factors about members of the family—age, sex, income, education, and so on—so that demographic characteristics could be matched with the programs people watched.

recordimeters

The SIAs and diaries were used for the national ratings of network programs. For local station ratings, Nielsen used only diaries. The expense of installing enough SIAs to determine accurate ratings in every market in the United States was too great.

In 1987 Nielsen introduced an entirely new type of machine called the **peoplemeter** for its national ratings. This machine includes a hand keypad that looks something like the remote control for a TV. Each member of the family is assigned a particular button on the keypad. They are to turn this button on when they begin watching TV, to push it periodically while they are watching, and to turn it off when they are not watching. The peoplemeter gathers both the information that previously had to be obtained from audimeters (is the set on and to which channel is it tuned) and diaries (exactly who is watching what programs). Therefore, Nielsen stopped using audimeters and diaries for national ratings as of fall 1987. Diaries are still used for many local ratings, although peoplemeters are used in large markets.

peoplemeters

The peoplemeter has been criticized because sometimes people just do not bother to push the buttons. Nielsen is now working on a passive meter—one that will determine which people are in the room without anyone needing to push any buttons.[2]

Obviously, there is more to the ratings process than mechanical devices attached to sets. Throughout the years, the techniques by which Nielsen gathers and analyzes data have changed. At first everything was analyzed by hand, perhaps with the aid of an old-fashioned adding machine. It took months for the reports to be issued after the data had been collected. With the advent of the computer, this changed. Data sent over phone lines from the SIAs or peoplemeters could be analyzed almost instantly. This led to **overnight reports** on the previous night's programming that could be delivered to executives by noon the next day. The data from diaries took longer to analyze, but it too was aided by the speed of computers and could be used to issue reports within weeks of the time the information was collected. As more companies obtained their own computer systems, they could interconnect their computers with Nielsen's and obtain the information on screens in their own offices.

overnights

FIGURE 10.5

A portion of the computer room at Nielsen's Florida facility. *(Courtesy of A. C. Nielsen Company)*

The methods of determining the sample homes has changed somewhat during the years. Throughout its history, Nielsen has tried to reflect in its sample the demographics of the entire United States. This is no easy task and involves using a great deal of census data, which, of course, has become easier because of the availability of computers. Nielsen uses the census data to identify a core of counties throughout the country that will provide the main demographics of the United States. Through a **random sampling** technique Nielsen identifies specific households that would be ideal to participate as sample homes. Random sampling involves chance. In essence, the addresses of all the homes that qualify are put in a box and one is drawn out. Of course, a box is not actually used. Most random sampling is now done by computers that are programmed to select every eighty-third (or some other number) household.

samples

Representatives then try to solicit the cooperation of the households selected with about a 55 percent success rate. Where they are not successful, alternate homes, which supposedly have similar demographic characteristics are chosen. Once the households are selected, a machine is attached to each TV set and the people in the home are paid, in cash or gifts, to cooperate and do what is needed to obtain the ratings data.

The demographics of the country change fairly rapidly; for example, the divorce rate goes up or down, the number of Spanish-speaking people changes, the average income takes a dip. For this reason, Nielsen must be constantly changing its sample in relationship to how fast the demographics change. Most people do not remain in the peoplemeter sample for more than two years.

multiple meters

The number of machines needed per household has also changed. When Nielsen began gathering data, most families had one radio set in the parlor and the whole family gathered around to listen. Throughout the years this pattern changed greatly, and now many homes have multiple TV sets viewed by different members of the family at differing times. Nielsen has had to connect meters to all of the household sets and gather data accordingly.

unlisted phones

Nielsen originally used only people with listed phone numbers, but as the number of unlisted phones increased significantly, the company began soliciting samples from people with both listed and unlisted numbers. To obtain

Business Aspects of Telecommunications

people with unlisted phone numbers, Nielsen programs all listed phone numbers for an area into a computer. The computer is then used to print out the numbers not included in the list that might be in the local area. Nielsen people then call some of those numbers—ones that are picked by random sampling.

The sample size has changed over the years as the number of households sample size have increased. The original sample of 1,000 was increased to 1,200 during the 1950s and 1,700 during the 1980s. The peoplemeter now samples 4,000 homes, which is about .004 percent of the 92 million American homes. The local markets that Nielsen samples with diaries require a total of about 100,000 diaries. The number of diaries per market is dependent upon the size of the market.

Changing and increasing samples raises the cost of doing the ratings. cost This cost is paid for by the customers who subscribe to the ratings services— networks, advertising agencies, stations, cable TV networks, and other companies concerned with advertising in the electronic media. Nielsen has raised its prices from time to time, but it has to be sensitive to what the market will bear. Nielsen also charges differing amounts to different entities. While networks may pay millions of dollars per year for the reports they desire, stations in small markets may only pay thousands per year.[3]

The rise of all the newer media has also changed the nature of ratings. newer media Originally, ratings were used only for commercial and public TV stations and networks because any other uses of the TV set were insignificant. But as the newer media, especially cable TV and videocassettes, began reaching significant penetration levels, Nielsen had to begin taking them into consideration. Its meters had to accommodate many more channels and had to be able to determine when programs were being recorded off the air and whether or not they were ever played back.

Even after Nielsen technology has determined that a set is on and tuned to a particular channel, it must determine what program is on that channel. program verification This is no easy chore. Originally this was done through network lineups and a large number of phone calls to local stations to determine what they actually aired. Then in 1982, Nielsen introduced the **Automated Measurement of Lineups (AMOL).** This involves equipment that codes each network and syndicated show in a manner similar to grocery store bar codes. Then, as each program is broadcast, the Nielsen equipment picks up the code and verifies whether or not the program is what was expected. If not, phone calls must still be made, but AMOL greatly simplifies the verification problem.[4]

The end result of all the data gathering is the reports that are compiled reports by Nielsen—these, too, have changed over the years. The first reports handled national radio and then local radio was added. Television followed the same pattern because at first nationally broadcast programs were the only ones large enough to measure. Local station measurement began during the 1960s. As cable TV became a viable force, some of the more popular pay and basic networks were measured and reported on. Videocassette viewing is measured mainly in terms of the number of people who are using their VCR and, therefore, are not watching what the networks are airing.

Nielsen's report of programs shown nationally is called **Nielsen Television Index (NTI).** When it had both SIAs and diaries, Nielsen actually issued two reports, the NTI and the NAC (National Audience Composition). The NTI gave the information from the SIAs in terms of total number of people viewing. The NAC came out later and broke the viewing down into the various demographics collected from the diaries. Peoplemeter data provides all of this information at once.

The NTI still emphasizes the networks—NBC, CBS, ABC and the newer Fox network. The data is used primarily by companies wishing to advertise nationally, and those advertisers still spend a majority of their money on the networks. However, summary data and some of the specific data includes cable TV and VCR usage. The NTI gives overall numbers regarding the number of people who watched the various programs, and it also breaks this data down by various subcategories including age, sex, education, part of the country, and minutes of the day.[5]

Reports on local stations are called the **Nielsen Station Index,** sometimes referred to as the **sweeps.** For these, Nielsen divides the country into more than two hundred **designated market areas (DMAs).** These geographic areas do not overlap but view the same general stations. All areas of the United States are surveyed at least four times a year; larger markets with peoplemeters are surveyed on a constant basis with reports issued as often as once a week. People in local areas are selected at random, but still represent the demographics of the particular local area being surveyed. These people are then asked to participate and are either given peoplemeters or shown how to fill out the diaries. If given diaries, participants mail them to Nielsen at the end of the survey week. Approximately half the people actually complete and return usable diaries. The data is fed into computers, and the reports generated give detailed breakdowns according to demographics and specific times of day.[6]

Nielsen has many specialized reports that it issues in addition to the NTI and the station index. Some of these deal with broadcasting's competing media. For example, "Cable TV: A Status Report" gives quarterly overviews of the impact of cable on television broadcasting, and "VCR Tracking Report" indicates the month-by-month trends of VCR usage. Others, such as "Television Viewing Among Blacks" deal with specialized audiences while still others, such as "The Sports Report" and "Primetime Feature Film Directory" give information about particular types of programs. In addition, Nielsen will do specialized reports for just about any entity that wishes to have the data programmed in a particular way to uncover particular trends or traits. This, of course, costs above and beyond the payment for the regular reports.[7]

Arbitron

Another rating service is **Arbitron** Ratings Company, formerly called American Research Bureau, or ARB, which measures both local television and radio audiences. It was originally formed in 1949 and in 1967 became a subsidiary of the computer firm Control Data Corporation.[8]

TIME	7:00	7:15	7:30	7:45	8:00	8:15	8:30	8:45	9:00	9:15	9:30	9:45	10:00	10:15	10:30	10:45	11:00	11:15
HUT	53.3	55.2	56.8	58.2	60.1	61.9	63.3	64.6	66.7	67.7	67.4	66.6	63.8	62.5	60.5	57.7	51.1	44.3

ABC TV ◄—LIFE GOES ON—► FREE SPIRIT HOMEROOM ◄—— ABC SUNDAY NIGHT MOVIE ——►
THE PREPPIE MURDER (PAE)

	7:00	7:15	7:30	7:45	8:00	8:15	8:30	8:45	9:00	9:15	9:30	9:45	10:00	10:15	10:30	10:45	11:00	11:15
AVERAGE AUDIENCE	9,210				9,580		8,010		14,000									
(Hhlds (000) & %)	10.0	9.3 *		10.8 *	10.4		8.7		15.2	13.7 *		15.8 *		15.7 *		15.5 *		
SHARE AUDIENCE %	18	17 *		19 *	17		14		24	20 *		24 *		25 *		26 *		
AVG. AUD. BY 1/4 HR %	8.7	9.8	10.6	11.1	10.2	10.6	8.5	8.9	13.6	13.8	15.7	15.8	15.9	15.6	15.6	15.5		

CBS TV ◄—— 60 MINUTES ——► ◄—MURDER, SHE WROTE—► ◄— ISLAND SON SPEC (PAE) —► ◄— WOLF SPEC —►

	7:00	7:15	7:30	7:45	8:00	8:15	8:30	8:45	9:00	9:15	9:30	9:45	10:00	10:15	10:30	10:45	11:00	11:15
AVERAGE AUDIENCE	17,040				19,060				14,640				11,700					
(Hhlds (000) & %)		16.9 *		20.0 *	20.7	20.2 *		21.2 *	15.9	15.5 *		16.2 *	12.7	13.1 *		12.3 *		
SHARE AUDIENCE %	33	31 *		35 *	33	33 *		33 *	24	23 *		24 *	21	21 *		21 *		
AVG. AUD. BY 1/4 HR %	15.7	19.0	20.1	20.0	20.0	20.5	21.4	21.0	15.4	15.7	16.5	16.0	13.2	12.9	12.5	12.1		

NBC TV (1) ◄ ALF TAKES NETWORK (7:25-8:00) (PAE) ► SISTER KATE MY TWO DADS ◄—— SNL 15TH ANN. SPC ——►

	7:00	7:15	7:30	7:45	8:00	8:15	8:30	8:45	9:00	9:15	9:30	9:45	10:00	10:15	10:30	10:45	11:00	11:15
AVERAGE AUDIENCE		5,990			8,200		10,870		18,700									
(Hhlds (000) & %)		6.5		6.4 *	8.9		11.8		20.8	19.3 *		21.1 *		22.1 *		21.1 *		17.7 *
SHARE AUDIENCE %		11		11 *	15		18		33	29 *		31 *		35 *		36 *		37 *
AVG. AUD. BY 1/4 HR %	13.6	6.9	6.1	6.7	8.6	9.3	10.9	12.8	18.5	20.0	20.6	21.6	22.4	21.8	21.4	20.9	19.8	15.6

INDEPENDENTS (INCL. SUPERSTATIONS)

AVERAGE AUDIENCE	11.5		12.6		14.2		14.8		15.7		11.4		9.6		7.7		6.0	
SHARE AUDIENCE %	21		22		23		23		23		17		15		13		13	

SUPERSTATIONS

AVERAGE AUDIENCE	2.3		2.3		1.9		2.2		2.5		2.8		2.4		1.9		1.7	
SHARE AUDIENCE %	4		4		3		3		4		4		4		3		3	

PBS

AVERAGE AUDIENCE	1.5		1.6		2.4		3.0		1.7		1.8		1.2		1.2		1.0	
SHARE AUDIENCE %	3		3		4		5		3		3		2		2		2	

CABLE ORIG.

AVERAGE AUDIENCE	6.8		8.6		7.2		6.5		5.6		5.4		4.9		4.4		3.8	
SHARE AUDIENCE %	12		15		12		10		8		8		8		8		8	

PAY SERVICES

AVERAGE AUDIENCE	2.7		2.5		3.7		4.1		4.1		3.5		3.5		3.1		2.5	
SHARE AUDIENCE %	5		4		6		6		6		5		5		5		5	

U.S. TV HOUSEHOLDS: 92,100,000

For explanation of symbols, see page B.

(1) NFL GAME 2, VARIOUS TEAMS AND TIMES, (PAE), NBC, (MULTI SEGMENT)

FIGURE 10.6

An example of a page from a Nielsen NTI report. *(Courtesy of A. C. Nielsen Company)*

WK1 11/01–11/07 WK2 11/08–11/14 WK3 11/15–11/21 WK4 11/22–11/28

WEDNESDAY
8.30PM–10.30PM

A sample Nielsen Station Index table is reproduced here showing DMA HH Ratings, Multi-Week Avg, Share Trend, and DMA Ratings (Persons, Women, Men, Teens, Child) for stations KABC, KBSC, KCBS, KCET, KCOP, KHJ, KMEX, KNBC, KTLA, KTTV, KJHY and their programs across four weeks in November 1984.

TIME PERIOD

WEDNESDAY
8.30PM–10.30PM

NOVEMBER 1984

For explanation of symbols, see page 3.
For RSE explanations, see page 2.

FIGURE 10.7

A sample Nielsen Station Index. *(Courtesy of A. C. Nielsen Company)*

Like Nielsen, Arbitron publishes a television station index at least four times a year—February, May, July, and November. Larger markets are surveyed more frequently.

Arbitron calls the geographic areas it surveys **areas of dominant influence (ADIs)**. To determine ADIs, Arbitron has divided the nation into 210 nonoverlapping viewing areas. Counties are placed in an ADI based on the stations most listened to in that county. Population size does not determine ADIs. In other words, a large area like Houston is considered one ADI while a small area like Lafayette, Indiana, is another. Sometimes very small areas are combined, such as Albany-Schenectady-Troy, New York. The largest ADI is New York City, with over 7 million households, and the smallest is Alpena, Michigan, with about 15 thousand households.[9]

Arbitron uses diaries for most of its television station ratings. The number of households given diaries depends on the size of the ADI, but the average is about 700 households per market. Each household gets a separate diary for each TV set and each VCR.

Like Nielsen, homes are selected to conform to demographics, then letters and phone calls ask participants to cooperate for a small monetary incentive, usually fifty cents. Arbitron has found that people feel more obligated to fill out the diaries if they are being paid but that they don't feel anymore obligated if they are given a large amount of money than if they are given the token fifty cents.

The diaries ask for demographic information about each family member and details regarding which family members watched which programs at what times on which sets. People mail the diaries to Arbitron's Beltsville, Maryland, headquarters, where data is analyzed. Arbitron, like Nielsen, has only a 50 percent return rate of useable diaries. Arbitron has a specific method for determining whether or not diaries are useable. They check for such things as whether or not the diary was mailed back before the programs listed were actually on the air, whether or not the demographic information about various people in the household is consistent throughout the report, whether the programs listed were actually on the air at the time they were listed, and whether the stations listed can actually be received in the area. If any of these conditions exist, the diary usually is not used.

However, Arbitron often accepts diaries with minor discrepancies or omissions and uses computer programs to fill in missing data. For example, if the name of a program is listed, but its station or cable channel is not listed, Arbitron can use its computer information to fill in the right channel. If there is no indication in the diary as to which people in the family watched a particular program, the Arbitron computer looks at data from other diaries in the same market and "guesses" who was watching in that particular household. Sometimes Arbitron even adds data if certain segments of the population do not return diaries in proportion to other segments. For example, if only 30 percent of diaries given to Chicano families are returned, whereas 50 percent of diaries given to black families are returned, Arbitron will multiply the Chicano data slightly so that it represents what might have been the case had 50 percent of the Chicano families returned the diaries.

ADIs

diaries

diary useability

diary changes

Shortly after the survey is completed, reports are sent to the clients who have ordered them. These reports are broken down into audience composition for each quarter hour and are organized by age, sex, and other demographic factors. The Arbitron reports come out at approximately the same time as the Nielsen reports.[10] For many years, most stations subscribed to both services so that they could select the data that showed them in the most favorable light, and they could present this data to advertisers. But recently, in part because of tightening budgets and in part because of discontent with the results that Nielsen and Arbitron have been reporting, stations have begun to subscribe to only one service.[11]

meters

Some of Arbitron's larger markets are measured with a meter called an Arbitron that is similar to the Nielsen SIAs. Arbitron is also experimenting with a peoplemeter called **ScanAmerica.** In addition to pushing buttons to indicate whether or not they are watching TV, people would run scanners past all the groceries they purchased. In this way, Arbitron would be able to correlate program watching with types and brands of products purchased.[12]

Unlike Nielsen, Arbitron does not have a service emphasizing network programming, although network overnight results can be gleaned from looking at the data compiled for the large metered markets.

radio surveys

An Arbitron-covered area not covered by Nielsen is radio. For many years Arbitron has been the leading company in this area. It measures 250 individual radio markets at least once a year. Larger markets are measured more often; in fact, in several very large markets the measurement is essentially year-round.

MSAs and TSAs

For each market, Arbitron measures two areas, **metro survey area (MSA)** and **total survey area (TSA).** The TSAs are geographically larger than the MSAs and will not receive the low-power stations as clearly as the MSAs. That is why Arbitron has two areas—the low-power stations will not compare favorably with the powerhouses in the TSAs, but can hold their own in the MSAs.

samples

For each market, Arbitron determines the number of homes needed for a valid sample. It then selects these homes from a telephone list that includes unlisted numbers. Arbitron employees send letters to people asking for their cooperation and then follow these up with phone calls. Diaries are then sent to the homes that have agreed to cooperate, and each person over the age of twelve is asked to keep a separate diary of radio listening. Diaries are mailed back to the Arbitron headquarters where the details are analyzed.

reports

Printed reports are then sent to subscribers—primarily radio stations and advertisers. These reports, like the television reports, give overall statistics and a breakdown of the data into various demographics and other pertinent categories. In addition, Arbitron will provide, for a fee, specialized reports for any customers who want them.[13]

TUESDAY

TIME			STATION			PLACE			
			Fill in station "call letters" (If you don't know them, fill in program name or dial setting)	Check One (✔)		Check One (✔)			
							Away From Home		
	From	To		AM	FM	At Home	In a Car	Some Other Place	
Early Morning (5AM to 10AM)									
Midday (10AM to 3PM)									
Late Afternoon (3PM to 7PM)									
Night (7PM to 5AM)									

IF YOU DID NOT LISTEN TO RADIO TODAY PLEASE CHECK ✔ HERE ➡ ☐

Please review each day's listening to be sure all your entries are complete.

FIGURE 10.8

A sample page from an Arbitron diary. *(Courtesy of © Arbitron Ratings Company)*

Other Ratings Services

Nielsen and Arbitron are the largest and best-known rating services in the United States, but there are many other companies that provide services regarding who is watching or listening to various programs.

For example, **Birch Radio** does radio research that uses the telephone recall method. Birch employees, using randomly generated phone numbers, call people in about 250 markets and ask them what they listened to yesterday. Birch then compiles the data and issues reports. Its service is much cheaper than the diary-based Arbitron service, so it is being subscribed to by an increasing number of radio stations.[14]

There are also many measurement companies that specialize their services to appeal to particular needs. For example, there is one company that specializes in data concerning news programs and another that tracks Hispanic viewing. Others concentrate on gathering ratings data for public radio

Birch

specialized reports

Specific Audience
MONDAY-FRIDAY 6AM-10AM

	Persons 12+	Men 18+	Men 18-24	Men 25-34	Men 35-44	Men 45-54	Men 55-64	Women 18+	Women 18-24	Women 25-34	Women 35-44	Women 45-54	Women 55-64	Teens 12-17
+WAAA WRRR METRO	8.6	7.2	1.2	10.1	13.1	5.4	4.3	9.8	11.2	10.4	11.1	3.1	14.9	8.6
WBBB METRO	.9	.7			3.6			.8	1.0				6.4	2.9
WCCC METRO	6.4	5.0		.8	6.0	5.4	14.9	8.5	3.1	3.2	4.0	12.5	12.8	
WDDD METRO	.4							.6	1.0	1.6				1.4
WDDD-FM METRO	8.7	9.8	23.8	12.4	4.8	1.8		6.2	14.3	10.4	1.0	3.1		18.6
TOTAL METRO	9.1	9.8	23.8	12.4	4.8	1.8		6.9	15.3	12.0	1.0	3.1		20.0
WEEE METRO	6.8	5.7	17.9	3.9	3.6		2.1	5.0	10.2	7.2	2.0		6.4	25.7
WFFF METRO	6.6	9.1	4.8	3.9	8.3	16.1	19.1	5.4	1.0	3.2	7.1	7.8	8.5	
WGGG METRO	1.3	2.1		2.3		3.6	6.4	.8		.8	2.0		2.1	
WGGG-FM METRO	20.6	23.6	19.0	25.6	21.4	37.5	21.3	20.6	6.1	19.2	28.3	37.5	25.5	2.9
TOTAL METRO	22.0	25.8	19.0	27.9	21.4	41.1	27.7	21.4	6.1	20.0	30.3	37.5	27.7	2.9
WHHH METRO	3.9	3.8	2.4	8.5			6.4	4.0	8.2	4.8	3.0	3.1		4.3
WIII METRO	11.3	10.7	26.2	8.5	6.0	12.5		10.2	21.4	12.8	4.0	4.7	4.3	22.9
WJJJ METRO	4.3	3.8	2.4	3.9	4.8	1.8	8.5	4.4	4.1	2.4	6.1	3.1	4.3	7.1
WKKK METRO	2.4	3.8	1.2	3.1	11.9	1.8		1.5		3.2	3.0			
WLLL METRO	7.3	4.3		2.3	7.1	5.4	4.3	10.8	7.1	10.4	15.2	9.4	4.3	1.4
WMMM METRO	2.7	.5					2.1	4.8	6.1	5.6	4.0	3.1	6.4	1.4
WZZZ METRO	.5	.7		2.3				.4	1.0	.8				
TOTALS AQH RTG	26.3	26.2	23.5	28.2	27.5	29.0	29.9	28.3	29.0	28.3	31.4	31.5	26.1	17.8

Footnote Symbols: ✱ Audience estimates adjusted for actual broadcast schedule. ✛ Station(s) reported with different call letters in prior surveys - see Page 5B.
\# Both of the previous footnotes apply.

ARBITRON RATINGS

WINTER 1987

FIGURE 10.9

A sample of the type of report Arbitron furnishes to subscribing radio stations. *(Courtesy of © Arbitron Ratings Company)*

or TV stations, medium market radio stations, or low-power TV stations. A few design their services primarily for cable TV. Some of these companies analyze the audiences of the various cable networks while others have methods for determining how many people are watching local origination or public access channels, neither of which are included in Nielsen or Arbitron surveys. Numerous companies also take ratings data that a station or network has obtained from Nielsen, Arbitron, or other companies and analyze it to find particular strengths that can be pitched to advertisers. They might find, for example, that a station's 4:00 to 6:00 P.M. slot has the second highest ratings if you add together women and children. Often these companies also offer (for a fee) to make suggestions as to how a particular station's ratings can be improved.[15]

Measurement Calculation

rating

Several different types of statistics are reported by audience research companies. The main one, of course, is the **rating**, which is basically a percentage of the households or people watching a particular TV program or listening to a particular radio station. Rating percentages take into consideration the total number of households or people having TV sets or radios.

Assume that the pie in figure 10.10 represents a sample of five hundred television households drawn from 100,000 TV households in the market being surveyed. The rating is the percentage of the total sample. Thus, the rating for station WAAA is 80/500, or 16 percent; the rating for WAAB is 50/500, or 10 percent; and the rating for WAAC is 70/500, or 14 percent. Usually when ratings are reported, the percentage sign is eliminated; thus, WAAA has a rating of 16. Sometimes ratings are reported for certain stations and sometimes for certain programs. If WAAA aired network evening news at the particular time of this rating pie, then this news would have a rating of 16 in this particular city. Of course, national ratings are drawn from a sample of more than just one market. There are approximately 92.1 million households in the United States, so the number of households watching national programs can be determined by multiplying the rating times 92.1 million.[16] For example, a rating of 20 is 20 percent times 92.1 million, or 18.4 million. Another way of looking at this is that each rating point is worth 921,000 households (921,000 times 20 equals 18,420,000).

A different measurement that is based on a universe, defined as all TV households or people using TV at the time, is called **share** of audience. In the pie shown in figure 10.10, 300 homes had the sets on—80 to WAAA, 50 to WAAB, 70 to WAAC, 100 to all others. In the other 200 homes, either no one was at home or the TV was off, so they do not count in the share of audience total. Therefore, WAAA's share of audience would be 80/300, or 26.7; WAAB's share would be 50/300, or 16.7; and WAAC's share would be 70/300, or 23.4. A share-of-audience calculation will always be higher than a rating unless 100 percent of the people are watching TV—an unlikely phenomenon.

share

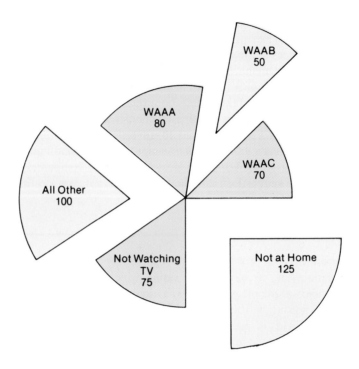

FIGURE 10.10
Ratings pie.

WAAB
50

WAAA
80

WAAC
70

All Other
100

Not Watching
TV
75

Not at Home
125

HUT

Another calculation involves **homes using TV (HUT).** This is the percentage of TV households that have the set tuned to anything. In the ratings pie example, the HUT figure is 60; 300 out of 500 households had the sets tuned to something (300/500 = 60 percent).

Another way to look at these three statistics is by their formulas:

Rating = homes tuned to station/total TV homes
Share = homes tuned to station/homes using TV
HUT = sets turned on/total TV homes

All the examples given here use households rather than individual people. That is still the way Nielsen and Arbitron figure most of their television measurements. However, as TV viewing becomes more fractionalized with different people in the same household watching different programs, this is gradually changing. The trend now is to also report the **PUT (people using television).** In the future, the PUT may become a more valuable measurement calculation than the HUT. Individual people rather than households have formed the basis for radio ratings for many years. In radio, the term used is **PUR (people using radio).**

PUT

PUR

Rating, share and HUT (or PUT or PUR) are the most common of the measurement statistics reported, but ratings companies do give other statistics. For example, for radio, **average quarter-hour ratings (AQRs)** are often given. This calculation is based on the average number of persons listening to a particular station for at least five minutes during a fifteen-minute period. Another statistic is the **cume**—the number of different persons who tune in to

AQR

cume

Business Aspects of Telecommunications

a station over a period of time. This is used primarily by advertisers who want to know how many different people will hear their message if it is aired at different times, say 9:05, 10:05 and 11:05.[17]

worth of rating point

Audience measurement calculations are used in a number of different ways. One of the most important is for selling advertising time. The higher the rating for a particular program or a particular time period, the more the station or network can charge for placing a commercial within that program or time period. Nielsen estimates that each of its rating points for prime-time programming is worth about $30,000. That means that a program with a rating of 20 can charge about $600,000 for a commercial while a program with a ratio of 15 can only charge $450,000. It is no wonder that networks and stations strive for high ratings.

CPM

Advertisers use ratings to make sure they are getting a good deal for their money. They need the ratings data in order to determine **cost per thousand,** or **CPM** (M is Latin for thousand). This is an indication to the advertiser as to how much it is costing to reach one thousand households. For example, if an advertiser pays $30 for a radio spot and the ratings show that spot is reaching five thousand households, then the advertiser is paying $6 for each one thousand households. Generally advertisers want CPMs close to $5 for general shows, such as network programs, that reach large audiences. However, with the recent erosion of the audience, network CPMs sometimes reach double digits.

industry health

Audience measurement is also used to determine the overall health of the electronic media industry. As of late, HUT levels have been diminishing. There are those in the industry who blame this on the peoplemeters. They say as many people are watching as before, but they just don't want to bother pushing the buttons. However, there are other indications that people just are not spending as much time watching television as they have in the past. To advertisers, this means they may want not only to bargain for lower prices from networks and stations, but they may want to invest some of their advertising dollars in other media or other forms of promotion.[18]

Ratings and shares also show the relative health of different segments of the electronic media business. The three commercial networks used to garner about 90 percent of the share of audience. Network programs with ratings of less than 20 usually did not last long. Now the three-network share is down to about 60 percent, and programs with ratings of about 10 are considered acceptable. Much of the difference has been taken up by independent stations, the Fox network, cable programming, and videocassette viewing. These aspects of the industry have been gaining in health at the expense of the networks' declining health.[19]

comparison

Measurement is also used as a basis of comparison. The share tells a station or network how it fares in relation to its competition. Although the HUT and ratings may be going down, a station can consider itself successful if its share of the viewing public is increasing. Ratings of the competition can be used as a bellwether. A 5 might sound like a low rating, but it is perfectly acceptable for most radio stations because radio is so diverse and specialized that a 5 is about the most many of the stations in a market can hope for. A station or network can also use ratings to compare against its past. If a station's rating or share decreases from month to month over a period of a year, the station has reason to worry.[20]

program changes

Then, of course, ratings are used extensively to make programming decisions. Programs with high ratings are kept on the air while those with low ratings are often quickly erased without the chance to grow or to develop an audience slowly. Similarly, people often disappear from the airwaves due to ratings. If a radio station's ratings slip during a certain time of day, the disc jockey on at that time can expect a pink slip. TV station news anchors are often changed in an attempt to bolster ratings.

Measurement calculations are varied in terms of types and uses because of the needs of different clientele. One advertiser might be more interested in the cumulative number of people hearing an ad than the number hearing at any particular time; another advertiser might be interested primarily in finding a station that delivers the largest share of the audience.

Other Forms of Research

Telecommunications organizations are often interested in other aspects of research other than the size and scope of the audience. For example, radio stations want to play music that people will listen to so that they do not turn the dial. To undertake this kind of **music preference research** the station must find people who like the overall format provided by the station. It would be futile to test individual country songs on people who are avid rock fans and would not listen to country. This means that a random sample selected from phone numbers would not be very effective. Usually stations take their sample from people already known to listen to the station, such as those who have won contests or have called in with requests.

music preference

techniques

The music preference research is conducted in several different ways—one method is over the phone. Researchers call selected people and play a short section of music to them and ask their opinion according to a prepared questionnaire. Another research method is to invite a group of people to a large auditorium and have them listen to music and give reactions, usually on a questionnaire form. The auditorium testing has an advantage in that the fidelity can be better than it is over the phone, and more songs can be played. This form of testing is more expensive than phone testing, however. A third research method involves inviting small groups of people to what are called **focus groups.** Again, the music can be played, but instead of reacting on questionnaires, the people discuss their likes and dislikes. This gives more in-depth information, but it is the most expensive method of all.

pretesting

The auditorium testing and the focus groups are also used for TV programs and commercials, often in the form of **pretesting.** Both advertisers and programmers like to know the success probability of an idea before they invest in it heavily. Organizations such as **Audience Studies Institute (ASI)** gather people into an auditorium and show them both proposed programs and commercials. They elicit the reactions of the audience members by questionnaires, various button-pushing techniques, or electronic techniques that measure perspiration. Sometimes audience members are given products that might be associated with certain programs in order to obtain reactions. If the pretesting is done with a focus group, it involves more discussion. Sometimes focus groups are used for testing ideas before any program material has even been taped.[21]

Other companies test commercials while the participants remain in their homes. These companies see that the ads are placed in a set period of programming on cable TV, a UHF station, or a willing VHF station. Potential testees are then telephoned and invited to watch the programming of a particular channel. The following day the same people are called again and asked their opinions of the commercials. If the results are positive, the commercial is launched on many other stations and/or networks.

testing at home

The pay-cable channels, such as HBO and Showtime, are not concerned with advertising but are concerned about people disconnecting from the service. Therefore, their research is directed more toward finding the right combination of movies and other programs to retain the largest possible number of subscribers. This can be done through phone surveys or focus groups.[22]

research to retain

A different kind of measurement conducted by **TVQ** researches attitudes and opinions about specific television programs and personalities. A questionnaire is mailed or different people are telephoned each month to determine their degree of awareness of particular shows or people and to determine how much they like the shows or personalities. The Q-score of each person or program—a number that indicates popularity—is then calculated. Having a low Q can mean unemployment lines for talent.[23]

TVQ

Industrial and educational organizations often undertake effectiveness research. They want to know if the programming they have produced has accomplished its goal in terms of such things as teaching certain information or establishing certain attitudes. Often this research is undertaken by giving the people who are going to be watching a particular program a pretest to determine what they know or how they feel. Then after participants view the program, they are once again given the same test to see if there has been a significant change.

effectiveness research

All of these forms of research are more qualitative than the ratings research, which tends to be quantitative. In other words, this research deals with concepts, cause and effect, and underlying feelings as opposed to numbers that indicate how many people have watched or listened to a particular program.

Ratings under Fire

Criticism of ratings does not attack their existence. Ratings are needed in order for broadcasters and advertisers to make programming and advertising decisions. Criticisms of ratings stem more from their methodology and interpretation.

All ratings are based on samples, rather than on the entire population. Therefore, it is the viewing and listening tastes of a very small percentage of people who are representing the tastes of the entire audience. At present it is simply too expensive and impractical to survey the whole population; therefore, samples must be used for audience measurement, just as they are for scientific research. A frequent criticism of audience research methodology is that the sample size is not large enough. The placement of peoplemeters in four thousand homes out of a total of 92.1 million is often decried, even by those who revere the ratings, as a sample size too small to prove anything.

Rating companies provide critics with a wealth of authoritative "proof" that their sample, though small, is statistically sound. These companies also point to their continuing attempt to make their sample representative of the entire population, with the same percentages of households with particular demographic characteristics in the sample group as there is in the population of the entire United States. For example, a rating sample would attempt to have the same percentage of households headed by a thirty-year-old high school educated black woman with three children living in the city and earning $25,000 per year as in the population of the whole United States. But just which of these characteristics are important to audience measurement can be avidly debated. Is a twenty-five-year-old high school educated black woman with no children who earns $20,000 a year a valid substitute?

Another problem with sampling, especially for television-based research, is that the samples are primarily households or people living in homes. Households is an out-of-date concept that assumes everyone watches the same programs on the same set. But even when the research does involve people, it is still based on in-home TV sets. TV sets located in college dorms, hospitals, bars, and hotel rooms are not included in the surveys. Networks and stations claim the rating services are missing data on millions of viewers because these types of places are not metered.[24]

Even after determining the ideal sample, companies are faced with the problem of uncooperative potential samples. About 30 percent of the people contacted refuse to have machines installed on their sets and about 45 percent do not want to keep diaries. Substitutions must be made for uncooperative people, which may bias the sample in two ways—the substitutes may have characteristics that unbalance the sample, and uncooperative people, as a group, may have particular traits that bias the sample.

Problems with ratings methodology do not cease once the size and composition of the sample have been decided and the people selected. There is also the problem of receiving accurate information from these people. Each rating method has its drawbacks in this regard.

Any system that employs meters is dealing with a technology that can fail. A certain number of machines do develop mechanical malfunctions, further reducing the sample size in an unscientific manner. The older machines, such as the SIA, recorded only that the set was on; they did not reveal whether anyone was watching it. People could turn on the TV set to entertain the baby—or the dog. This problem has been solved by the peoplemeters, but people can watch TV with the peoplemeters and not bother to push the button. This deflates the ratings. Some people attempt to appear more intellectual by turning on cultural programs but not actually watching them, or they may react in other ways contrary to their normal behavior simply because they know they are being monitored.

Diaries are subject to human deceit also, for people can lie about what they actually watched in order to appear more intellectual or in order to help some program they would like to see high in the ratings. Some people simply forget to mark down their viewing or listening and then attempt to fill in a week's worth of programming from memory—perhaps aided by *TV Guide.* In addition, only about half the diaries sent out are returned in a form that can be used for analysis.

diary problems

The telephone interview techniques can also contain bias—one form being the bias of the interviewer. If the person doing the interviewing has certain preferences toward programs or stations, he or she may influence a person who is unsure of what actually was seen or heard to respond toward the interviewer's preferences. Many interviewers are inexperienced, unskillful, and temporary employees, so they have not developed good interviewing skills. Lack of memory or purposeful inaccuracies can affect interviews based on recall, just as with diaries. Calls made to determine what a person is viewing or hearing at the moment are unable to supply data involving late night or early morning programs because phone calls are inappropriate at those times.

interview problems

Stations and networks, too, can influence ratings. During rating periods, many stations engage in the practice called **hypoing.** They broadcast their most popular programming, hold contests, give away prizes, and generally attempt to increase the size of their audience—usually a temporary measure. During sweeps periods, networks try to help their affiliates by canceling regular shows and programming blockbuster specials or movies. In some ways this defeats the purpose of ratings because it does not indicate how many people regularly watch the network-affiliated channel. It just tells how many watch the one-shot big-gun programs.

hypoing

Another overall criticism of ratings has been that they are too secretive and that the rating companies keep methodology and sampling techniques too close to the vest. This opens the door for corruption and incompetence.

secretiveness

Even after rating material has been collected, it is still subject to computer error and printing error. The fact that computer programs are sometimes used to fill in missing data on diaries strikes some as invalidating the results. With the sample rate already small and the return rate low, how can the computer accurately predict the "right" answers?

computer problems

Another important problem is differences in results. Arbitron and Nielsen have often differed as much as several rating points on certain programs or time periods.[25] Even within one service, there are discrepancies. For example, Nielsen's overnights from machines often differed from its later diary reports. With the advent of the peoplemeter, the differences have been even more glaring. Ratings among children and teens are significantly lower for the peoplemeters as compared to diaries and other meters, ostensibly because young

discrepancies

people do not push the buttons. On the other hand, sports programming rates higher with peoplemeters. The speculation here is that men, watching a sports game intently, will push the buttons while in the past most women kept the family diaries and did not always record the husband's sports viewing. At any rate, the discrepancies among services makes skeptics question the accuracy of any of the ratings.[26]

Although rating company methodology is often criticized, management interpretation is the area most criticized. Rating companies publish results and really cannot be held responsible for how they are used. This area of error is the domain of broadcasting and advertising executives.

emphasis

The main criticism is that too much emphasis is placed on ratings. Even though rating companies themselves acknowledge that their sampling techniques and methodology yield imperfect results (usually not accurate to one percentage point), programs are sometimes removed from the air when they slip one or two rating points. Actors and actresses whose careers have been stunted by such action harbor resentment. TV history is full of programs that scored poor ratings initially (e.g., "Hill Street Blues," "Cheers") but, for one reason or another, were left on the air and went on to acquire large, loyal audiences. Trade journals will headline the ratings lead of one network over the others when that lead, for all programs totaled, may be only half a rating point. As TV diversifies and rating numbers become smaller, accuracy becomes even more of a problem. The extent of error is greater with numbers such as 3 or 5 than it is for ratings of 20 or 25. Ratings should be an indication of comparative size and nothing more, but in reality their shadow extends much further.

The overdependence on ratings often leads to programming concepts deplored by the critics. In a popularity contest designed to gain the highest numbers, stations and networks neglect programming for special interest groups. All programming tends to become similar, geared toward the audience that

quantity only

will deliver the largest numbers. Programmers place emphasis on viewer quantity, often to the neglect of creativity, station image, public access, flexibility, availability to the community, and station services to advertisers.

imitation

Dependence on ratings tends to perpetuate the imitative quality of programming. When one show receives a high rating, many similar shows are spawned. This imitative quality has become rampant even in the cable TV business, which many hoped would be the place for creative new ideas. But cable, too, has become quite dependent on advertising, and hence ratings.

Advertisers are also guilty of being slaves to quantity. Often a small but select audience might be best for a specific purpose—for example, estate planning insurance might be better advertised on a classical music station with a low rating than a Top 40 station with a high rating. Advertisers pay very little

qualitative factors

attention to qualitative factors such as opinions and attitudes of people toward programs and products, brand name recognition in relation to program identification and purchasing decisions, and level of audience attention to selected material such as commercials. Advertisers are not prone to demand such research. They, like broadcasters, seem content to assume that quantity is the primary goal and that cost-per-thousand—regardless of the composition, attitude, or attention of the thousand—will move goods from the shelf. With the present audience measurement structure, ratings should be used as an aid, not an end. All too often this is not the case.

In Defense of Ratings

The intangibles associated with electronic media listening and viewing are enormous. Unlike newspapers, the consumption of which is spread out over time, broadcasting vanishes as soon as it has been presented. In addition, radio waves refuse to obey political or geographic boundaries, making the concept of a market a muddled one at best. Networks and stations cannot survey their own audience totals, for the numbers would be suspect. Then, too, the desired result is thrice removed from the original stimulus. Programs are produced to attract a large audience to watch the commercials, which theoretically will induce people to buy products. No wonder, then, that supplying meaningful ratings is difficult. And yet something is needed because millions of dollars are involved. Given these parameters, ratings are the best method devised as yet. intangibles

Rating companies can defend their methodology by declaring that conscious manipulation of the truth by sample participants will average out. While some people inflate their diaries to improve the ratings of their favorite rock station, others exaggerate to an interviewer the number of listening hours for a country-western station. averaging out

To the critics of sampling procedures, the rating companies can reply, "All right, you come up with a better idea." No two people in the country are exactly alike, so sampling procedures must do the best they can. In general, the methods used by rating companies are as good as any yet devised. As times and conditions change, ratings companies change their methodology. For example, they changed from household to people ratings for radio and are in the process of doing so for TV. Out-of-home viewing will probably eventually be included also. samples

The size of the samples is such that no one claims they are accurate to the exact percentage point. Larger, more refined sample sizes could be easily accommodated if subscribers were willing to pay the cost.

Likewise, more qualitative data could be gathered, interviewing techniques could be improved, education about ratings could be more widespread—but someone must pay. There has been no demand for these improvements because subscribers—electronic media companies and advertisers—have not been willing to foot the bill. qualitative data

Hypoing can affect ratings, but it is outlawed by the FCC. The FCC has never prosecuted any stations for this violation, but on occasion, rating companies have left stations out of the report because of blatant hypoing. hypoing

Rating companies claim that they must be somewhat secretive, particularly in regard to identification of the sample households. If any significant secretiveness

number of households were bribed, this could severely endanger rating validity. The rating companies are overseen by an organization called the **Electronic Media Rating Council (EMRC),** formed by broadcasters during the 1960s to accredit the various rating companies. These companies feel having the EMRC check such procedures as sampling techniques and interviewing procedures serves to keep rating companies honest.

democratic process

The effects that ratings have on program content are simply the result of the democratic process. Audience members get what they vote for. If stations were to use another criterion, say creativity, as the basis for advertising rates, the situation would be far more unjust than is the present quantitative rating system. Creativity is an abstract that really has not been defined, let alone counted.

dependence

As for the great dependence put on ratings, audience measurement companies claim that is not their problem. Ratings companies do not cancel programs or fire stars. If executives are using ratings to make those kinds of decisions, then that is all the more reason to realize that ratings are an extremely important part of the electronic media world.

Audience measurement technology can stand improvement. However, this improvement needs to be conducted through cooperative efforts on the part of rating companies and electronic media professionals.[27]

Conclusion

Audience measurement has become an important part of the telecommunications business because it enables advertisers and broadcasting executives to estimate how many people are viewing or listening to their programs or commercials.

Many different methodologies have been used or are being used to determine audience size and characteristics, but they all have flaws. Fan mail and prizes were used in early days, although not very scientifically. Telephone recall was used by the Crossleys, but this method suffers from the flaw that people do not always remember what they heard or saw in the past. Coincidental telephone technique, such as that used by Hooper, solves the problem of recall but has the additional problems of people lying about what they are doing. The face-to-face roster recall interview method can aid the interviewee's memory, but the interviewer can have biases that influence the answers of the interviewee. The postcard questionnaires, such as those used by BMB, did not yield a high return rate. All mechanical devices can break down. The SIA was accurate in indicating whether or not the set was on but could not tell if anyone was watching it. Peoplemeters indicate whether or not people are watching, provided those people remember to push their buttons. Diary methods, such as those that have been used by Arbitron and Nielsen for their station indexes, are only as accurate as the memories of the people who fill them out and are subject to both intentional and unintentional human inaccuracies. Obviously, no perfect method of audience measurement has yet been devised.

The two largest audience measurement companies are Nielsen and Arbitron, which, working with ADIs or DMAs, publish numerous reports for advertisers, networks, and stations, such as overnights, sweeps, and NTIs. The ratings companies use sampling techniques, which are often criticized in terms of their accuracy. Nielsen reports on local and national television, while Arbitron handles radio and local television. The fact that the two services do not always agree creates skepticism of the measurement system.

Numerous other ratings companies exist, some of which specialize in the newer media such as cable TV, low-power TV, and videocassettes. Still other companies handle more qualitative types of research. This is often conducted through phone interviews, auditorium testing, or focus groups. Music preference research and program and commercial pretesting are examples. TVQ measures opinions about people and programs; corporations and educators often undertake effectiveness research.

The figures the companies collect can be calculated in many ways: ratings, shares, HUTs, PUTs, PURs, AQRs, cumes. By using these different calculations creatively, stations can make themselves more attractive to advertisers than if only ratings were calculated. Audience measurement is also used to determine the health of the industry or parts of it, to calculate CPMs, and to make changes in programs and talent.

Measurement information, with all its flaws, is still better than no information at all. Ratings may improve as both measurement and media change.

Thought Questions

1. To what extent should ratings affect whether or not a program is taken off the air?
2. Suggest better methodology for conducting ratings than that presently used by Nielsen and Arbitron.
3. Could broadcasting exist without ratings? Explain your answer.

Programming

The product of the electronic media is programming. It is the antics and actions of people and objects that constitute the reason for existence. Ranging from the ridiculous to the sublime, this programming is executed by human beings with ordinary intelligence and ordinary prejudices. A select few program for the many, but the many, by a simple flick of a switch, affect the decisions and even the careers of the few. Programs inform, teach, entertain, or merely occupy time, but they are often the basis for conversation, interaction, or even decision making.

PART 4

Entertainment Programming

Introduction

Entertainment programming is the predominant form of programming in both radio and television. And yet, it is often difficult to distinguish between entertainment and information. A talk show contains information, but people watch it mainly for the enjoyment of seeing celebrities. Sports programming shows real events as they are happening, and yet many people watch to be entertained or exasperated. Children's programming can be entertaining, informational, or both.

A further difficulty arises when one tries to divide entertainment programming into different genres—as this chapter does. Again many programs cross lines. Is "Twin Peaks" a drama or a soap opera? Is a program like "The Days and Nights of Molly Dodd" a drama or a comedy? (For awhile it, and similar shows, were referred to as "dramedies.") Is "Bambi" a movie or a children's program? How is a cop show with musical numbers categorized? Nevertheless, programs are divided into categories, in part so that the public knows what to expect, at least in a generalized way.

CHAPTER 11

Watching television is like making love—not a reasoning activity.

Television Quarterly

Music

Music is the mainstay of radio, with disc jockeys' chatter and platters filling the airwaves. A listener who is patient enough can uncover just about every type of music imaginable.

Because of the impact that radio airings have on record and CD sales, large stations are usually deluged with free promotional copies. Smaller stations do buy records and CDs, but often at reduced rates.

ASCAP, BMI, SESAC

All stations, however, do have to pay for the right to air the music through arrangements with the **American Society of Composers, Authors, and Publishers (ASCAP), Broadcast Music, Inc. (BMI),** and the **Society of European Stage Artists and Composers (SESAC).** These are music licensing organizations that collect fees from stations in one of two ways.

blanket and per-program fees

One way is called **blanket licensing**—for one yearly fee, a station can play whatever music it wants from the license organization without having to negotiate for each piece of music. The other way is called a **per-program fee** whereby the station pays a set amount for each program that utilized music from the licensing organization.

radio

All of these fees have been historically controversial, both in terms of amount and how they are determined. Radio has been happier with the situation than television. Because they air so much music, radio stations generally opt for the blanket license, which is based on a percentage of the station's revenue. Having paid that, the station can air whatever music it wants that is controlled by the music licensing company without further negotiations. The only controversial part is the percentage of revenue the stations must pay. This fee is frequently renegotiated between the music licensing organizations and a group called the All-Industry Radio Music License Committee, which represents radio stations. Presently the rate is about 1.4 percent.[1]

The per-program rate has a more complicated base of pay determination. Generally a station has to pay about fifty times its highest one-minute advertising rate plus a percentage of the revenue it receives for the particular program on which the music is to be aired. This amounts to about three times what the blanket license fee would be for the same period of time. However, for stations such as all-news stations, which use very little music, it is less expensive than a blanket license. Generally the radio stations that choose per-program licenses use so little music that they do not find the fee offensive.

TV

Such is not the case with television. TV stations do not like blanket licenses because they use very little music in the local programs they produce. They often see their fees for music constantly increasing, even though their use of music does not. This is because the fees are tied to the revenues of stations. When station income increases, fees for music increase even though the total use of music may go down. In addition, most of the music heard on TV stations is on programs produced by production companies who have already paid a one-time, **needle-drop** fee, which enables them to distribute and exhibit the programming with the music in it. The stations feel they have, in one way or another, paid for the programming and should not need to pay again for the music contained in it.

At one point the TV stations took ASCAP and BMI to court over the legality of blanket licensing, but in 1985 the Supreme Court ruled that blanket licensing was legal.[2] It did not say that all licensing had to be blanket, however, so the feud continued. In 1987 another court case ruled that ASCAP had to make available to broadcasters an economically viable per-program license in addition to the blanket license. The logistics of this are still being worked out, but in the meantime stations are paying both blanket licenses and per-program fees on an interim fee structure worked out by the courts.[3]

other entities

Other entities besides commercial radio and TV stations pay money to music licensing organizations. These include bars, restaurants, concert halls, public TV stations, and cable TV networks. Usually the licensing organizations do not bother a new entity while it is in the formative stages. For example, low-power TV stations have only recently been asked to pay licensing fees.[4] Cable networks did not pay when they were first started, but once they became successful, ASCAP and BMI came knocking. After much hassling, a federal judge decided pay-cable networks should pay a blanket fee of fifteen cents per subscriber and basic networks should pay 0.3 percent of gross revenues.[5] ASCAP, BMI, and SESAC distribute the money they collect to com-

distribution of money

posers and publishers in accordance with the number of times the music has been aired. The top hits, naturally, gain the largest percentage of income. In order to determine which pieces are aired most frequently, each licensing organization periodically surveys a representative sample of stations and asks them for their **play lists.** The information from these lists of musical selections and the number of times they were broadcast is then entered into a computer, which determines the pay rate for each piece.

history of licensing organizations

Obviously, one licensing agency rather than three could handle this chore, but three have evolved, mainly to ensure proper competitive practices. In the early days of radio, only ASCAP existed. When it raised its fees to an extent that radio stations considered exorbitant, the broadcasters countered by forming BMI. The idea behind BMI was that stations would play only music by composers and publishers represented by BMI, circumventing the need for ASCAP. However, ASCAP ceased its high rate demands, and most stations now play music represented by both ASCAP and BMI and pay licensing fees to both. ASCAP and BMI try to woo successful composers through special financially rewarding contract provisions, so musicians profit from having the two competitive organizations. SESAC primarily represents foreign and religious music; therefore, many stations do not bother to contract with it. In recent years, SESAC has captured a few hits, and it is now seen more frequently on station expense records.

ASCAP, BMI, and SESAC pay only composers and publishers, not performers and record companies. Exposure on radio is assumed to increase record demand, from which both performers and record companies profit. However, record companies and performers generally do not feel this way, and sometimes performers compose their own music in order to reap the benefit of the licensing payments. Licensing agencies are nonprofit by nature, so after paying

company structure

expenses connected with surveys, computer calculations, and personnel, they distribute the rest of the money. The largest number of employees are field representatives who handle problems related to collections and monitor non-subscribing entities to make sure they are not playing music represented by the field representative's licensing organization.

rock lyrics

One of the controversies surrounding music on radio involves the lyrics of rock music. During the mid-1980s, a group of Washington parents became particularly concerned about the sexual and violent nature of the lyrics. Some stations became somewhat careful about the lyrics they aired, but most stations said they did not broadcast the offensive lyrics and that attempts to police what was aired amounted to censorship. This continues to be a controversy.[6]

music on early TV

Historically, music as an entity in itself has had a more minor role in TV than in radio. Most thirty-minute or hour shows by prominent musicians had to become variety shows in order to maintain the attention of viewers. "Your Hit Parade," performances of the top-ten tunes of the week, lasted from 1950 to 1956 but was eliminated by rock 'n' roll, music that did not fit the program's talent or format. Dick Clark's "American Bandstand," a glorified disc jockey show on which teenagers danced to the current hits, has had the greatest longevity of any music show, having been shown in some form or other for about forty years.[7]

classical music

Public broadcasting regularly airs concerts of classical music from various concert halls around the country. Leonard Bernstein was successful with his "Young People's Concerts," which were a combination of music performances and low-key instructions about music. "Voice of Firestone," a classical music program, maintained a small but appreciative audience from 1949 to 1963, when its cancellation caused a great deal of controversy. Firestone Tire Company wanted to continue sponsoring the program and was willing to pay all bills, but the network canceled the show anyway because it could not deliver a large enough audience to serve as a lead-in to other network programs.

The minuscule role of classical music on both radio and TV is decried by broadcasting critics. Only about a dozen classical music stations are left on the AM band. On the FM band, classical music is found primarily in the public radio portion of the spectrum.

In several instances, a proposed change in classical music format has brought citizen unrest. During the late 1960s the sole classical music station in Atlanta, Georgia, announced that it was going to be sold to new owners who planned to change its format. Shortly thereafter, WEFM, a Chicago station that had been playing classical music since 1940 made a similar announcement. In both instances, citizens protested the transfers, and the issue wound its way through the FCC and the courts, picking up a similar case involving WNCN in New York along the way. Conflicting judgments were issued at various judicial levels concerning whether or not format change should be considered in licensing decisions. In 1981 the U.S. Supreme Court ruled that the FCC need not consider the uniqueness of a radio station's format before granting a license renewal or transfer; it should let the marketplace determine format decisions. This was a setback for cultural organizations and classical music lovers.[8]

FIGURE 11.1

MTV's taping of the British rock group, Squeeze. *(Courtesy of Warner-AMEX Satellite Entertainment Company)*

The music concept that has had the most success on recent TV is **music videos,** started in 1981 by MTV, a twenty-four-hour-a-day, advertiser-supported cable network. These three-to-four-minute minifilms made to accompany rock music quickly became a big hit with teenagers, enabling MTV to become one of the few cable services making a profit. With MTV's success, rock music videos began to appear everywhere—on networks, stations, other cable channels, videocassettes, and at dance clubs. The videos were criticized, however, for their sameness, sexism, and violence. The music video business also spread to other forms of music. The Nashville Network initiated country-western music videos and the MTV-owned cable network, VH-1, makes music videos for oldies.[9] None of these ever reached the fervor-pitched success of MTV, and the popularity of rock videos began to wane by the late 1980s as the novelty wore off. Even MTV stopped programming music videos twenty-four hours a day and began experimenting with other forms of programming such as comedies and talk shows.[10]

music videos

Drama

Dramatic programs have changed greatly over the decades. Radio was replete with them until TV took over, at which time radio drama essentially disappeared. From time to time various radio programs initiate a revival in an attempt to reawaken public interest and acceptance.

The TV **anthology dramas** of the 1950s, such as *Marty* and *Requiem for a Heavyweight,* probed character and motivation and emphasized the complexity of life. Although these plays were popular with the public, they became less acceptable to the advertisers, who were trying to sell instant solutions to problems through a new pill, toothpaste, deodorant, or coffee. The sometimes depressing, drawn-out relationships and problems of the dramas were inconsistent with advertiser philosophy and largely led to their demise by the 1960s.

1950's anthology drama

What replaced these anthology dramas were **episodic serialized dramas** with set characters and problems that could be solved within sixty minutes. With series such as "Gunsmoke," "Route 66," "Marcus Welby, M.D.," and "Miami Vice," plot dominated character, and adventure, excitement, tension, and resolution became key factors. Westerns, detective stories, mysteries, science fiction thrillers, and medical shows all tended to have good guys and bad guys. Although the main characters were the same week after week, they rarely seemed to profit from lessons learned on previous programs and were as pristine at the end of the episode as at the beginning. Problems of individual episodes could be solved, but never the overall motivation for the series because that would mean the series itself would have to end.

Various forms of dramatic programs cycle in popularity, with doctor shows being big one year, police shows dominating the next, and lawyers holding the limelight a year later. Most of these forms had precursors in style and content within novels, films, and radio, where danger, panic, pursuit, and climax had held sway for years. However, television called for changes in concept because the small screen demanded intimacy rather than spectacle, few characters rather than many, and reliance on close shots rather than the long shots of movies or the imagination-induced shots of radio.

The more probing type of drama has surfaced occasionally with network single presentations, such as *Death of a Salesman* and *The Glass Menagerie,* and public broadcasting series, such as "Hollywood Television Theater" and "American Playhouse." Dramas of longer duration, known as **miniseries,** became popular beginning with the 1975 serialization of Irwin Shaw's *Rich Man, Poor Man.* Two years later the miniseries *Roots,* Alex Haley's saga of his slave ancestors, aired eight straight nights to the largest TV audience up to that time. *Roots* was also one of the first series to be called a **docudrama**— a program that presents material that has a factual base but includes fictionalized events. Docudramas became controversial because, by combining fact and fiction, they could lead the viewer to believe events happened that, in actuality, were fictional inventions by the scriptwriter. Both the miniseries and docudrama forms still exist, but are not aired as often or as successfully as they were during the late 1970s and early 1980s.

Dramas with more complexity, such as "Hill Street Blues," a police series, and "St. Elsewhere," a medical series, were introduced during the 1980s. These had to be watched carefully in order to follow the plot; they were less pap to the mind than many of the previous series had been.

The drama show that was touted as the model for the new TV of the 1990s was "Twin Peaks." This stylized, shocking saga, directed by David Lynch, was a serialized cliffhanger that dealt each week with the question of "Who killed Laura Palmer?" It did not become a model, however, and was canceled in 1991, a victim of the vagaries of programming.[11]

One of the main problems encountered by TV writers and networks revolves around the fact that, of all forms of entertainment, drama has the greatest capacity to evoke strong and even disturbing emotional responses in its audience. For this reason, TV drama is particularly susceptible to criticism

FIGURE 11.2
A scene from the drama *Roots*, a David L. Wolper Production depicting one of Alex Haley's ancestors who was sold into slavery in the New World.
(Courtesy of The Wolper Organization, Inc.)

and censorship within and outside the industry. Examples of network or sponsor management deleting or rejecting controversial content are numerous, but probably even more numerous are outcries from pressure groups, government agencies, and the public at large. Although presentations regarding politics, bigotry, religion, and other controversial topics have come under fire, the subjects of sex and violence in TV drama conjure up the hottest arguments.

Sexual permissiveness, both in society and on the TV screen, has increased over the years. The low-cut dresses that raised eyebrows during the 1950s are considered modest by today's standards. TV drama has broached sexually sensitive subjects, such as homosexuality, premarital intercourse, and incest, usually to the almost instant outcry of critics. In the end, the heat passes, and the networks go on to conquer another sexual taboo.

sex

The newer media are now receiving most of the criticism regarding sexual programming. R- and X-rated movies are available over cable TV and on cassettes. The producers of this material argue that the programs are not available to anyone who happens to tune in but are distributed in a restricted way to those who wish to pay for them. Opponents argue that such pornographic material is tasteless and should not be available because it leads to greater promiscuity, particularly among the young. Critics also feel it desensitizes those who watch it to many acts considered socially unacceptable. The battle lines have been drawn, and future developments will be interesting.[12]

Violence, even more so than sex, appears to recycle in a predictably unpredictable manner. A hue and cry will emerge from various segments of society followed by a TV impoundment of guns, crashing cars, knives, and fists. But to some viewers, nonviolent programming seems bland, and it does not draw the audience that its more violent counterpart does. So gradually the guns and knives reemerge until they are so prevalent that a hue and cry once again arises.

violence cycles

Entertainment Programming   291

As far back as 1950 Senator Estes Kefauver asked the U.S. Senate if there was too much violence on TV. The first major outcry arose in 1963 after the assassination of President Kennedy. Claims were made that all the violence on TV had led to the possibility of assassination. In 1967 an antiviolence crusade led to an investigation by the Senate communications subcommittee.

In 1972 the Surgeon General issued a report that stated the causal relationship between violent TV and antisocial behavior is sufficient enough to merit immediate attention.[13] Hardly anyone paid immediate attention, but a growing protest against violence reached a crescendo in 1976 and 1977, which led to network program changes that led to diminished protest. Part of the 1977 outburst against violence was the result of a court case in which it was alleged that a fifteen-year-old boy killed his elderly neighbor because he watched too much violent TV.[14]

During the early 1980s the subject of violence, coupled with sex and profanity, was resurrected by Reverend Jerry Falwell, head of the Moral Majority, and Reverend Donald Wildmon, who organized the Coalition for Better Television. They took credit for the fact that several advertisers canceled sponsorship of violent programs.[15] More Congressional hearings took place during the mid-1980s, and Senator Paul Simon introduced several bills to curb violence on TV.[16]

Both quantitative and qualitative problems plague the violence debate. Measuring violence is not like measuring cups of sugar. Is pushing someone in front of a runaway cactus the same violent act as pushing someone in front of a car? Should the humorous "pie-in-the-face" slapstick comedy be considered violent? Is it violent for one cartoon character to push another off a cliff when the one pushed soars through the air and arrives at the bottom with nothing injured but pride? Should a heated argument be treated the same as a murder? Is it worse to sock a poor old lady than a young virile man? Should a gunfight be considered one act of violence, or should each shot of the gun be counted?

violence indexes

Despite all these measurement pitfalls, indexes abound in an attempt to tell whether violence on TV is increasing or decreasing. One of the oldest violence-measuring systems was developed by Professor George Gerbner of the University of Pennsylvania. For more than ten years he has had trained observers watching one week of TV fare a year to count acts of violence according to his complicated formula. Because his count includes just about everything remotely violent, the networks take offense at his calculations, and CBS and ABC have developed their own violence indexes.[17] The National Citizens' Committee for Broadcasting, somewhat with tongue in cheek, developed a "violence index" that calculated how many years each network would have to spend in jail if convicted of all the crimes it portrayed in one week—the range was from 1,063 to 1,485 years.[18]

violence effects

Measurement is not the only pitfall connected with violence. The effect of TV violence on society is also debated and hard to determine. Many research projects and surveys have been conducted, the findings of which have generally been severely (perhaps even violently) challenged by both friends and foes of TV fare. Some of the results that have surfaced are enlightening. (1) People who watch killings and woundings on TV show a greater immediate tendency toward aggressive behavior than do those who watch chase scenes and arguments. (2) People who watch violence on a large screen show greater tendencies toward aggressive behavior than do those who watch on a small screen. (3) The inclusion of humor in a program dampens the tendency toward aggressive behavior on the part of the viewer. (4) There is little difference between news programs that contain only nonviolent news and ones that contain both violent and nonviolent items in terms of increased inclination toward aggression. (5) The more children identify with violent characters in a program, the greater is their inclination toward aggression. (6) There is little correlation between what children consider to be violent acts and what mothers consider to be violent acts. (7) People think there are more bloody scenes on TV than there actually are. (8) Four out of ten people feel that violence is harmful to the general public and to children in particular. (9) Four out of ten people say that they avoid watching violent shows. (10) Hardly anyone believes that watching violence hurts him or her personally.[19]

Innumerable organizations have joined the battle to curb violence. The PTA (Parent-Teacher Association) held hearings on the issue in eight cities. The American Medical Association wrote a letter to advertisers urging them to refrain from advertising on violent programs. The National Citizens' Committee for Broadcasting (NCCB) distributed a list of the advertisers who most frequently advertise on violent shows. Consumers' groups have boycotted products advertised on violent programs. Even interindustry groups, such as the National Association of Broadcasters and the Screen Actors Guild, have at times called for a halt to violence.

The entire violence issue tends to generate heat and may remain unresolved in future generations.[20]

Situation Comedy

Situation comedy shows are perhaps the purest form of entertainment in that their aim is to make people laugh. This is not an easy task. It takes strong-penned writers and strong-willed actors and actresses to crank out humorous lines and actions week after week.

The grande dame of situation comedy is Lucille Ball, whose antics will probably live forever in reruns. Others who have made their mark in this form of programming are Henry Winkler as the Fonz in "Happy Days," Robert Young in "Father Knows Best," Bill Cosby on "The Cosby Show," Alan Alda of "M*A*S*H," and Dick Van Dyke and Mary Tyler Moore, first together and then on separate shows.

The general successful format for a comedy show is the development of characters who are placed in a situation that has infinite plot possibilities, the creation of complication, the reign of confusion, and the alleviation of the confusion. The problems encountered are usually the result of misunderstanding rather than evil, and the audience can relax because it knows the problem will be solved.

format

FIGURE 11.3

The cast of the long-running and highly-popular "M*A*S*H."
(Suero Orlando, © Gamma-Liaison)

early comedies

The early situation comedies made an attempt to be believable, but the necessity to crank out programs accelerated a trend toward paper-thin characters and canned laughter. One of the mainstays became the idiotic father ruling over his patient and understanding wife and children. The advent of "The Beverly Hillbillies," a family of nouveau riche hillbillies who moved to Beverly Hills where their uncultured life-style clashed with the accepted standards, was cited as evidence of the decadence of TV and perhaps the depth of silly exaggerated situations, slapstick corny plots, and unbelievable characters. But these early comedies live on because reruns of them, and later ones, fill enormous blocks of time on independent stations and some cable networks.

A breakthrough in comedy series occurred during the 1970s with the debut of Norman Lear's "All in the Family," whose bigot lead, Archie Bunker, harbored a long list of prejudices. This series, unlike any previous comedy series, dealt with contemporary, relevant social problems and even with politics, heretofore taboo for comedy series. This program and subsequent similar ones raised the status of situation comedy in the eyes of critics and the public alike.

social relevance

antistereotype

During the early 1980s two comedy programs stole the show, mainly because they defied stereotypes. "The Cosby Show," starring Bill Cosby, dealt with a black family, but provided situations that anyone could relate to. The other program was "The Golden Girls," which dealt in a very wholesome, energetic way with four older women who shared a house in Florida.

caustic comedy

The late 1980s saw a rise of caustic comedy. The Fox network had two hit shows in this category: "Married . . . With Children" dealt irreverently with marriage, and "The Simpsons" glorified the underachiever. On ABC "Roseanne" featured an overweight family that often acted outrageously. Although these programs were often criticized for their negative approaches, they received audience loyalty.

Situation comedy, in general, has been criticized because of the emphasis on sex, the improper portrayal of members of minorities, too slavish obedience to ratings, and the outlandish financial demands of the stars. Situation comedy has also been criticized for making it appear that all problems can be solved in thirty minutes. This concept was particularly pursued during the 1960s when the first television-weaned generation began demanding wholesale reforms. There were those who felt these young people had been exposed to so much TV that their view of reality was a simplistic one in which all problems could be solved easily and quickly, not allowing for real-world complexity.

But most of the producers of situation comedy shows are concerned with making people happy. As Norman Lear has said, "I would hope . . . that people turn off these shows and feel a little better for having seen them."[21]

Variety and Reality Shows

Variety shows are a hard act to follow—especially for the variety performers themselves. In the days of vaudeville, a stand-up comedian, juggler, or musical group could survive for years by keeping on the move and performing for new audiences in each town—not so with television. In one prime-time hour, a comedian will have exhausted his or her supply of jokes before an unseen audience of 20 million. What staves off unemployment? Because variety shows absorb jokes faster than writers can write them, juggling acts faster than performers can learn new ones, and musical numbers faster than singers and orchestras can rehearse them, few variety shows have enjoyed longevity.

The longest running variety show was Ed Sullivan's "Toast of the Town," which was seen on CBS every Sunday at 8:00 P.M. for sixteen years. Different talent appeared each week, with Mr. Sullivan giving straight-laced and straight-faced introductions to each. No one ever accused him of being a stand-up comedian, so he did not run out of funny material, but he did know how to put together a first-rate weekly show with wide audience appeal that included such coups as The Beatles, Elvis Presley, and the Singing Nun.

One of the most controversial variety shows was hosted by the Smothers Brothers, who rose from obscurity during the 1960s to become a top-rated team in their first season. In a period of political division caused by the conflict in Vietnam, the brothers leaned heavily on political satire. Constant battles raged over the censorship that the two brothers felt CBS used to stifle their creativity. Amid arguments over edited segments of programs and lawsuits, the brothers were relieved of their show.

The list of entertainers who were on prime-time network variety shows includes Jackie Gleason, Sid Caesar, Dinah Shore, Andy Williams, Carol Burnett, and Donny and Marie Osmond.

Nowadays, most of the true variety shows have disappeared and are usually only one-time specials. "Saturday Night Live" might be considered a variety show, but it has its own unique form. Talent shows are a type of variety show, but they, too, are rare.

FIGURE 11.4
A 1987–88 attempt at
a full-scale musical-
variety show was
ABC's "Dolly"
starring Dolly Parton.
However, it did not last
the year. *(Courtesy of
Sandollar Television, Inc.)*

pay-cable

Early pay-cable adopted a version of the variety show by taping stand-up comics appearing at clubs or engaging in a friendly competition to determine who was the best comic. This was fare that could be taped even though union agreements had not been signed. Later pay-cable continued with stand-up comics, some taped especially for the services.

reality-based programs

A newer version of variety shows are the types of programs that are often referred to as reality-based programs. These involve a host or hostess introducing real people who perform unusual acts. In fact, one such show was called "Real People" and featured ordinary Americans who had unusual abilities in terms of strength, dexterity, hobbies, or other accomplishments. The "Candid Camera" type programs that show people reacting to unusual circumstances are also reality-based shows. The latest craze in these types of shows was spawned by "America's Funniest Home Videos." People send in tapes they have recorded, and the best are shown on the air—accompanied by humorous introductions and dialogue.

criticism

Networks are sometimes criticized because they don't foster the development of true variety shows anymore. But when the shows were aired, they were often criticized for sameness; they all had singers who performed similar silly skits. The stand-up comics and the reality-based shows are also criticized for sameness and silliness. However, most variety and reality shows provide wholesome family entertainment, and they do offer variety within the television diet.[22]

Movies

Movies on TV is another evolving area in which the emphasis is shifting from local TV stations to networks to pay-cable to videocassettes. Many of the regular drama series are produced on film, and as such could be considered movies for TV, but they are not generally placed in this category. All movies that are first shown in the theater and then released to TV fall into the movie category, as do films without continuing characters that are made specifically for TV.

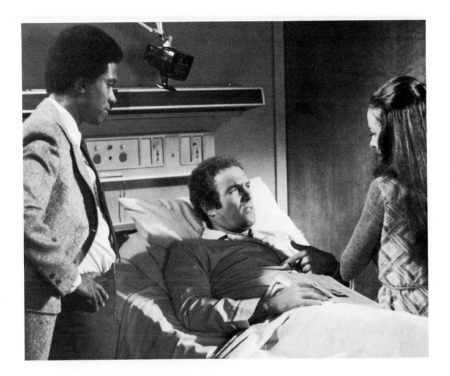

In the early days of television, theatrical films were the mainstay of local independent stations. With a twenty-year backlog of films just sitting on the shelf, the film studios were happy for this new source of revenue. No union contract had envisioned this bonanza, so at first there were no **residuals** to be paid. But as the use of movies on TV became popular, both the guilds and unions negotiated contracts with producers calling for the payment of residuals. With costs thus greatly increased, the producers turned to the networks, which obviously had larger pockets than independent TV stations. The phenomenon of movies on TV caught hold in a big way, and by 1968 there were movies on at least one network each night of the week. Soon the twenty-year backlog of movies was depleted.

local

networks

The networks then began contracting for movies made especially for TV. These are still being produced in fair abundance. Many of these made-for-TV movies are low-budget, quickly produced, grade-B movies, but occasionally one emerges that achieves critical and popular acclaim.

made-fors

Of course, big box office movies still find their way onto TV, but the cost per film to a network is generally in the millions. Such film contracts usually stipulate that the films cannot be shown on TV for a specific period of time after their release—to try to ensure that the TV showings will not divert revenue from the movie theaters. This period of time is referred to as a **window.** A film with a two-year window cannot be shown on TV until two years after it is released to theaters.

Entertainment Programming

At present, three categories of movies are seen on local and network broadcast TV: the oldies that are making the rounds for the umpteenth time, the made-for-TV movies, and the recent releases that are being shown over the airwaves for the first time. Today movies do not account for as large a percentage of broadcast time as they did during the 1960s. Nevertheless, they still make their presence known.

pay-TV

During the 1970s several new outlets for movies arose in the form of pay-TV, subscription TV, videocassettes, and video discs. The early local pay-TV services featured movies but usually were not able to afford the big blockbusters. When Home Box Office's national satellite service began to reap success, HBO began negotiating for the more glamorous films. Once again, union contracts had not envisioned a booming pay-TV market, and the fledgling pay-cable TV services were able to obtain movie product fairly inexpensively during their initial years.

windows

They were also able to negotiate a shorter window than for commercial TV. As a result, films were sometimes seen on pay-TV within months rather than years of their theatrical release. The home video services—cassettes and discs—eager to distribute movies, were able to negotiate an even shorter window. The general release pattern became movie theaters, then pay-per-view, then cassettes and discs, then pay-cable networks, then commercial network TV, then basic cable networks, and then local stations. As the success of movies on pay-TV increased and as union negotiations solved payment and residual problems, the major pay-TV services began ordering made-for-pay-TV movies.

controversies

Controversies have arisen concerning the content and presentation of movies on both pay- and commercial TV. One recent controversy involves the **colorization** of films. Old black-and-white films are being made into color films through computerized processes. The distributors ordering the colorizing (led by Ted Turner) are doing so because they believe color films have more audience appeal than black-and-white films. Directors, talent, set designers, and other creative people disagree, saying they created the films for black-and-white and do not want their artistic integrity tampered with by color.

A complaint directed against commercial TV is that the content of movies is so heavily censored to eliminate sex and violence that the films are edited beyond recognition or sensibility. On the other hand, there are complaints that the violence and sexual material present in movies is much too explicit and that films dealing with homosexuality, rape, prostitution, and similar subjects should be kept off both commercial and pay-TV.

This criticism generally does not apply to the home video market, which distributes primarily unedited theatrical films. Because they are rented or purchased, the content of films in this form do not raise as much ire as when they travel through open airwaves. Occasional cries are heard from theater owners who say that movies shown on any other distribution form hurt their business. In an "if you can't lick them, join them" style, some theaters have set up videotape movie rental facilities in their lobbies. But movies, in some form or another, will no doubt continue to be a staple of television.[23]

FIGURE 11.6
Oprah Winfrey
interacting with her
audience. *(Courtesy of
Harpo, Inc./Paul Natkin,
photographer)*

Talk Shows

Most TV talk shows capitalize on the average person's desire to know what
makes celebrities and unusual people tick. The shows constantly parade people
past hosts or hostesses who attempt to bring out the unique or peculiar in the
guest. The late-night talk shows are probably the best known, with Johnny
Carson being the dean of talk show hosts. David Letterman is another of the
most popular hosts, primarily because of his unique brand of humor. The
afternoon features talk show rivals Phil Donahue and Oprah Winfrey. Their
shows utilize not only guests, but reactions and questions from members of
the audience.

on TV

Radio talk shows often involve celebrities, too, but many are much more
audience oriented. Some are devoted entirely to members of the public who
call to express opinions or ideas on specific or general subjects. Others feature
experts on various subjects who then answer listeners' questions.

on radio

The cost of talk shows depends primarily on the quality and demand of
the guests and host. Some talk shows are virtually free, for they are beset with
requests from aspiring authors, dog acts, one-man bands, and the like, who
wish to appear on the program for the free publicity. Other shows pay top price
to obtain "hot properties" of the show business and political worlds.

cost

Public access cable TV has also become a fertile area for talk shows.
Hosts and hostesses interview friends, acquaintances, business associates, and
sometimes celebrities or would-be celebrities in an effort to further their causes
or their careers.

on public access

Naturally, not all guests on talk shows turn out to have scintillating personalities, so the shows are occasionally criticized for being boring. Some hosts
capitalize on abrasiveness in order to provoke responses from guests, and this,

criticism

too, is criticized. Some talk shows have been dubbed "tabloid TV" because subject matters, such as sex changes, Satanic worship, and wife beating, seem beyond the bounds of propriety to some elements of the public.

Talk shows run the gamut of network produced, station produced, and independently produced. Some radio talk shows draw top ratings in particular markets, and TV talk shows, although they are not the largest of drawing cards, tend to draw consistent audiences.[24]

Audience Participation Shows

Audience participation shows have always been popular. Early radio had its quiz shows for both children and adults, and today quiz and game shows abound on TV.

early shows

TV took over the quiz-game program idea early in its history with a 1942 simulcast on radio and TV of "Truth or Consequences," a program on which contestants who could not answer questions had to participate in generally silly activities.

quiz scandals

Most of these early shows had modest prizes for the winning contestants, but during the mid-1950s, the stakes began to increase as such programs as "The $64,000 Question," "The $100,000 Surprise," and "The $64,000 Challenge" made their debut. Of course, the 1958 **quiz scandals** gave the quiz-game show area a temporary blow. For a while, no chance-oriented shows dared touch the airwaves, but gradually additional low-stakes programs referred to as game shows emerged during daytime hours. The prize for "The Dating Game" was an expense-paid date for the contestant and the person he or she selected. "The Newlywed Game" featured household items for recently married couples who agreed on answers to questions about each other.

present-day shows

Gradually money and expensive prizes crept back in, and the game shows dallied into the evening, particularly the early evening hours when local stations have control of the programming fare. The biggest hit in game shows became "Wheel of Fortune," which is the highest rated syndicated show in television history. The show brought fame to its card turner, Vanna White.

The gamut of opinion regarding game-quiz shows varies. Some people think the games are educational because of the information contained in the questions. Others think the games feed on avarice and gambling instincts, and still others believe the games make fools of all the contestants who participate and waste the time of those who watch.

cost

Game shows are among the least expensive to produce. All talent except for the host or hostess is free, the set can be used over and over, and the prizes are donated by companies in exchange for mention on the show.

criticisms

The degree of commercialization inherent in these programs is often questioned. Some programs appear to be one long commercial as the merits of the various prizes are revealed. The games themselves are criticized for being inane and childish and for a sameness that seems to permeate most of them. However, many viewers compete or empathize with both winners and losers, and there is never a lack of people lined up to try their luck or skill on big-time TV.[25]

FIGURE 11.7
Hostess Vanna White and host Pat Sajak show a winning contestant his new car on "Wheel of Fortune." *(Courtesy of Wheel of Fortune)*

Soap Operas

Soap operas arose during the heyday of radio and succeeded in dominating the afternoon hours with stories dealing mostly with the homemaker struggling against overwhelming adversity—sick and dying children, ne'er-do-well relatives, weak husbands.

Television adopted the soaps, often referred to as "daytime TV," at about the same time other programs switched from radio to the new medium. Many of the original traits were retained: Each program is serialized in such a way that it entices the viewer to "tune in tomorrow"; the plot lines trail on for weeks; music is used to designate transition; very little humor is included in the dialogue, as adversity is the common thread. Soap opera characters, unlike their evening dramatic and comedy counterparts, live with their mistakes and are constantly affected by events that happened on previous programs. They also grow old and have children who grow older.

What has changed from the old radio soap opera days is the program content. Although there are still some homemakers struggling against overwhelming adversity, the emphasis is now much more on male-female sexual relationships. Infidelity, premarital sex, artificial insemination, mate swapping, impotence, incest, venereal disease, frigidity, and abortion have been added to nervous breakdowns, sudden surgery, and missing wills. For a period of time, subject matter was often tried first on soap operas to determine if it would be fit for evening drama and movies. Soap operas then became part of evening programming with successful programs such as "Dallas" and "Dynasty." In fact, there are those who feel "Twin Peaks" was a glorified soap opera.[26]

traits

changing content

Sometimes soaps are produced by the networks, and sometimes they are produced by independent companies. The daytime soaps are among the most profitable TV ventures, for production is cheap and ads are plentiful. The same paper-thin scenery is used day after day, and because soaps are a world of words and close-ups with very little action, hardly anything is consumed or destroyed. In recent years, some soaps have taped on location, but most are still studio bound. The nighttime soaps have budgets and sets more akin to prime-time drama.

Daytime soap opera stars are paid much less than prime-time talent, a fact that the former often decry because they work at a much more hectic pace. While the nighttime stars are working to crank out one program a week, the talent of the soaps must produce one program a day. Understandably, this leads to some production sloppiness where blown lines are left intact in the aired product. Such incidents are remarkably rare, however, if one considers the time pressures under which the actors are performing.

Soap opera regulars, if they can take the pace, can be fairly sure of long-term employment, for many soaps have survived while there have been dozens of turnovers in the prime-time area. Some of the longest running soaps include "Search for Tomorrow," "Days of Our Lives," "General Hospital," and "The Young and the Restless."

For the most part, daytime soaps are put in a second-class stepchild position by critics and the broadcasting industry alike—mainly because of their airtime, cheap production, and maudlin story lines. However, there are those who feel that from a literary point of view, soaps are superior to nighttime dramas and comedies. Soap writers are relieved of the chore of solving all problems in thirty minutes, so they can explore character and probe motivation in a way that provides viewers with more realistic, albeit exaggerated, situations.

It was assumed for many years that only middle-class homemakers and shut-ins made up the audience for soap operas. But in recent years many "closet-case fans" have emerged, including baseball players, nighttime TV stars, college students, politicians, and many men and women who work nights. In fact, a small weekly magazine, *Soap Opera Digest,* which prints capsule plots of each soap on the air that week, has been very successful marketing its product to those who must miss an episode of their favorite soap and do not want to fall behind the story line.[27]

So, despite the fact that Heather has been jilted at the altar by John, who has discovered that his father is impotent and he is the love-child of an affair between his mother and Dr. Winton, thus making him a first cousin of Sharon, who is in love with Tom, the husband of Tricia, who has just had an abortion in order to cover up her affair with Richard while Tom was out of the country searching for his child of a previous affair who had been put up for adoption—soap operas will no doubt continue.

costs

long runs

status

audience

FIGURE 11.8
A wedding scene from
"General Hospital."
(© 1987 American
Broadcasting Companies,
Inc.)

Children's Programs

Never could Sky King, The Lone Ranger, The Green Hornet, Howdy Doody,
Kukla, or Mickey Mouse envision the furor that has arisen over children's
programming.

The airtime for children's programming has not changed since the 1930s.
Saturday morning and after-school hours were the domain of the young in
early radio days and still are with TV today. However, radio, by virtue of its
aural nature, emphasized imagination and sound effects, whereas TV emphasizes sight and action. Modern-day children's radio programming is essentially nonexistent except for a few programs on public radio.

TV networks started children's programming early with an emphasis on   early programs
puppets, such as Howdy Doody and his real-life friends Clarabell the Clown
and Buffalo Bob, and Kukla and Ollie with their real-life friend Fran Allison.
The longest running kiddie show on network TV was "Captain Kangaroo,"
starring Bob Keeshan. Beginning in 1955, it ran Monday through Friday until
1982.

Children's programming was important on early local TV stations, too.
Most programs consisted of a host or hostess whose main job was to introduce
cartoons and sell commercial products. During the 1960s, networks overwhelmingly adopted the likes of "Felix the Cat," "The Roadrunner,"
"The Flintstones," "Popeye," and "Tom and Jerry"—and therein began the
controversy.

For many years these cartoons dominated Saturday morning TV, making for one of the most profitable areas of network programming. The cartoons were relatively inexpensive to produce, and advertisers had learned that children can be very persuasive in convincing their parents to buy certain cereals, candies, and toys. The result was profits in the neighborhood of $16 million per network just from Saturday morning TV.[28]

But gradually the situation changed. Parents who managed to awaken for a cup of coffee by 7:00 A.M. Saturday noticed the boom-bang violent, non-educative content of the shows along with the obviously cheap mouth-open/mouth-close animation techniques. A group of Boston parents became upset enough to form an organization called **Action for Children's Television (ACT),** which began demanding changes in children's programs and commercials. Scholastic Aptitude Test scores started going down as the first television generation took the tests. Researchers realized that children under five were watching twenty-three and a half hours of TV a week and that by the time they graduated from high school, they would have spent 15,000 hours in front of the tube. Government agencies also discovered that those nutritious cereals weren't so nutritious after all. The Children's Television Workshop developed "Sesame Street," and its successful airing on public TV proved that education and entertainment could mix.

All of this led to a long, hard look at children's TV. Various independent researchers, most of them at universities, undertook studies to determine the effects of TV on children. In 1969 the U.S. Surgeon General appointed a committee of twelve prestigious researchers to investigate the effects of television crime and violence on children. These twelve worked for two and a half years on the project and came to the conclusion that a modest relationship exists between viewing violence on TV and aggressive tendencies.[29]

The overall body of research on children and TV yielded some conflicting results, but overall seemed to indicate that changes did need to be made in children's TV. Studies have shown that children do learn reading and vocabulary from TV, but the children who watch TV the most are the ones who do poorly in reading in school. Watching TV generally cuts down on book reading, but certain TV programs that refer to books actually increase book reading. Children three and younger understand very little of what they watch on TV, and yet they will sit mesmerized before the set. Nine out of ten children between the ages of seven and eleven understand social messages when present in programs. Some studies show that children predisposed to violence are more apt to increase violent behavior after seeing it on TV than are so-called normal children; other studies show exactly the opposite. It has been determined that watching TV is an activity involving mainly the brain's right hemisphere, which contains nonverbal, nonlogical, visual, and spatial components of thought. From this research it is theorized that watching TV may hamper development of verbal and logical abilities. One study conducted on highly creative children found that their creativity dropped significantly after three weeks of intensive TV viewing.[30]

ACT

research effects

FIGURE 11.9

A scene from "Fraggle Rock." This animated series, which aired on network TV, originated from the live action "Fraggle Rock" created by Jim Henson and shown on cable TV. *(Courtesy of Henson Association, Inc., and Marvel Productions, Ltd.)*

Led and cajoled by ACT, a number of other organizations began demanding reforms in children's TV. They took note of the fact that there were some fine children's programs on TV—"The Wonderful World of Disney," "Lassie," and "Mr. Rogers' Neighborhood,"—but they were after changes in the cartoons, slapstick comedy, and deceptive commercials. In 1974 the FCC issued guidelines for children's television. It stated, among other things, that stations would be expected to present a reasonable number of children's programs to educate and inform, not simply entertain. It also stated that broadcasters should use imaginative and exciting ways to further a child's understanding of areas such as history, science, literature, the environment, drama, music, fine arts, human relations, other cultures and languages, and basic skills such as reading and mathematics.

By the mid-1970s most stations and networks had acquiesced, at least in part, to the reform demands. Programs with names such as "Kids' News Conference," "What's It All About?," "Let's Get Growing," and "Villa Allegre" hit the airwaves. Many of these shows attempted to teach both information and social values. These were more expensive to produce than cartoons, so the amount networks and stations spent on children's programs increased.

program changes

ABC developed after-school specials that dealt dramatically with socially significant problems faced by children, such as divorce and the death of a friend. In the area of cable TV, Warner-Amex established a satellite service called Nickelodeon that consisted solely of nonviolent children's programming. Unfortunately, most of the socially relevant programs did not receive high ratings.

With the 1980s, the tide turned again and the FCC, in the spirit of deregulation, dismissed the idea of maintaining or adopting standards for children's programming. The 1974 rules were ignored.[31]

The networks, eyeing their sinking children's TV profits with fear, canceled many of the expensive education-oriented programs and returned primarily to the world of cartoons. Some of these cartoons were based on toys and became known as **toy-based programming.** A toy was developed with the idea in mind that a TV series would be developed around it. Some people viewed this programming as a thirty-minute commercial for the toy.[32]

Citizens groups and Congress complained about all of this,[33] but broadcasters pointed out that children, like adults, need entertainment as well as education and that, given a choice, they will still choose "The Flintstones" over "Young People's News" in the same way that adults choose to watch an old movie rather than a sterling documentary. This phenomenon, they say, is a problem beyond their realm. Parents are the ones who must control the set, and as long as parents use it as a cheap baby-sitter, children's viewing hours will not be curtailed and their habits will not be changed.

By the 1990s, it looked like the pendulum was swinging back. Congress passed a bill saying that when stations came up for license renewal, they would have to prove that they had served the educational needs of children.[34]

The conflict over children's television is likely to continue. Parents and citizens groups will argue that broadcasters should provide quality children's programming, even if it is not profitable because such programming meets the nation's needs. Those in the television industry will defend their programming and their bottom line.[35]

Conclusion

Entertainment programming has both friends and foes. The music available on radio is often applauded because stations, overall, provide variety to meet the tastes of everyone. Music on TV has experienced high points, such as the Bernstein concerts and the Firestone hour. MTV has been a trend setter and a financial success.

Dramatic programming is sometimes applauded because it provides the escapism that many people need in their lives. There have been probing, humanistic high points in the drama area, such as *Roots* and *Rich Man, Poor Man.*

Situation comedies are healthy in that they make people laugh and have the ability to highlight social problems in a humorous manner.

Variety shows and reality-based shows generally provide wholesome family entertainment and lend a change of pace to the television fare.

Movies are often high quality because they have been produced initially for theaters. Some made-for-TV movies have also received critical acclaim.

Talk shows can be informative and allow audience members to get to know what makes celebrities tick.

Audience participation shows can be educational, and the networks like them because they are inexpensive to produce.

The characters in soap operas can be well developed because the story line lasts almost indefinitely. People on these programs are affected by their mistakes, making this genre somewhat realistic. Soap operas also give networks a chance to try out ideas to see if they are potentially acceptable for nighttime TV.

Children's programming provides both entertainment and baby-sitting for children. Some of this programming has been of an excellent nature, such as "Captain Kangaroo," "Sesame Street," and after-school specials.

Criticisms expressed toward entertainment programming are more abundant than the applause, but it is always easier to criticize than it is to create.

Music programming is criticized because both radio and TV have essentially abandoned classical music. Music lyrics are criticized for the sex and violence they encompass. Within the industry, performers and record companies are dissatisfied with the method of music licensing payment.

The glow of nostalgia affects drama in that people criticize current dramatic presentations because they are not as creative or "deep" as the radio and TV drama of the 1950s. The recurring waves of sex and violence are deplored by many organized groups and individuals.

Situation comedy also comes under attack for allowing too much sexual permissiveness. Other criticisms revolve around misportrayal of minorities, oversimplification of solutions to problems, exaggerated situations, and slapstick approaches for getting a laugh.

Freshness is hard to maintain on variety shows and reality-based programs. Soon they succumb to a sameness that many people criticize.

Movies are often relatively old by the time they are broadcast because of the window used to protect theater owners. Pay services and home video have access to movies sooner, but at a cost to the consumer. The battle over too much or too little censorship becomes heated on occasion. The colorization of movies has created an internal battle within the movie industry.

Talk programs are criticized for being dull and sometimes abrasive or sensationalized.

To many, audience participation shows are an inane, childish, and a waste of time that capitalizes on the gambling instincts of the human race.

Soap operas, like many other forms of programming, are criticized for their high degree of sexual content and their often morose nature. The work needed to produce soaps on a daily basis is often hard on the actors and others involved in their production.

Slapstick, violent children's programming is considered not only a waste of children's time but a detriment to their growth and development. Both organized groups and experimental studies provide continual pressure for the upgrading of children's programming.

Entertainment programming will continue to evolve as society itself evolves. Ironically, radio, which pioneered most of the forms of programming that now appear on TV, uses very few of them. Subject matter once taboo on soap operas is now handled in children's programming. Organizations influential in entertainment, such as ASCAP, BMI, ACT, and the numerous production companies, will continue to interact. The forms and quantity of entertainment programming available to the public will no doubt multiply, but the public will still determine success or failure as it sits in judgment.

Thought Questions

1. Does violence on TV adversely affect society? Support your answer.
2. Should performers and record companies receive money from ASCAP and BMI? Explain.
3. How can children be encouraged to watch quality programs?

News and Information Programming

Introduction

As more and more people turn to television and radio as their major source of information about what is happening in the world, the programming that generally falls within the news and public affairs area develops more social significance. However, it is often an uphill fight for those committed to news and information programming to see their program ideas reach fruition. Not only is such material likely to envelope a station or network in controversy, it is also generally expensive to produce.

News and information programming is not watched by nearly as many people as entertainment programming and, hence, cannot sell high-priced ads—in fact, sometimes it cannot sell ads at all. Because public service programming generally costs three times as much as entertainment programming per rating point, it must depend on the profits of its more glamorous sisters to support it. Yet this area is of great importance in the realm of social significance.

CHAPTER 12

Television has learned to amuse well; to inform up to a point; to instruct up to a nearer point; to inspire rarely. The great literature, the great art, the great thoughts of past and present make only guest appearances. This can change.

*Eric Sevareid,
longtime CBS commentator*

News

speed of news

Americans declared their independence on July 4, 1776, but it wasn't until many months later that the British learned of the declaration. Likewise, during the War of 1812, the Battle of New Orleans was fought weeks after the war was actually over, for word had not gotten to New Orleans. Today the slightest little rift between countries can be reported, analyzed, and even blown out of proportion within a matter of minutes. More people today are aware of what is happening in the world than ever before, and basically radio and television can take credit for this. The greatly increased worldwide communication of the past few decades is also painless to the viewer or listener. A mere flick of a dial can bring one up to date on current events or at least ensure that no great disaster has occurred.

In times of disaster the electronic media have become the main source of help and information—directing victims to sheltered areas, seeking help from outside sources, communicating vital health and safety information, and calming jangled nerves. The advent of portable equipment and satellite transmission for both radio and TV allows news to be reported rapidly and allows for firsthand reports through **actuality** interviews— question-and-answer sessions conducted with those involved in the news story.

News Sources

wire services

Gathering news is generally a complex process, with stations and networks depending on a variety of sources for news. The news **wire services,** Associated Press (AP) and United Press International (UPI), have been a traditional source of national and international news from the early days of newspapers. They have specialized services for both radio and TV that include copy written in electronic media style.

Wire service news is collected by a bevy of reporters stationed at strategic points around the world. These reporters have regular beats, such as government offices and police stations, which they cover to gather news. Sometimes reporters actually go to the places to cover the news; other times they can get all the information they need by making phone calls. The wire services also use stories from **stringers**—people who are not on the wire service payroll but who are paid for stories they supply that are actually used by the wire service.

The stories gathered by reporters and stringers are sent by satellite or wire to central AP or UPI offices where they are assembled and often rewritten. Then the stories are sent to stations and networks, where they are printed out on machines specially made to receive the signals. Stations and networks, along with newspapers and magazines, pay the wire services subscription fees for both the machine and the news.

Many other countries besides the United States have wire services, and some networks and stations choose to subscribe to them in order to obtain more international news and different slants on the news being reported. In addition, major U.S. cities have local wire services used by local stations and cable systems to keep them abreast of local news.

The use made of wire service news is largely up to the organization that receives it. Some radio stations simply rip the news from the wire service machines each half hour or hour and read the writing on the paper verbatim over the air. Other stations select news items and rewrite them in ways they feel will appeal to their audiences. Still others use the wire services information only as leads and dig for the facts themselves.

In addition to written copy, many wire services and other organizations provide visual material, both still pictures and video footage. These organizations, such as Associated Press' TVDirect, Independent News Network, and Spacenet, collect visuals and footage from stringers, from their own reporters, from local stations, and make it available to various stations and networks. They pay the sources for the news and charge the organizations who use the material. In this way, these companies act as middlemen or brokers of news.

Those in the business of gathering news also subscribe to various **data banks.** Sometimes they use these for news tips, but more often they use them to supply background information needed for stories. News organizations often have satellite or radar equipment to monitor weather conditions, and they have **scanners**—devices that listen in on police and fire radio communications. In this way they can get leads on stories, sometimes arriving at the scene of a crime before the police. Some news organizations have their own airplanes and helicopters so that they can gather news from the air.

The three major broadcast networks—ABC, CBS, and NBC—have their own national and international reporters who cover stories for their network news. They also make footage not used in their national newscasts available to their affiliates. This works in reverse, too. Affiliated stations often supply footage to networks, particularly when a major story occurs in their city.

network correspondents

Cable News Network has an elaborate news-gathering structure that makes news available to others besides CNN. CNN has been particularly aggressive in selling (or sometimes giving) its news to foreign countries. As a result, CNN news is seen in abundance throughout the world.

Another increasingly important source of news is the general public. People call stations or networks with tips on news stories. These, of course, must be checked out very carefully. Now that many people have **camcorders,** ordinary citizens sometimes supply news organizations with footage just because they happened to have their camcorder with them when a major event occurred where they happened to be.

News organizations often make liberal use of each other in order to keep up to date on happenings. People at TV stations read newspapers; radio stations listen to news on TV and on other radio stations; ABC personnel listen to the news on CBS and NBC and vice versa. More and more CNN is setting the agenda for newscasts throughout the country and the world. The *New York Times* used to be *the* standard for news, but CNN is quickly taking over that role.[1]

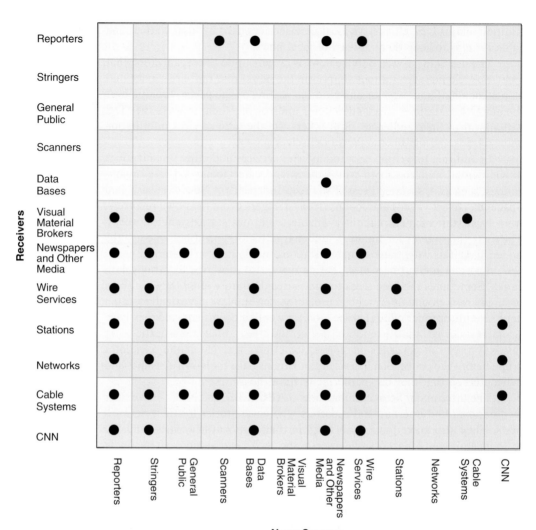

Receivers / **News Sources**

FIGURE 12.1
Interrelationship of
various news sources
and receivers.

The News Process

While reporters are gathering worldwide, national, and local news, other people
remain at the station or network undertaking other duties that will lead to
newscasts. The number of people needed depends on the emphases given to
news at the particular entity. A small radio station may have a news staff of
one person who decides what stories to broadcast, makes minor revisions in
wire service copy, makes a few phone calls to confirm facts or gather material
for stories, writes the script for a three-minute news update once an hour, and
also reads the news over the air.

At a large news organization, a news director has overall responsibility
for the news operation, hiring and firing people and setting the general guidelines for the approach. Each individual newscast is overseen by a news producer who organizes the program and decides what will and will not be

writers and producers

312   Programming

included. If a newscast is to include special types of information such as consumer affairs or entertainment news, segment producers may oversee those particular items. Assignment editors keep track of the stories that need to be covered and send reporters to cover these stories. Writers rewrite wire service copy and stories sent in by reporters and prepare the intros and outros for the total newscast. In some TV facilities tape editors put together the news footage that has been gathered in the field.

Most newsrooms are now computerized so that stories and ideas can be typed into a computer by a reporter or writer, recalled from the computer onto a screen in another location by the producer, who can then rewrite or in other ways act upon a story. Computers also aid the assignment editor in keeping track of what may need to be covered in the future. Computers can be used to help avoid overlooked details, to keep track of crew status and location, to distribute messages to the staff, and to file material used in the past.[2]

computers

Satellite news gathering (SNG) has changed the news process at many local stations. This concept was pioneered by **Conus,** a Hubbard Broadcasting-owned company, in 1984. Initially Conus gathered news with trucks equipped with satellite uplinks and sold the information to specific stations equipped with downlinks. Stations could send one reporter to a news site to get the local slant on a story, and Conus would transmit this back to the local station. For example, if a plane crash occurred in Chicago, a station in Atlanta could send one reporter to Chicago to report on surviving passengers from Atlanta. Conus encouraged stations to buy uplinks to make the news stories they covered available to other stations through the Conus structure.

SNG

local news

Other entities set up regional or national cooperative arrangements. For example, stations in the state of Florida established an SNG network called Florida News Network. With uplinks in five different cities, it provides coverage of the state primarily for stations throughout the state.

As more and more stations purchased **satellite news vehicles (SNVs),** they were able to interchange more news and depend less on the national news-gathering organizations, including the major networks. Stations were also able to go further afield to gather news. The older microwave trucks that stations used to gather news could only transmit about thirty miles, so stations could not cover stories further away than that. Satellite signals can travel throughout the country, so station crews can cover stories in more remote or distant areas.[3]

At most radio stations, the process of selecting the stories that will be aired is ongoing because news is broadcast many times throughout the day. Radio networks generally feed stories to their stations as they occur and also provide several news programs throughout the day. Thus, the news producer of a station affiliated with an ABC network might decide that the 8:00 A.M. news should consist of five minutes of ABC news as broadcast by the network; a story on a fire in Delhi that is on the UPI wire but not covered by ABC; an update of the condition of a hospitalized city official that the producer obtained by phoning the hospital; the report on a liquor store robbery gathered by one of the station's reporters; a report of a murder received from the local wire service; and the weather as received over the phone from the weather

radio station decision process

bureau. For the 9:00 A.M. news, the stories of the Delhi fire and liquor store robbery may be included again, slightly rewritten and accompanied by two stories excerpted from the 8:00 A.M. ABC news broadcast, a report phoned in live by a station reporter covering a school board meeting, an update on the hospitalized official, and the weather.

There are a fair number of all-news radio stations that broadcast news continuously throughout the day. They have a large number of crews, a variety of news sources, and a large staff of people working at the station. Needless to say, an all-news format is very expensive.

Most television news has a more defined countdown because the major effort is devoted to the evening news. There are exceptions, of course, including late night and early morning newscasts incorporated from time to time in both network and station schedules. Cable News Network provides twenty-four-hour news, so its procedures are closely akin to an all-news radio station.

News producers at most local stations and the major networks spend the day assessing the multitude of news items received to decide which twenty or thirty will be included on the evening news and in what order. Usually the producers of the various newscasts meet early in the morning to assess what they feel will be major stories. They convey this to the assignment editor who then assigns news crews to cover these events.

The crew may consist of one person recording shots to be included in a story delivered totally by the anchorperson, or the crew may be two or three people, including a reporter and equipment operators. Once the story is covered, the reporter may return to the home base, write the story, and work with an editor to cut the footage. Or the reporter may turn everything over to a writer and go out on another story. Or the footage may be microwaved or sent by satellite while the reporter and crew move on to another story as writers and editors complete the first story. Sometimes the reporter does a **standupper** live from the location during the newscast.

Changing events of the day often alter how the news is gathered. Communication transpires frequently between producers and reporters in the field, for example:

"There's a hurricane warning in Florida. Should I cover it?" "Yes, forget the Georgia peanut contest and get there right away."

"Forget the vice-president's luncheon and cover the French embassy picketing. Try to get an interview with the ambassador."

"The prime minister just resigned and I lined up an interview with his son. Can you give me five minutes of the show?" "Not five, but maybe two."

"Got the interview with the ambassador." "Send it on the satellite. We'll use it."

"The Senate just confirmed the president's nominee for the FAA. Should I try to interview him?" "Don't bother, we have too much other news. Get over to the secretary of state's news conference."

FIGURE 12.2
CBS evening news
anchor, Dan Rather.
(UPI/Bettmann Newsphotos)

"Rewrite this story on the earthquake prediction so it only takes thirty
    seconds. We have to make time for the ambassador's interview."

This type of decision exchange goes on at TV networks and local stations until *the newscast*
close to airtime when the "final" stories are collected, written, and timed. The
anchorpeople get into position, the director and crew prepare for the broadcast, the edited tapes to be rolled in to the newscast are put in order, the tele-
prompter copy is readied, the computer graphics are polished, and the newscast
begins.

    Last-minute changes can still occur, for the news is sometimes changed
even as it is being aired—"Just got a satellite damage report on the Florida
hurricane. Substitute that for the earthquake prediction."[4]

    Presentation of news is another important area. For radio, the days of *presentation*
the dulcet-toned announcers are gone; news is generally read by disc jockeys
or reporters. However, stations do vary both content and presentation of news
broadcasts in relation to their audience. A rock station will have a more fast-
paced presentation than an oldies station. Because TV news is actually a money-
maker for local stations, they are particularly anxious to lead the ratings;
therefore, local stations attempt to find newscasters who will appeal to viewers.
The networks hope to establish trustworthy, congenial newscasters who will
maintain a loyal audience.[5]

    Special elements, in addition to the daily news, fall under the jurisdiction *special news elements*
of the news department. Some of these—such as presidential news confer-
ences, congressional hearings, and astronaut launchings—can be predicted in
advance and adequate preparations can be made. Others—such as riots,
earthquakes, and assassinations—must be handled as well as possible by reporters and crews who happen to be close at hand covering other assignments.
Often stations interrupt their regular programming to broadcast special events.

Accolades and Criticisms of News

service function

Broadcast journalism is proud of its service function. As early as 1932 New York area stations conducted around-the-clock operations to cover the Lindbergh kidnapping case. In 1937 radio provided the main communications for the flood-stricken Ohio and Mississippi Valleys, and in 1938 it did likewise for southern California. Many times since then radio has acted as the main communication conduit during disaster. Radio and TV have brought the world the stories of the Hindenberg explosion, the declaration of World War II, the invasion of Normandy, the surrender of Germany, presidential news conferences, deaths of presidents, political conventions, the visit by Soviet Premier Khrushchev, space orbits, riots, President Nixon's trip to China, Watergate hearings, Iran-Contra hearings, the war in the Persian Gulf, and most other major events. For most of this coverage, broadcasting has been praised for its decorum and service to the nation.

And yet broadcast journalism is one of the most criticized institutions in our country today. It is blasted by government officials, liberals, conservatives, middle Americans, and members of its own fraternity.

The criticism centers primarily on what news the electronic media present and how they present it. Generally conceded is the fact that radio and TV bring important events to their audiences, but many critics claim that those media have not yet learned how to make the prime issues of our time understandable. The broadcast media are good at covering wars and fires, but inadequate in dealing with subjects such as inflation, unemployment, and economics.

visual emphasis

Television, particularly, is obsessed with visual stories and sometimes may downplay an important story simply on the grounds that there are no exciting visuals to accompany it. Sometimes stories are so highly visual that the facts become obscured. The weather, for example, can contain so many computer graphics, satellite feeds, and digital effects that viewers are left wondering whether or not it will rain.

capsulized news

Both media must provide capsulized news. The thirty-minute evening network news programs, when all commercials have been added, boil down to twenty-two minutes of news—approximately the equivalent of three newspaper columns. Obviously this cannot be the day's news in-depth, explained and analyzed. And yet the news that is chosen is often trivial. Greater coverage is given to the president eating a taco than to his views on the nation's economy. If a station or network has achieved a scoop, it will dwell on that story even though the story itself is relatively unimportant.

creating news

publicity

TV is sometimes held responsible for the evil effects of publicity given to violent lawbreakers. Murderers, kidnappers, skyjackers, and the like receive so much news coverage that they are boosted almost to a celebrity status. Stations have particularly difficult decisions to make when someone who is holding hostages demands access to the airwaves in order to denounce some particular group or institution. At times broadcasters have been encouraged simply to ignore such acts in hopes that the perpetrators will stop when they see that no one is paying attention to them. But when hostages' lives are in jeopardy, the decision is a difficult one.

Tasteless coverage of the victims of violent acts is also a problem. There are reporters who attempt and often succeed in interviewing people who have just seen a close relative killed or have just lost all their possessions in a disastrous flood. The interview answers given by victims under these circumstances are often not rational and further sensationalize the news event. In their sensationalizing, news media are often accused of judging the guilt of a suspect before he or she has had a chance to receive a court trial. "Trial by the press" is a frequently used ploy of lawyers who think that their clients do not have a chance because of adverse radio and TV publicity. Violence shown in newscasts is often criticized as tasteless and unnecessary—just a ploy to get high ratings.

In many types of stories, broadcast journalists accentuate the negative— accent on negative unemployment going up is reported more frequently and with more fervor than unemployment going down. Because of the speed of radio in particular, news is sometimes reported inaccurately just to be first. With the advent of minicams and satellites, pictures of events can be shown before the reporter has had time to gather the facts.

Another objection that some individuals have raised, particularly against network broadcast journalism, is that it is a biased product of the "liberal bias eastern establishment." The criticisms range from charges of outright purposeful distortion of facts to editorializing by exclusion. Politicians and businesspeople, who are often targets of news darts, are particularly prone to cry bias. They bemoan not only what is reported, but how it is reported and what is left out of what is reported.

Some criticisms of the news process come from those working within the news industry itself. Newscasters are constantly pleading for more airtime for time news so that they can correct the flaws brought about by the short, capsulized treatments. One-hour evening newscasts rather than half-hour newscasts are proposed frequently, but these proposals have been largely unheeded—mainly because entertainment programming supplies more dollars.

As networks, stations, and cable systems tighten their belts, many of the financial cutbacks functions and people associated with the news departments are disappearing. Foreign bureaus are being cut, and fewer reporters are being sent to cover a single story. The total number of stories being covered has also been cut down because stories covered and not aired are money down the drain. The people within the news departments decry this and claim that the quality of news will suffer. They also worry that the gathering of the news will become more centralized and fewer interpretations will surface.

Tension also exists between local stations and networks, mainly because station-network tension of SNG. With the satellite possibilities, the importance of the network is diminished. This could portend major changes in the status of network evening news.

Criticism is bound to be a permanent irritant to broadcast journalism. For one thing, most of the critics are members of the upper-middle class who do not get most of their information from television and are, therefore, not part of the mass audience to which broadcasting is attempting to relate. Much of the criticism of broadcast news is closely tied to news events themselves; however, broadcasters can in no way control world affairs. Perhaps most important, broadcast news is caught squarely between the two purposes of the broadcast industry. The first, as stated in the 1934 Communications Act, is to serve the public interest, convenience, and necessity. The second, as stated by broadcasting's management and stockholders, is to make a profit.[6]

Documentaries

Documentaries are designed to give in-depth coverage of subjects that can be dealt with only superficially in news programs. They require extensive research and expensive production.

types of documentaries

Radio documentaries were not uncommon on early radio and occasionally are produced today by public radio networks or local stations. Local TV stations also produce documentaries of local issues—sometimes run as **minidocs** within the news. They will air three to five minutes about a certain subject each day of the week, and later they may edit the segments into one unified documentary to air as a program by itself.

However, historically the best known and most controversial documentaries are produced by the TV networks. Ed Murrow and Fred Friendly invented the TV news documentary during the early 1950s with "See It Now," which presented bold, strong programs on controversial issues. The series was canceled in 1958 because it lost its sponsor, its production costs increased, CBS was tired of fighting the problems it caused, and Murrow and Friendly

changes in documentaries

got tired of fighting CBS. Since then documentaries have gone through phases of varying emphasis, largely depending on the degree to which the networks are pressured into presenting public interest programming. Sometimes documentaries appear only during the Sunday afternoon "intellectual ghetto" hours, and sometimes they enter prime time. The CBS weekly series "60 Minutes," a compilation of several subjects, has often been tops in the weekly prime-time ratings, but there are those who question whether or not it should be labeled as documentary.

hard and soft

Documentaries are usually divided into **hard** and **soft** on the basis of their subject matter. The hard documentaries, by far the more controversial, are usually the result of investigative research of current topics. Examples include: "The Uncounted Enemy: A Vietnam Deception," a program indicating that the number of enemy troops in Vietnam was purposefully underreported; "The Selling of the Pentagon," a program about the money allegedly spent by the Department of Defense for public relations efforts aimed at selling the American public on defense projects; and "Boys and Girls Together," an exploration of teenage sexuality. Soft documentaries give in-depth information regarding less controversial subject matter, such as "The Louvre," "The White House Tour with Jacqueline Kennedy," and "Lyndon Johnson's

FIGURE 12.3

A scene from the "Frontline" documentary dealing with the Nicaraguan Contras. *(Courtesy of a consortium of public television stations: KCTS Seattle, WGBH Boston, WNET New York, WPBT Miami, and WTVS Detroit/ Gustavo Sagastume, photographer)*

Texas." National Geographic nature specials and some of the scientific programs on the cable TV Discovery Channel could also qualify as soft documentaries.

Documentaries cause innumerable problems for the networks with respect to both expense and content. If a dramatic program set in the 1800s employs words not in the vernacular of the time, it may be criticized, but with none of the severity that occurs when a public official is misquoted in an unfavorable light. Documentaries also have traditionally low ratings and often do not recover their costs. Stations have been required by the fairness doctrine to give varying points of view for all controversial issues; therefore, if some individual or group could prove its point of view was not fairly represented, it could demand free time to present information. Obviously, this adds further cost.

problems created by documentaries

A network frequently finds itself in a difficult situation after airing a controversial documentary. After CBS aired "The Uncounted Enemy," it found itself in a libel case against General William Westmoreland. When "The Selling of the Pentagon" was aired, CBS network executives were called to testify before congressional committees regarding the content of the show. Documentaries have been known to cause internal dissension within the network family. Producers argue with network executives who want to censor material, and stations within the affiliate family sometimes refuse to carry a program because of the subject matter.

Documentaries are subject to some of the same criticisms leveled at news—a liberal bias, an emphasis on the negative, editorializing by exclusion, and unnecessary sensationalizing. In addition, they are criticized for appearing too infrequently on the program schedule. The only regularly scheduled prime-time documentary, "Frontline," appears on PBS. Network

editorial ideas

executives, of course, counter these arguments with dollars and cents and an appeal that perhaps what is needed in this country is less advocacy and less material that will divide the nation into fragmented groups.[7]

Editorials

Radio and TV stations are not mandated to editorialize, and many choose not to. This is due in part to fear of the controversies that may ensue and in part to the difficulty of complying with the present strings attached to editorializing. If a station is going to endorse a political candidate, it must notify all of the candidate's opponents. If it is going to say something negative about a person, it must give the person advance notice and an opportunity to reply. Small stations generally do not have the staff to handle these requirements and sometimes do not even have the staff to write the original editorials.

Ideas for editorials are usually conceived by a station management team and then are presented by a member of top management. Some editorial material comes from networks or syndicators in the form of **commentaries.** Occasionally these consist of a series of programs designed to present a spectrum of opinion on a particular subject, covering all points of view that might be considered in the controversy. Commentaries are personal viewpoints, while editorials express the viewpoint of the station management.

pros and cons

Editorializing is itself a controversial subject. Some critics feel that because the number of frequencies is limited, radio and TV stations should not be allowed to editorialize at all because doing so gives these broadcasters an unfair advantage over others in the community. Other critics feel that editorializing is guaranteed by the first amendment and that, in fact, broadcasters who do not editorialize are shirking their public duty. Methods of editorializing are also debated. If editorials are presented within news programs, they may be mistaken as news, but if they are presented at other times, they can be a jarring interruption to program continuity.

blandness

Editorials that are aired are often criticized for their blandness. The subjects covered—such as public parks, automobile safety, and school crosswalks—are often so noncontroversial as to be hardly worth the status of editorial. Presentations are generally dry and nonvisual and are considered boring if they last very long. As a result, editorials are generally the least glamorous of radio and TV programming elements.[8]

Sports

Sports programming is a hybrid of information and entertainment programs. It involves material broadcast from the scene, generally live and of real events, but most people watch it for entertainment purposes.

sports producers

The most highly publicized of sports programming is America's major sports—football, basketball, baseball—seen on the broadcast and cable networks. However, archery, badminton, caterpillar tractor pulling, and every other conceivable type of sport can be found somewhere on radio or TV. So much sports is consumed by television that independent networks, such as Hughes Sports Network and Television Sports Network, have prospered by selling commercial time nationally and paying independent stations to carry the shows. Throughout the country, local radio, TV, and cable systems have their own sports programs, usually in conjunction with a local college or university. Regional cable TV sports networks that show local teams on a number of cable systems have become particularly popular.[9]

Throughout their short history, sports and broadcasting have had an unusual symbiotic relationship. The Dempsey-Carpentier boxing match gave early radio its first big boost; early TV had its wrestling matches; ESPN was one of the first successful cable networks; and many of the people who have installed backyard satellite dishes are sports junkies who delight in the thirty or more feeds they can watch at any time during the weekend.

interrelationship

As sports proved its value to broadcasting, the **rights fees** television had to pay to air the games skyrocketed. Between 1970 and 1990, National Football League rights climbed from $50 million to $468 million, the National Basketball Association went from $10 million to $132 million, and major league baseball rights rose from $18 million to $365 million.[10] The money was used to pay the ever-increasing sports stars' salaries and to build profit into the whole sports structure.

rights fees

For many years the networks, cablecasters, and stations broadcasting sports continued to profit from them despite the rising rights fees. Advertisers were eager to access the affluent males regularly attracted to sports programming and, on occasion, paid over a million dollars for a single spot in a sports event. By the 1990s, the spiraling balloon was beginning to lose air, however. Networks were sometimes losing money on sports events; they had to give money back to advertisers because they could not deliver the size of audience they promised. As a result, networks were becoming less inclined to pay high rights fees. In fact, in some cases the broadcast networks, which had tried to keep the cable networks out of major sports for years, seemed eager to sell off some of their rights to ESPN. A leveling off of rights fees could pinch the sports organizations somewhat because they have become dependent on television fees in order to remain solvent.[11]

Of course, a setback of this nature would not mean an end to the sports-broadcasting relationship, just a readjustment of terms. It would not be the first time that the relationship had hit rocky times. During the early days of television, sports had been negatively affected by overexposure on TV. One of the first sports to suffer was boxing, which was one of the most popular events on early television, with fights virtually every night of the week. While everyone was watching boxing on TV, no one was supporting club boxing, so about 250 of the 300 small boxing clubs in the United States closed up shop between 1952 and 1959. The result was no fresh talent and a boxing lethargy that was not restored until the coming of Muhammad Ali. Baseball was affected too—while baseball club owners were greedily grabbing every golden nugget TV offered for the rights to their games, attendance at these games fell 32 percent between 1948 and 1953. Similarly, attendance at college football games dropped almost 3 million between 1949 and 1953.

overexposure

Eventually, teams began restricting the number of telecast games, but it was a long hard pull to coax sports fans back into the stadium. Once there, however, the fans seemed to enjoy the media coverage. Sometimes the fact that a game is to be telecast seems to draw fans rather than drive them away.

changing sports

One of the other sore points that arose between the sports and broadcasting worlds involved the degree to which sports were changed to accommodate television. For example, there were two kickoffs during the 1967 Super Bowl because NBC was in the middle of a commercial during the first kickoff. Likewise, the 1978 tennis competition time was changed at the last moment from 1:00 P.M. to 12:30 P.M. in order to accommodate taping requirements of TV. As a result, many paying fans missed several games of the match. These and other similar events caused indignation. But as the years passed, it was increasingly difficult for people in sports to register moral indignation because sports itself became largely show biz. The athletes seemed to adjust their adrenaline levels to fit the needs of the television world. Besides, it seemed a small price to pay when the result was that millions of spectators who otherwise would not have had the opportunity could enjoy the game.

athlete commercials

Another selling-of-the-soul aspect of the sports-broadcasting relationship involves the frequent commercials that feature top athletes. Part of the controversy revolves around whether sports stars should stoop to peddling, but the money earned for such work is certainly tempting. Large sports salaries, plus money from commercials, have turned many athletes into company presidents and entrepreneurs. Another aspect of the controversy revolves around the probable duping of the American public when they see sports idols extol the virtues of a product without, in most instances, any expertise or credentials in the area. In order to limit this phenomenon somewhat, athletes and others engaged in testimonials must now at least use the product they advertise.

Even if athletes do not perform for commercials, TV still tends to require that they be actors as well as athletes. The number of interviews athletes must submit to is enough to make the shiest ones glib, and those who want to keep their private lives private find themselves hounded just like movie stars.

blackouts

Another controversial aspect of sports and broadcasting involves **blackouts.** Sports owners realized that they were cutting their own throats when they televised all their games. As a result, a law was passed by Congress in 1973 that allows football, baseball, basketball, and hockey games to be blacked out (not broadcast) up to ninety miles from the origination point unless the game is sold out seventy-two hours in advance of game time. The biggest critics of this policy are the fans who are deprived of seeing the game on TV.

equipment sophistication

There are occasional outcries from the sports world against TV equipment that is so sophisticated it out-umpires umpires and out-referees referees. On the other hand, the sophistication of modern TV equipment often enhances the viewing with slow motion, split-screens of different parts of the game at the same time, and cameras placed at reverse angles to give a different point of reference to a play. Some stadiums now have large-screen TVs that show parts of the game in the stadium while it is in progress.

FIGURE 12.4
ESPN coverage of a soccer match. *(Courtesy of © ESPN/Thomas F. Maguire, Jr., photographer)*

costs

The sophistication of TV equipment is another factor that adds to the cost of sports. Aside from paying for the rights, broadcast and cable networks must pay production and transmission costs, making sports broadcasting one of the most expensive forms of programming that stations or networks can undertake. Obviously, the sport cannot be brought to the TV station's studio, so the studio must go to the sport. It is not uncommon for a network to gather 20 cameras, 30 microphones, 2 remote trucks, and 100 technicians to cover a sporting event. Even local stations and cable TV systems find it expensive and complicated to cover the local high school football games. The 1984 Summer Olympics utilized 216 cameras, 3,500 people, 26 mobile units, 4 helicopters, and 3 houseboats.[12]

production

The actual broadcast of a game can produce ulcers for those involved. The announcer must attempt to be clever and articulate about plays while he or she listens through a headphone to instructions being barked by the producer and director—"After this play remember to do the promo for next week's game and mention the sponsor's name. Tell the audience that Governor Flupadup is here because we want to get a shot of him"—and trying to comprehend messages being passed under his or her nose—"That was Schlocks's fourth time for hitting three triples in twelve games. Attendance is 27,982. Station break time." The director must choose the best picture from among the twenty or so displayed before him or her, usually with the help of assistant directors watching particular monitors—for example, one assistant director watching only the isolated instant replay cameras and another watching for interesting crowd scenes. Electronic chalkboards, super slo mo, reverse angle cameras, and a host of special effects add to production values.

Whatever the problems are with the sports-electronic media relationship, it is not likely that either party will initiate divorce proceedings. Broadcasters like the audience numbers that accompany sports events. So many sports organizations have built their entire budgets around television, that if television were to withdraw the money, the sports structure would collapse. The fans would not approve of that.

Magazine Shows

A magazine show is really defined more by its form than its content. It consists of short segments on a variety of subjects, similar to a printed magazine. Also, like its printed counterpart, its segments can cover virtually unrelated subjects, or they can deal with material of a common theme.

Magazine shows first came to the fore during the late 1970s when the **prime-time access rule** was instituted. Under this FCC rule, networks were allowed to program only three hours of the 7:00–11:00 P.M. time period, and local affiliates were to program the rest. Although most of the local stations opted for game shows and other inexpensive fare, Westinghouse tried a concept called "P. M. Magazine." The parent company produced a number of segments for the show and distributed them to all its stations. These stations then added a few segments of their own local material. In some instances, the material produced locally was incorporated into the nationally distributed material. Westinghouse first put this program on its stations in 1977, and although it was not an instant hit, the company stayed with the concept and eventually it became very popular.

As often happens with successful shows, imitations began to appear. Other stations began producing magazine shows to fit into their prime-time access periods. Most of these dealt with events in the local community and featured a host and hostess who acted as host/interviewers.

The concept spread beyond prime-time access, and networks and syndicators began distributing magazine shows. These were varied in nature—some serious and some frivolous. Some dealt with a unified theme (e.g., unusual hobbies), and others covered a vast array of subjects.

The word "magazine" has been connected to a number of types of programs because it does deal with form rather than content. For example, "60 Minutes," "20/20," and similar concepts have been labeled as magazine shows, although some would argue that they should be called documentaries. The fact that they contain segments, some of which are rather light-hearted, makes them a hybrid.[13]

Educational Programming

Educational programming is not an area America excels in. Most countries exceed the United States in both the quantity and quality of such programs. In many countries radio and television stations broadcast direct instructional material to schoolchildren as a matter of policy, including programs to

FIGURE 12.5

Shooting a segment of "PM Magazine." This segment dealt with Kathy Swenson, a competitive dog musher who raced in the Yukon Quest. *(Courtesy of KFMB-TV, San Diego/Geary Buyaos, photographer)*

supplement what is occurring in the classroom—such as science experiments or programs about foreign places that students might be studying—as well as course work in which the total instruction is from radio or TV. Although programs of this nature do appear on many American public TV stations and some cable networks, they are generally not utilized to the extent they are in many foreign countries.

 In other countries, television is also used as a primary source of adult education. Some of the underdeveloped countries have full-scale TV courses to reduce illiteracy. In some totalitarian countries, watching certain television courses is one of the prerequisites for promotion. England has an open university that enables people to obtain college degrees through a combination of television and correspondence. In the United States, college credit and adult education courses are programmed in various sections of the country over public and commercial stations and through cable networks, but the extent and use of these is less than that of some other countries.

 Other educational types of American programs are about the activities of the schools and self-help, how-to programs. However, these are generally on stations at the least popular time—6:00 A.M. on weekdays or on Sunday mornings. The reason for this is that educational programming is rarely sponsored—even if it were on at a better time, it would not draw a significant audience. Part of this is due to the subject matter of the programs, and part is due to the fact that these programs are generally low-budget and do not include all the production elements of more expensive entertainment shows.

in-school programming

adult learning

scheduling

News and Information Programming   325

FIGURE 12.6

Galileo, played by
Aaron Fletcher,
contemplates the law
of inertia in this
program for a college
credit science series
called "The
Mechanical Universe."
*(Courtesy of the Southern
California Consortium)*

radio

Very little is done educationally on radio, even though many of the concepts of educational TV programs could be presented quite adequately using audio only.

cable TV

Some original cable TV franchises required that channels be set aside for education, but many of these requirements have been ignored, and the channels in operation are not overly successful.

One thing that has been promoted rather successfully is the preparation of study guides for regular broadcast and cable programs. Teachers can assign viewing as homework and then follow up with educational activities in the classroom.

Probably the most successful instructional programming is produced in the corporate realm. Programs intended to train employees or to enlighten some element of the company about products or services are often well produced and effective.[14]

Religious Programming

For many years, religious programming was another broadcast stepchild usually aired during the early morning or Sunday morning time slots on both commercial radio and TV stations. This form of programming was generally quite inexpensive for the stations because it was supplied, on film or tape, free of charge by the various religious sects. Some stations charged religious groups for airing the programs; others gave free airtime.

FIGURE 12.7
TV minister Jerry
Falwell in the control
room of his nationally
telecast program "The
Old Time Gospel
Hour." *(Courtesy of The
Old Time Gospel Hour/Les
Schofer, photographer)*

The networks paid very little attention to religious programming, although one of the earliest "hits" on network TV was Bishop Fulton J. Sheen, who preached each Tuesday night opposite Milton Berle's "Texaco Star Theater." This caused Sheen to comment that he and Berle worked for the same sponsor—Sky Chief.

early programming

A type of religious programming that had some impact was that broadcast on radio or TV stations dedicated exclusively to religion. At first these were few in number and very local in nature. A great deal of the airtime was spent soliciting contributions from listeners, a procedure that was quite successful.

religious stations

The advent of satellites and cable TV made religious television a much more powerful force. Programs that were local could now be distributed nationwide and receive contributions nationwide. Religious cable networks such as TBN (Trinity Broadcast Network) and PTL (Praise the Lord) were established. The religions that used television the most tended to be evangelistic, leading to the coining of the term **televangelism.** This televangelism became a billion dollar a year business, and ministers such as Jimmy Swaggart, Robert Schuller, Pat Robertson, Jerry Falwell, and Jim and Tammy Bakker became known nationwide and launched political as well as religious movements.

national coverage

A major blow was dealt to TV evangelism in 1987 when Jim and Tammy Bakker of the PTL network were involved in a sex and hush-money scandal. The ensuing investigation revealed a number of financial improprieties of the Bakkers and the PTL, which had been subject to very little auditing as a nonprofit organization.

scandal

Congress investigated the tax-exempt status of all televised evangelism; then matters seemed to simmer down, and the religious forces went about trying to repair their tarnished image.

Religious programming of a more traditional nature still appears in its Sunday morning time slots, but the leaders of these religious groups are generally angered by televangelism, which they feel may be replacing the church.[15]

Public Affairs

types of programs

Public affairs programming is a general term for information programming that does not fall into other categories, primarily consisting of interview, discussion, and on-site programs that deal with issues of concern to the citizenry. Some long-running network programs like "Meet the Press" and "Face the Nation" have featured prominent names in the news being interviewed (or grilled) by top journalists. The network morning shows and Ted Koppel's "Nightline" could also be considered public affairs, although some would put them in the news category.

Most public affairs programs are local productions that deal with community problems. Some local radio and TV stations are very community oriented and program significant, timely, fairly frequent public affairs programs, but most try to skimp on both the quantity and quality of these programs because costs generally cannot be recouped through commercials.

Although stations are criticized by the public for their lack of good public affairs programming, they find that when they do air quality material, very few people watch, making that programming an unattractive buy for advertisers.

Public affairs programs sometimes necessitate that a station give time and facilities to a group whose point of view was not represented when a particular controversial topic was discussed. This, of course, adds to the cost.

On the positive side, there are many instances in which public affairs radio or TV programs have been instrumental in helping correct injustices or problems within the community, thus constituting a worthwhile public service.

A station whose personnel become involved in the community is usually held in higher esteem in that community than a station whose personnel remain aloof. Therefore, the community-involved station may receive overall higher ratings than the aloof competitor and can afford to put money back into public affairs programs that aid the community. In fact, all stations profit from the community they serve and, thereby, have economic justification for providing public affairs programs.

issues

Finding adequate issues and program participants is generally not a problem faced by stations—all communities seem to have adequate spokespeople with opinions to express. Local cable TV operators have begun to capitalize on this phenomenon by encouraging "do-it-yourself" public affairs programming. Members of the community are trained in the rudiments of TV production and are able to use the cable TV facilities and equipment to produce their own public affairs programs, which are then cablecast over access channels.

The overall future of public affairs programming is uncertain. Economic and social fluctuations affect both the content and availability of these programs. When stations are economically healthy, they are more inclined to program public affairs. When social issues are of burning significance, both larger demand and larger audiences surface for public affairs programming.[16]

Politics

Sometimes all of a station manager's other problems seem dwarfed when he or she enters the area of political broadcasting. The biggest problems come during the years when there are major elections. Political candidates then fill the airwaves in their attempts to become elected or reelected. According to **Section 315** of the Communications Act, stations must give equal opportunity (commonly called **equal time**) to all political candidates who are running for the same office. (See chapter 17 for more details on Section 315.) If any candidates feel that they have not been given equal treatment, they can appeal to the FCC, and the station then becomes involved in a hearing that has the potential of being appealed all the way to the U.S. Supreme Court.

Section 315

Issues other than equal time also arise at election time. One is the overriding role of broadcasting in the election process. Television has become the most potent force a candidate has at his or her disposal, especially if the candidate is running for a national office. No whistle-stop campaign can get a face and views in front of as many people as can one television commercial. There are those who complain that for a candidate to win an election, he or she must project as a TV personality and that perhaps the country should not be run only by the glib and the glamorous. Candidates must rely more on their public relations firms and media advisors than on their political views and philosophies.

strong role of television

Broadcasting's coverage of political events is also criticized for giving too much emphasis to frilly baby-kissing events and too little to the issues that divide the candidates. Most of the spots that stations sell to potential office holders are thirty seconds long—hardly time enough to tell how one would run major aspects of the government. There are also complaints that too much money is spent on broadcast advertising, making running for office a game for the rich. Devious methods used for obtaining campaign money have led to some public financing of presidential campaigns and other reforms, but it is still extremely expensive to seek office.

Coverage of political conventions is also a thorny topic because network coverage of the Republican and Democratic nominating conventions has changed the political structure. In some ways this has been an advantage—the delegates now tend more to business than to partying because they know they are under a watchful eye. But because the conventions are televised, they have become show business, with major actions being programmed for prime time and with contrived suspense building so viewers won't flip the dial to a cop show. The actual cost to the networks for covering a convention is approximately $3 million, leading them to pool resources in order to cut costs.

Despite its imperfections, broadcasting brings to the people more information about candidates than they could gain by themselves, while at the same time it simplifies the campaigning process for the candidates.

Coverage of election results, whether they be national, statewide, or local, also provokes controversies. The fact that networks have reported election results in some parts of the country while people are still voting in other parts can distort election results. For example, Democrats in the West charged that the reporting of Reagan's obvious landslides in both the 1980 and the 1984 presidential elections before the polls closed in their time zone meant that many Democrats did not bother to vote, and as a result, state and local Democrats

suffered. Also decried is the fact that networks predict winners early by conducting **exit polls**—asking voters who are leaving the polls how they voted. Broadcasters do not want to stop this; they are trying, instead, to encourage Congress to remedy the time difference situation by instituting uniform poll closings throughout the country.

Between election campaigns, politics is still evident in broadcasting as the media air press conferences and intramural political disputes that may arise. Reporters try not to come too close to the politicians they cover, but there have been complaints that broadcasters do have their political darlings and are in a position to foster the careers of those they like. Likewise, reporters have often been accused of unjustly causing the downfall of politicians through personal revelations that would not affect their ability to govern.[17]

Special Interest Programming

Special interest programming is primarily a product of the cable TV age. When the three networks dominated programming, they geared their material to the masses and had little room for programming that would appeal to only a small

FIGURE 12.9
The 1988 Democratic candidate debates—an example of political broadcasting. *(© Bill Gillette, Gamma Liaison)*

segment of the population. As the number of channels and viewing options increased through the addition of many independent TV stations, as well as through cable and videocassettes, the concept of **narrowcasting** took hold, and programming geared toward special audiences became fairly popular.

changing emphasis

Within cable TV, whole networks are devoted to bringing people financial news, the weather, health information, and travel facts. These channels air up to twenty-four hours a day, and people watch them when they need specialized information.

C-Span is another type of special interest programming that covers the House of Representatives and the Senate. It is watched from time to time by a fairly large audience, but most of the time the material is far from prime-time quality.

C-Span

Ethnic programming is also of a special interest nature featured by several cable TV networks. It is also fairly common on independent TV stations. Many markets have one or more TV stations designated to showing programming, usually in a foreign language, of a dominant ethnic group. Some stations show several different types of ethnic programming at different times of the day.

ethnic programming

The home shopping channels, stations, and programs that have sprung up are another example of a new type of special interest programming. People who like to shop watch for long periods of time; others tune in trying to find a good buy.

home shopping

Some types of special interest programs have been on both radio and television for years, but they are more abundant now. These include shows about gardening, cooking, farming, and consumer education, as well as movie reviews and how-to programs. Some of the how-to programs have been particularly successful on videocassette.[18]

other shows

FIGURE 12.10
Off-monitor shot of
C-SPAN's coverage of
the senate. *(Courtesy of
C-SPAN)*

Overall, radio and TV are dynamic information sources that have captured public attention in a short period of time. Because of their influence, they have great responsibility to provide accurate and thorough information.

Conclusion

Preproducing, producing, and airing information programming can be a complex process. The myriad of sources available for news—wire services, picture services, network reporters, station reporters, data bases, and individual citizens—must all be constantly checked to supply viewers and listeners with up-to-date newscasts on an ongoing basis.

The research that must precede the airing of either a hard or soft documentary so that it is accurate is also enormous. Likewise, conceiving and researching editorials is a time-consuming process.

In sports programming, a crucial preproduction activity is the negotiation of rights with sports organizations. The logistics of transporting equipment and people to the scene of the sporting event involves expense, personnel, and time.

Producing magazine shows involves gathering information on a variety of subjects and then tying it together, usually with some sort of loosely connected theme.

Preplanning and consultation must be a part of educational programming so that the material presented meets the intended educational goals.

Religious programming is generally underwritten and produced by the religion involved and tends to be of the evangelistic nature.

Public affairs programs are generally the cheapest and easiest to produce, but even at that they generally do not pay their way.

Those overseeing political broadcasting must study equal-time regulations carefully and, of course, the complicated logistics of setting up election-night coverage.

Special interest programs vary greatly in their level of complexity and came about largely because of the greater number of channels available to the public.

Although the media are praised for the service they provide with news and information programming, criticism of this type of programming also abounds. News is lambasted for being biased, sensationalized, capsulized, tasteless, and trite. Broadcast news rarely undertakes complex issues and when it does, it seems to fail in making the issues comprehensible. More attention is paid to the looks and demeanor of the newscasters than to the news itself. The media can bring glory to lawbreakers.

Documentaries share many of the criticisms lodged against the news. In addition, networks are criticized for the very limited quantity of documentaries produced and the unfavorable times at which they are usually scheduled.

Many stations that do not editorialize are criticized for shirking their public duty, and those that do editorialize encounter conflicts because their editorials are considered unfair or dull.

The hoopla and high financing that accompany sports events sometimes alter the games. Athletes who become admired and revered are then criticized when they plug commercial products. Blackout rules bother fans but are part of the economic base of the sports-media connection. The sophisticated equipment sometimes bothers the referees but the production resulting from the equipment is enjoyed by the fans.

Magazine shows can be expensive and repetitive. The major criticisms of educational programming is the lack of quantity and, to some extent, the dullness.

Some people question the degree to which televangelism has replaced the church. Scandals have tended to lower the credibility of this type of broadcasting.

Public affairs programs suffer from dullness and poor airtime and can bring conflict and turmoil to stations.

Many of the problems with political broadcasting stem from Section 315 of the Communications Act. In addition, broadcasters are criticized for emphasizing the frills rather than the issues of political campaigns, for reporting election results before the polls are closed throughout the country, and for fostering the careers of politicians they favor.

Special interest programming can be so narrow it does not attract a viable audience.

News and information programming have become very important to the structure of our society. Regulations and customs that change this type of programming usually indicate major changes in society.

Thought Questions

1. Do you feel that broadcast news is biased and/or tasteless? Explain.
2. Should radio and TV stations be mandated to editorialize? Why or why not?
3. Is it proper to rearrange sports events in order to accommodate broadcasting? Explain.
4. Should political candidates be given free advertising time by stations? Why or why not?

Programming Practices

Introduction

The acquiring of programming for the various electronic media has become quite complex during recent years. In the "old days," networks supplied programs to stations. When the stations weren't airing network programs, they broadcast programs produced by themselves or programs they had bought from a third party, usually **syndicators** (companies that sold programming they produced or acquired from a variety of sources).

But now, with the proliferation of media forms and with the tightening of budgets, it is often hard to tell the buyer from the seller. A station looking to buy from a syndicator may also be trying to sell one of its programs to the syndicator. Syndicators have as potential customers a vast array of media—radio stations, TV stations, cable networks, low-power TV, MMDS.

All of those involved with programming have a common goal—attracting an audience. But for some of the players that means attracting as large an audience as possible and for others it means attracting a more specialized audience. These different goals require different programming practices. Companies involved with programming also want to make a profit (or at least not lose money). Often that is difficult to do and leads to some of the complex arrangements that have sprung up in the programming field as everyone tries to think of a better way to make a buck.

Probably the best way to understand the major programming practices is to examine the buyer and seller roles of the main players.

CHAPTER 13

Imitation is the sincerest form of television.

*Fred Allen
performer*

Commercial Radio Networks and Syndicators

networks

Radio networks have changed greatly over the years. During the 1930s and 1940s they supplied most of the popular programming of the time—soap operas, children's programs, drama, and comedy. When this programming moved over to television, the radio networks became basically news services. Gradually radio networks increased their services to include features and total programs, and new networks arose to supply music.

For the most part radio networks create their own programming. Each network has news reporters throughout the world who gather information that the network supplies to the stations. Networks also select and program music that is sent over satellites to stations.

Sometimes stations, especially those owned by the network, supply some of the network programming. A talk show, for example, may be produced by a local station and sent to the network that then sends it out to other stations.[1]

syndicators

Radio syndicators also supply programming to stations in much the same way that networks do. In radio, the line between a syndicator and a network is rather blurry. Syndicators usually supply less programming than networks, often specializing in certain types of material such as commentaries, self-help programs, or jingles. Although syndicators sometimes acquire programming from individuals or small audio production companies, most of them create their own programming. Some of the present radio networks started out as syndicators and then expanded their programming to the extent that they began calling themselves networks.[2]

Radio networks and syndicators are not very noticeable in radio. Most stations want to sound local, so they do not emphasize the fact that they are playing network or syndicated programming. Some stations affiliate with more than one network, so they do not have the network identity that is present in the TV world.[3]

Commercial Radio Stations

formats and features

Almost all radio stations have a particular **format**—a particular type of material that they program. Some of the overall formats used by radio stations include adult contemporary, beautiful music, CHR/Top 40, country, middle-of-the road, news, oldies, religious, and talk.[4] Within the context of the format, stations may program special features such as business reports, comedy, drama, real estate reports, editorials, children's programs, local sports, live concerts, boating reports, farm reports, personality interviews, public affairs, health information, and science updates.[5]

The decisions regarding format and special features will generally be made by a management team. In some small stations this "team" may consist of one person. In larger stations it may include the general manager, program manager, sales manager, business manager, and chief engineer. Inputs from all are necessary to determine whether the costs of the programming decided upon can legitimately be borne by the revenue the station will generate. Some stations, especially those that are doing poorly and want to make a change, bring in consultants to advise them.

Many factors affect these format and feature decisions. One consideration is the programming already available in the station's listening area. For example, an overall format of country-western music may be decided upon because no other station in town offers that type of music. However, if there are only a few country-western music fans in the local area, this format will probably not be adequate for drawing an audience. Therefore, the composition of the listening audience is another important factor. A rural audience will probably be more interested in frequent detailed farm reports than will a city audience. The interests of the community and the interests of the station management can also affect programming. If the town has a popular football team, the station might choose to broadcast football games. If the station manager is particularly interested in boating, he or she might promote boating reports. decision factors

Some stations rely heavily on networks and syndicators to supply format and feature material and have very few local employees. They will have local commercials and often a local person who makes announcements between and within network segments, but they depend primarily on material sent to them on tape or by satellite.

However, most radio stations produce most of their own programming locally. To undertake this, one of the first things the program director does is design an hourly **clock,** showing all the segments that appear within an hour's worth of programming. Generally each time segment within an hour resembles that same time at other hours. For example, if news comes on at 9:15, it will also come on at 10:15, 11:15, 12:15, and so on. In this way, the listener knows what to expect of a station at each segmented time of the hour. the clock

Because most radio stations opt for some type of music format to constitute the bulk of their programming, the program manager must find disc jockeys whose talents fit the station's format. For example, rock 'n' roll music disc jockeys must be capable of fast, lively chatter, and a classical music host must be more subdued and capable of pronouncing foreign works. For an audience call-in show, the on-the-air talent must have good rapport with people. disc jockeys

The program manager must also begin building a station record and CD library. This is usually just a matter of making sure that the station is on the right mailing lists. Because record companies are eager to have their selections aired, the program manager sometimes finds that he or she must fend off salespeople wanting to give records they want the station to plug. The program director and disc jockeys must decide not only what music to play but how often and how long to play it. Most songs wear out in terms of popularity after a few weeks, and someone has to decide when a particular selection should no longer be played. music library

Special features must be handled by the program manager as the need dictates. If local college news is to be aired, a communication system for obtaining the news must be set up. If public affairs discussions are to be held, participants must be contacted and coordinated. If a local fair is to be covered, the details for a remote coverage must be cleared. If an editorial is to be aired, special features

FIGURE 13.1

A typical radio station
clock.

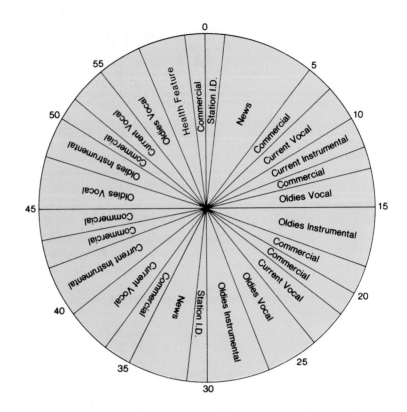

people and organizations with opposing viewpoints must be allowed to present
their points of view. If city council meetings are to be covered, arrangements
must be made with the proper government officials.

news

Most stations program some news, which is usually obtained from a network or read from wire service copy. Stations with all-news formats, of course,
have very elaborate methods for collecting and broadcasting news.

Compared to television programming practices, radio practices are fairly
simple. Stations are in the driver's seat and can choose what elements they
wish to produce locally and what they wish to obtain from networks or syndicators.

Public Radio Networks and Stations

Public radio is somewhat self-contained in that public stations produce their
own programming and receive programming from public radio networks. Occasionally they obtain material from an outside source, especially the BBC,
but for the most part public stations and networks interact with each other.

networks

Most of the larger, high-powered public radio stations affiliate with
National Public Radio and/or American Public Radio. These two networks
distribute mostly cultural and classical programming, and the public radio
stations do not stray very far from that image in their local programming.

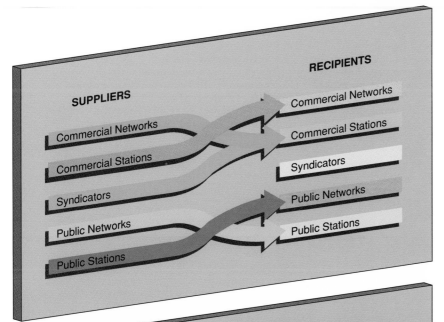

PLATE 9
Suppliers and recipients of radio programming.

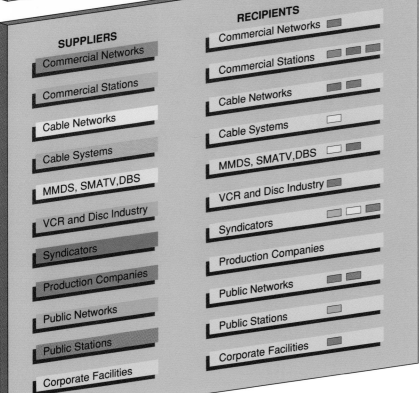

PLATE 10
Suppliers and recipients of television programming. The color blocks in the recipients' column match the color of the suppliers.

Programs, especially long-running ones, often go through an evolution. For example, "Cheers," which started in 1982, has changed several cast members and added others. These changes give room for growth and alterations in the story lines. For the first five years Ted Danson (Sam) owned the Boston bar and the female lead, Shelley Long (Diane), worked as a waitress. When Long decided to leave the show, the story line changed. The main female lead became Kirstie Alley (Rebecca), who bought out Sam, making him the employee. Shown in Plate 11 is the original cast. Top row: John Ratzenberger (Cliff); Middle row: George Wendt (Norm); Rhea Pearlman (Carla); Nicholas Colasanta (Coach); Bottom row: Shelley Long (Diane); and Ted Danson (Sam). Shown in Plate 12 is the current cast. Standing left to right: George Wendt (Norm); Rhea Pearlman (Carla); Woody Harrelson (Woody); Kelsey Grammer (Frasier); Bebe Neuwirth (Lilith); John Ratzenberger (Cliff); Seated: Kirstie Alley (Rebecca); and Ted Danson (Sam).

PLATE 11

PLATE 12

Often successful programs with a multitude of characters spinoff some of the characters into successful shows of their own. Such was the case with "Happy Days," started in 1974, featuring Ron Howard as Richie Cunningham and Henry Winkler as Fonzie (shown left to right in Plate 13). Two of the minor characters of "Happy Days" were Laverne De Fazio and Shirley Feeney. In 1976 they were spunoff into their own show "Laverne and Shirley," starring Penny Marshall and Cindy Williams (shown left to right in Plate 14).

PLATE 13

PLATE 14

PLATE 15

"Star Trek" has had an interesting history. The original NBC network series, produced by Gene Roddenberry in conjunction with Paramount, started airing in 1966. It had a very devoted, but small, audience, and NBC, upset with the low ratings, canceled "Star Trek" in 1969. It lived in reruns for many years and gathered a following of "Trekkies," fans who periodically begged for a new renewal of the series. An animated series and five "Star Trek" movies all did well. Then in 1987 Roddenberry and Paramount began producing "Star Trek: The Next Generation" for the syndication market. Eventually the number of shows produced for syndication surpassed the number produced for the first series. Both series followed the voyages of the Starship Enterprise (pictured in Plate 15), but the casts were entirely different. The mission of the first crew was "to boldly go where no man has gone before," while the mission of the second crew was to go "where no *one* has gone before."

Bob Hope (shown in Plate 16) has had a long-term association with NBC that actually dates back to the "Golden Age of Radio" days. His present contract calls for several specials every season, most of which attain high ratings.

Plates 11, 13, 14, 16 (Everett Collection)
Plates 12, 15 (© Sygma)

PLATE 16

Public radio station managers usually program classical or offbeat music, remotes of local concerts, and public affairs shows that seek out cultural and thought-provoking elements of the community.

The public radio networks differ somewhat in their programming philosophy. National Public Radio produces most of its own programs and emphasizes news and public affairs. American Public Radio redistributes programs produced by public radio stations and concentrates more on music than NPR. Both networks receive compensation from the local affiliated stations.

Some public radio stations do not affiliate with either NPR or APR. Most of them are run by college students who make the same kind of decisions that seasoned professionals in commercial radio make. They decide on formats and features and, although they do not need to concern themselves with commercials, they must keep an eye on finances so that production costs do not exceed the budget.

Public radio has only a small niche in the overall radio programming market, but it is one that usually reaches upscale, influential people.[6]

Commercial Television Networks

Network prime-time TV programming is the most viewed, most controversial, and most maligned of all broadcast programming. Ideas for network shows come from personnel within the networks, production companies, and on rare occasion, from the public at large.

The first material that is usually developed for a prime-time series is a written **treatment**—the general idea for the series and ideas for some of the specific programs. These treatments usually come from one of the many production companies, such as Universal Studios, Carsey-Werner, Paramount, and Aaron Spelling Productions, that regularly submit ideas to networks.

If network executives like the treatment, the production company is invited to submit a **script.** If the script is approved, it is produced into a **pilot**—one program taped or filmed as it might appear in the series. If the network likes the pilot, the series goes on the air. Networks generally look at 2,000 story lines a year and then order 200 scripts, which get whittled down to about 30 pilots. Six or eight of these usually become on-air shows, and, if the network is lucky, one of two will be a hit.

In addition to series, networks obtain movies from production companies. Sometimes these are movies that have already been produced for motion picture theaters and do not need to go through any sort of development process. Other times they are movies made specially for television. For these, production companies will pitch network personnel with a treatment. Accepted treatments will then be written as scripts, and approved scripts will be produced.

Producers and networks battle over the amount the network pays to the producing company in the form of **license fees** to air a series or movie. Originally, the amount paid covered all the costs of the production company and

college radio

series development

movies

license fees

left a small profit. However, as production costs have risen, the amount the network pays often leaves the producer with a hefty deficit. If the series or movie is successful, this money can be recouped through syndication or foreign sales, but often the producing company finds that the more it produces, the deeper in debt it falls.[7]

episodes

Another thorny element that arises with series involves the number of episodes the network will order from the production company. During the 1960s, if a series idea made it through the pilot stage, its producers could usually count on the network ordering about twenty-four programs. But now so many series are canceled shortly after initial airing that networks order far fewer programs, sometimes only three or four. If a series is canceled, this further affects the production company's expenses because it cannot average the costs of sets, costumes, and the like over a large number of programs.

seasons

The concept of a **season** has changed in the network-production company interrelationship. It used to be that new series and rescheduled old series would begin in September and, for the most part, continue until June. Summer would be devoted to **reruns** of some of the season episodes, and then the new season would begin again in September. Now, because so many series are canceled, new series are likely to begin at any time during the year. Some of the networks engage in year-round pilot development instead of having all the pilots produced at one time.[8]

decision factors

As the networks determine what to program, when to program it, and what to cancel, they consider numerous factors. Networks look at ratings of present programs, fads of the time, production costs of the series, the overall program mix the network hopes to attain, ideas that have worked well or poorly in the past, the kind of audience to which the program will appeal, and the type of programs that the other networks may be scheduling.

programming strategies

Often networks attempt **block programming**—the programming of one type of program, such as situation comedies, for an entire evening. This is done in an attempt to ensure **audience flow**—the ability to hold an audience from one program to another. If networks are introducing a new series, they may program it between two successful existing series with the idea that people will stay tuned through the whole block. This is usually referred to as **hammocking.** Another concept, called **tentpoling,** involves scheduling a new or weak series before and after a very successful program. The hope here is that the audience will tune in early and stay tuned after the popular show. Networks often try to **counterprogram** what is on the other networks. If ABC is airing dramas on Wednesday night, CBS may decide to make Wednesday night a comedy night in order to attract a different audience. Counterprogramming is less effective than it used to be because networks must consider not only the other networks but all the programming on cable TV and videocassettes as well. It is impossible to counterprogram all that may be coming into a consumer's home. Networks also engage in **bridging**—programming a show so that it runs over the starting time of a new program on the competing network. The philosophy here is that people will stay with the network show until it is over and then will not want to change to the competing network because they

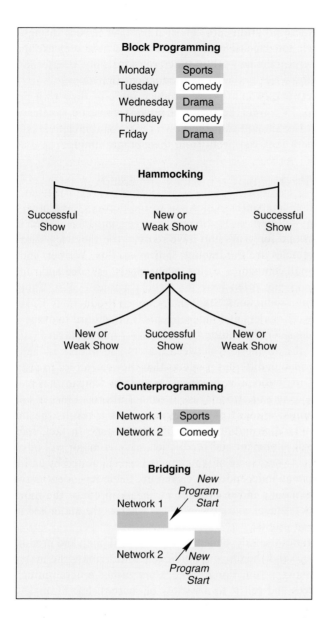

FIGURE 13.2
Diagrams that
illustrate several
programming
strategies.

will have missed the beginning of the show. However, again it is impossible to bridge for all the available channels. In addition many people use their remote control devices to switch back and forth between programs, sometimes watching several at the same time.

Networks program a great deal of material in addition to their prime-time series and movies, such as news programs, public affairs, documentaries, sports, soap operas, and children's programs. The ideas for and production of most of these programs come from the networks themselves, although soaps and children's programs are sometimes supplied by production companies.

other programs

Programming Practices   341

Sometimes these programs are canceled because of poor ratings, altered because they are too expensive, or keyed down because they are too controversial, but generally the non-prime-time programs enjoy greater longevity than the prime-time series provided by production companies.

network production

The networks would like to produce more of their own programming than they do. However, the **financial interest-domestic syndication (fin-syn)** rules restrict the amount that their own programming networks can produce and the amount they can profit from their programming.[9]

Television Production Companies and Syndicators

majors and independents

The production companies, which sell to the networks and others, are organized in many different ways. Some are large companies, called **majors,** that have been around for years and have roots in the theatrical film production business. Examples are Paramount, Universal, Fox, Warner, and Columbia. Others are small companies, called **independents,** founded and run by individuals. The companies often bear the owners' names—Witt/Thomas/Harris, Aaron Spelling Productions, Stephen J. Cannell Productions, Carsey/Werner. The line between majors and independents is often hard to draw. At one time Disney was considered a minor, but now it is a major player.

joint deals

One of the distinctions between majors and minors is that the majors usually have enough different products that they can afford to distribute them themselves. Distribution is very expensive, and if a company has made its name on one series or one or two movies, it cannot afford to hire a salesperson to push that limited amount of programming. As a result, the independents sometimes make distribution deals with the majors. In fact, sometimes the deals involve development and production, also. A major will pitch a number of ideas to a network, some of which have been suggested by an independent. If an idea from an independent is picked up, the independent may rent studio space on the major's movie production lot and produce the movie or series there. Contracts are, of course, worked out wherein the major and independent share costs and profits.

syndicators

Other times the independent companies will pitch and produce the products themselves and then hire a syndicator to distribute the material after it has had its network run. These syndicators gather programming from a variety of sources and sell it to a variety of sources. For example, one of the biggest syndication companies is Viacom, which distributes, among many other things, the "The Cosby Show," produced by Carsey-Werner.[10]

non-network programming

Syndicators and production companies are involved not only with old network shows (often called **off-net**) but also many programs that are never seen on the network. Production companies regularly pitch ideas to cable networks, public broadcasting, corporate facilities, and anyone else who might be able to pay them for their services. Some production companies specialize in types of programming such as sports, game shows, religious programming, or children's fare, but many others try a variety of genres until something hits.

Of course, many run out of money before becoming profitable and fall by the wayside.

Although syndicators sell off-net, they also sell many programs that have never been on the network and are not intended for the network. Examples are information shows such as "Entertainment Tonight" and "A Current Affair," and talk shows such as Oprah Winfrey and Phil Donahue. Syndicators often handle packages of movies that have been produced by small and/or large movie production companies. Nowadays' dramas and comedies are produced for **first-run syndication,** meaning they are not intended to be shown on networks first—they go straight to the syndication market. Magazine shows, sports, reality shows, how-to programs, and special interest programming are all grist for the syndication mill.[11] Sometimes these programs are produced by production companies, and sometimes they are produced by the syndicator itself.

Originally, syndicated programming was sold mainly to television stations. However, with the advent of cable TV and other media forms, the syndication market became much broader. But syndicators still do a great deal of business with local TV stations.

buyers

Commercial Television Stations

Local television stations are usually divided into two groups, **affiliates** and **independents.** Affiliates are those stations that receive programming from networks, and independents are those that are not affiliated with a network but rather acquire all their own programming. The line between these two has become blurred, however, with the advent of Fox. Although stations affiliate with Fox, they still consider themselves independents because Fox does not supply nearly as many hours of programming as do NBC, CBS, and ABC.[12]

affiliates and independents

Independent stations have the greater job in terms of acquiring programming because they have twenty-four hours a day to fill, whereas affiliated stations receive the bulk of their programming from the network.

The programming of some independent stations is somewhat akin to radio station formats—that is, they have one particular type of programming. Stations that air entirely religious material or foreign language programs fall into this category. They receive their material from specialized suppliers. In the case of foreign language programming, they may deal directly with one or more foreign countries. Some low-power TV stations buy programming from the cable networks, such as HBO and ESPN.

specialized programming

But most independents offer a potpourri of programming obtained from various sources. One very common form of programming is old movies, obtained either directly from the production companies such as Paramount and Columbia or through syndicators. By the time most movies get to independents, however, they have had abundant exposure in theaters, on videocassettes, on pay-cable, and on network TV. This is an advantage to independents because they know the track record of a movie and can promote it and sell ads for it based on its past. On the other hand, many people will have seen the film and may not be interested in viewing it again.

movies

When a station acquires these movies, it usually obtains the right to air them as often as it desires for a specific time, such as one year. The station tries to negotiate so that the price it pays for the movies is low enough that it can make money by selling commercial time during the initial run and reruns.

Some film companies offer independents **packages** of films that have not been shown on network TV because networks have been less interested in films in recent years. These films have been shown on pay-TV and, therefore, do not draw well on network TV. When film companies offer these films to independent stations, they try to set up **ad hoc networks**—groups of independent stations that agree to air the movies at approximately the same time so that the films can profit from at least some degree of national promotion.[13]

The second type of syndicated material, old network programming, was for many years the greatest source of program material for independent stations. The programs had been successful, and viewers were familiar with them. Because many people like to see their favorite programs over again, a steady audience could be obtained. The programs were also relatively inexpensive. Stations in different sized markets paid different amounts, but often series could be bought for several thousand dollars per episode.

Syndication companies would offer these off-net shows as they became available to stations in cities throughout the country. All independents would have the opportunity to bid on the series, and the station in each market that bid the highest received exclusive rights to air the series in its market at whatever time of day they wanted. So a station in Detroit might air "Happy Days" at 4:00 P.M., while a station in Milwaukee would air it at 10:00 A.M., and a station in St Louis would air it at 7:00 P.M.

Most independents would **strip** old network programs. Where the network aired a series on a once-a-week basis, the local station offered it on a strip schedule of Monday through Friday in a time period selected according to viewer interest and availability. In other words, a station might air reruns of "Gilligan's Island" five days a week at 4:00 P.M. to attract schoolchildren and might air reruns of "Star Trek" at 8:00 P.M. on Monday through Friday to attract adults who had been working during the day.

Because of this type of scheduling, a large number of programs, usually about one hundred, had to be available in order to make the airing schedule worthwhile. During the first several decades of TV this was no problem because network shows would run for several years. But as networks adopted a policy of canceling shows at a rapid rate, the number of programs produced shrank and, therefore, the number of shows available for syndication dwindled.

As a result, a number of different practices arose regarding the selling of old network shows. First, the prices rose dramatically. Top-rated series could command six figures per episode, and even series low in the ratings could commandeer $50,000 per episode. Independents complained about this, but there was little they could do because the market was responding to supply and demand.[14]

A practice known as **futures** was also developed. Some of the successful programs, such as "Happy Days," "Laverne and Shirley," and "Little House on the Prairie," not only set high prices, but also asked stations to commit to buying the programs long before they were actually available for airing on a rerun strip. In other words, an independent might buy a series in 1977 but not be able to begin showing it until 1980. Independents were not happy with this method of buying because they were not sure the programs would still be applicable several years hence. But if one station in a market would not buy this way, another would in order to try to ensure that it would have the highest ratings in the future.

futures

To sweeten the pie a bit, some production companies began promising independents they would continue producing a particular series to provide approximately one hundred programs even if the series was canceled by the network before it reached that number. The price at which programs were being sold in syndication made this economically feasible.

continued production

Distributors also tried creative packaging. They would acquire several series of similar style or content that had had short network runs and offer them together under a unifying theme, such as family humor or high adventure. The price to independents was reasonable, and the programs received extra, unexpected life.

packaging

For the long-running series, however, syndicators became even more aggressive about setting prices. Sometimes they would offer a particular series in a city and set a floor price of more than $100,000 per episode. They would then tell all interested stations that they could offer whatever they wanted above that price through a closed-bid process. The station submitting the highest bid over and above the floor price would get the series.

price setting

Independents generally felt maligned by these practices, but there was little they could do because off-network programming continued to be their most popular product.[15]

Eventually, they turned more and more to a third source of programming—programs that were specifically made to be syndicated. These were produced with the idea that the main distribution vehicle would be independent stations. Game shows and talk shows had been marketed this way for a number of years, but some independent station owners got together and decided to produce some sitcoms that would be aired on all their stations. The first of these was "Small Wonder," a comedy aired initially in 1984 on independent stations owned by Gannett, Hearst, Metromedia, Taft, and Storer. It was successful enough that quite a few more comedies and dramas were produced specifically for syndication.[16]

made-for-syndication

About this time, the market was growing for syndicated material, so other companies started producing made-for-syndication shows knowing they had not only independents but also cable TV networks as a potential market.

There are other sources of outside programming besides the syndication market from which an independent station can choose. For example, several **independent networks** offer live sports events, such as golf, football, basketball,

independent networks

and soccer, on an as-they-occur basis. Other networks supply specials such as beauty contests.

payment methods

Different methods of payment exist for syndicated programming. Sometimes a station pays a set **cash** amount for a program or series of programs, and then the station sells commercials to fill all the spots. Other times stations engage in a **barter** deal with a syndicator. The syndicator sells and inserts two minutes of ads in a half-hour show, leaves the station four minutes to sell, and the station gets the program for free. In essence, it is bartering its time in exchange for programming. Other deals involve **cash plus barter.** The syndicator sells one minute, leaving five minutes for the station, but the station pays the syndicator a small amount of cash. These types of payment arrangements, called **barter syndication,** are worked out on an individual basis between stations and syndicators.[17]

affiliated stations

Affiliated stations sometimes deal with syndication companies, too. In 1971, when the **prime-time access rule** was instituted to limit the number of evening hours a network could program, a large new market opened up for syndicated programming. Affiliated stations bought this programming to fill the time that networks were no longer allowed to fill.[18] In fact, there have been instances when a program originally aired on CBS has been purchased by an ABC affiliate from a syndicator and run during non-network time.

network-affiliate payment

However, the bulk of the programming aired by an affiliated station comes from the network. For this programming, the network and the station sign a **network-affiliate contract,** which is a rather complicated document. It states that the network will pay the station for the network programs that the station airs. This may at first seem backward because the station is the one receiving the goods. However, the network sells most of the ads connected with the program, receiving its revenue that way. The more stations that air the program, the larger the audience, and the higher the price the network can charge for the ads.

The amount the network pays the station is decided by a complicated formula that considers the amount the station charges for ads, the time of day, the length of the program, and the number of commercials that the network plans to include in the program.

The network pays the station at a percentage of what the station could have earned had it filled the time with local ads. Of course, the network will pay the station a larger total for a one-hour than for a half-hour program. Usually the network will not fill one or two commercial minutes in a program so that stations can sell local ads. The network then pays the station less than it would if all the commercial time had been filled. For example, a network pays a station more for airing a one-hour program with four minutes of network ads on Monday at 8:00 P.M. than it pays the station for airing a one-hour program with ten minutes of ads on Monday at 1:00 P.M.

The whole question of network payments to affiliates has been a sore point. Networks, with their shrinking audiences and shrinking dollars, feel that one way they can develop better bottom lines is to cut back on the amount of money they pay affiliates to run programs. Affiliates, of course, are unhappy with this concept. Discussions are held regularly on this subject.[19]

In addition to payment procedures, there are many other details in the network-affiliate contract. A station can refuse to clear a network program if it feels the program is unsuitable for its audience or if it wishes to air material of greater local importance. Lately station rejection has become more of a problem than it used to be, mainly because stations are unhappy with the quality of the programming put out by the network. They feel they can garner greater audiences and advertisers by programming other material, such as syndicated fare.[20]

program rejection

Other provisions stated in the contract are as follows: The network pays the station less if the program is aired on a delay basis—later than the network suggests it be aired; the network holds liability responsibility for lawsuits that may arise from any network program; and if a station changes ownership, the network has the right to decide if it wishes to offer continued affiliation to the new owners.[21]

other provisions

Some affiliated stations are owned and operated by the network and are often referred to as **O and Os.** The networks not only provide programming to the O and Os, but they also have financial control over the stations and can make decisions concerning the general operations of the station. Sometimes all of the O and Os will exchange programming among themselves. One station will produce the programming and the rest will use it. This is a cross between network and local programming.

O and Os

Both independent and affiliated stations usually program their own local material in addition to the programming received from the network, production companies, and syndicators. Sometimes they sell some of these local programs to syndicators, but mostly they just broadcast them locally.

local programming

A staff of people at the station make decisions regarding the type of local programming to be undertaken. The most common form of programming is local news, in part because it serves the community well, and in part because it is usually a money-maker for the station.

news

Some stations also produce local sports. This is an expensive proposition, but if the sports teams covered are an important part of the community life, this, too, can bring profit to the station. In addition, local stations often undertake public affairs programs, usually in the form of an inexpensive talk format. And, on occasion, a local station will produce a hard-hitting documentary on a local topic or an entertainment program starring local talent. However, as economics become tighter, stations are less likely to produce expensive local programming that does not pay for itself in terms of audience and ads.

other programming

At first glance it appears that the commercial affiliated station program manager has an easier job than his or her counterpart at the independent station. The network provides most entertainment shows and some news and public affairs. This means that the program manager needs to oversee only the local programs. Needing to find programs for a fraction of the program day permits the program manager of an affiliated station to be more selective. The added

affiliated and independent differences

advantage of viewer habit in watching the network station provides more money to bid for and obtain the best available talent in the local market.

Less obvious are the program manager's problems of finding good airtime for the required programs. A network may program every day from 7:00 A.M. until 10:00 P.M., with the exception of several hours in the late afternoon and early evening and some time on the weekend. This means that the program manager's selections must air at times when the audience is minimal or when children constitute the major portion of viewers. Such limitations tend to stifle the program manager's creativity and make the station vulnerable to complaints by local community groups.

Another problem of the affiliated station program manager is the need to constantly review network programs as they relate to the morals and mores of the community the station serves. Programs that seem perfectly innocent in Los Angeles, New York, and other larger markets may offend many people in smaller towns. This can be a costly and time-consuming problem at best and can cost the station its license at worst. In other words, network affiliation is not a cure-all for the program department.

Programming a local station, either independent or affiliated, is a difficult job. The program manager must satisfy both the community and the station owners. The station is licensed to serve, but it also must maintain economic stability.

Public Television

Within the public broadcasting realm, all stations are affiliates because they receive at least some of their programming from the Public Broadcasting Service. PBS stations have both similarities and differences with commercial stations that are affiliated with NBC, CBS, or ABC.

The public station program managers have the same problem of finding adequate airtime for non-PBS programs, but public stations, operating through the **Station Program Cooperative,** have a freer choice concerning which of the network shows they wish to air. This may change, however, as PBS begins to centralize its programming process. Stations may be encouraged, more than they are now, to all air the same programs at the same times.

Because commercials are not involved, the elaborate contracts of who pays whom how much and under what circumstances are not needed. The stations pay PBS for the programs they air, but they do so mostly with money that has come from the government through the Corporation for Public Broadcasting. PBS pays the stations that produce the programs, but it does so with money that comes from other stations. In this way, public broadcasting network money is rather circular and does not involve many outside commercial organizations.

PBS program ideas are station generated to a much greater degree than are those of the commercial networks. The stations propose ideas for the Station Program Cooperative, and the stations oversee the productions. Of course, there are programs that PBS acquires from sources other than local stations, particularly the programs from the BBC.[22]

PBS

station input

348   Programming

Because the livelihood of a PBS show or series is not dependent on rating points translated as advertising dollars, the scheduling process is not as frantic as that found at the commercial networks and stations.

local programming

The responsibilities of the local programming staff vary from station to station. For example, some stations like to emphasize formal education programs, and others prefer general public affairs programming. Public TV station program heads, like their commercial counterparts, are responsible for producing, acquiring, and scheduling whatever programming is needed.

For a typical public TV station, if such a thing exists, the daily programming might be as follows: From 7:00 to 10:00 A.M., the station airs children's programs obtained through PBS; from 10:00 A.M. until noon, it airs programs designed for use in schools as part of the curriculum. (Some of these programs might be produced locally, and others are produced at other stations or by production companies.) From noon to 1:00 P.M., the station programs local news (locally produced); and 1:00 to 3:00 P.M. again would be taken up with classroom instruction. Children are generally out of school by 3:00 P.M., so from then until 6:00 P.M., there is a repeat of the PBS children's programming aired in the morning. The hour from 6:00 to 7:00 P.M. is devoted to college credit courses, and the evening hours until sign-off consist of PBS programming that includes a cooking show, a documentary, a drama, and a symphony.

Cable TV and Other Media

Cable TV networks employ a myriad of programming practices. Pay services, such as HBO and Showtime, buy theatrical movies from production companies almost always before the broadcast networks. They also each make deals with the movie studios to obtain exclusive rights to theatrical films. These rights do not exclude broadcasters or other media; rather, they exclude other pay services. In other words, Home Box Office will make an exclusive deal with Paramount so that Paramount's films do not go to Showtime. Sometimes one of the pay services will even help pay for the theatrical movie in exchange for exclusive rights.

pay services

These same services hire production companies to produce movies, comedies, and other programming specially for them. The process is similar to that of commercial networks. Production companies pitch ideas, and cable network executives decide which of the ideas they want to go into production. There is no season for pay networks, however, so ideas are pitched throughout the year. Sometimes made-for-cable programs are sold into syndication, just like the off-net shows of the commercial networks.[23]

basic services

Some of the basic cable networks, such as ESPN, FNN, C-Span, and CNN, produce almost all of their own material. They send crews to various places around the nation and the world to gather the material that they need, or they produce in their own studios. Other basic networks, such as USA, The Family Channel, and The Nostalgia Channel, produce some of their own programming and acquire the rest, mainly from syndicators. They buy the same network reruns, movie packages, and specialized programs that the local TV stations buy. In fact, broadcasters sometimes fret because syndicators have occasionally offered movies and other programs to basic cable networks before offering them to TV stations.[24]

narrowcasting

Much of cable has adopted the concept of **narrowcasting** as opposed to broadcasting. With narrowcasting, the program producer assumes that only a limited number of people will be interested in the subject matter or program as opposed to the mass appeal sought by network TV. In this way, cable takes on some of the characteristics of the specialized magazines—a channel or program that emphasizes health, movies, sports, culture, or music.

networks and affiliates

Although the words "network" and "affiliates" are used in the cable business, they have a slightly different meaning than they do in broadcasting. Some companies own both satellite services and cable systems, so the organizational structure is similar to the O and Os of broadcasting. For example, Time-Life owns Home Box Office and ATC, a cable company. But the financial relationship does not give the cable network control over the cable system and does not mean that the cable system carries programming only from that cable network. Because cable systems have many channels, they affiliate with many networks. It is not at all uncommon for a ATC cable system to carry Home Box Office, ESPN, CNN, USA, MTV, and a myriad of other services. Cable systems do, however, usually carry those networks in which the parent company has a financial interest.

Since cable is subscriber based, the cable systems pay the cable networks for the programs. Cable systems give a small percentage of the money that they collect from subscribers to the cable networks.

local programming

In addition to carrying the networks, cable systems carry local TV and radio stations and produce some of their own local programming. Many have local origination channels that provide community news and other community programming. Most also have public access channels that allow members of the public to provide programming. Some participate in regional cable networks, usually to obtain local sports. A company will cover sports in a particular area and then sell the service to as many cable systems in the area as it can. One of the newest forms of cable system programming involves the Fox network. Cable systems in areas of the country that cannot receive Fox programming on a local broadcast station are affiliating with Fox and showing its programming on one of the system channels.[25]

other media

Patterns of program practices for other media such as MMDS, SMATV, and the proposed DBS services, vary and are not settled as yet. They try to obtain programming from the cable networks, but often have difficulty doing this because the cable networks do not want to give programming to media forms that will compete with the cable systems. Other than that, these media seem to be lining up at the same door where everyone else lines up—the motion picture production houses.

Hollywood only produces three hundred to four hundred movies a year, and many of these are far from blockbusters. As a result, consumers are often dissatisfied with what they think is inadequate programming on the part of the media. All of them seem to show the same movies—over and over.

Corporate and Personal TV

Almost all corporate TV is produced by the company that uses it. Some general programming, such as health and safety information, is initiated and produced by production companies who then sell it to as many businesses as they can. But most corporate video is so company-specific that preproduced programs do not work.

There are several different ways that corporate programs are conceived and produced. Often a company will have an **in-house** staff of about three people who will have at their disposal a studio and television equipment. They will be responsible for producing the company programs. The in-house staff may initiate ideas, or other people within the company may come to them with ideas. They are responsible for seeing that all the elements needed—performers, equipment, and props—come together at the appropriate time. They then act as crew during the taping, sometimes augmented with people who **freelance.** Some of the programs are taped in a studio, but many are taped at remote locations, particularly around the plant.

in-house production

Another way programs are produced is to have them taped **out-of-house** by a production company. The production company does not come up with the ideas; someone within the corporation does. But the production company executes the idea, and often both in-house and out-of-house production methods are used. Simple productions, which can be handled by a small crew and limited equipment, are done in-house; more complex productions are taped out-of-house. Someone within the company must find the proper production group and oversee the program so that the end result is what the organization wants.[26]

out-of-house production

The consumer aspects of the videocassette and video disc businesses thrive primarily on movies. The movie production companies release their films to the industry fairly soon after they are shown in theaters, copies are made, and the tapes and discs are delivered to stores by distribution companies. Some self-help videocassette tapes, such as exercise programs and hobby-oriented material, have also done quite well. These are usually produced by small production companies and then turned over to distribution companies who see that they get on store shelves. In some cases, small production companies or even individuals have managed to make a small profit by distributing narrowly-focused material themselves. Tapes about such subjects as boating or holistic health can be sold by advertising in magazines that appeal to people interested in those subjects.[27]

videocassettes

The interactive aspects of video discs and computers have tended to emphasize informational subject matter, except, perhaps, for video games. Future directions in this type of programming will be interesting to watch.

interactive

Programming for any of the electronic media is both a challenging and creative activity, be it a broadcast network blockbuster or an experimental home video. This year's hit can be next year's flop, and an idea that has languished for years can suddenly become a success.

Conclusion

Programming practices in the electronic media are quite complex. Buyers become sellers and recipients become suppliers. As belts tighten, every entity tries to squeeze as much as it can from its programming.

One of the key elements in the world of programming practices is syndication. Companies engaged in TV syndication produce material themselves or obtain programming from a variety of sources—movie studios, major production companies, minor production companies, broadcast stations, cable TV pay networks, and cable basic networks. Then they sell the programming to many of the same types of organizations—broadcast stations (especially independents) and cable networks. Syndicators can sell on a barter, cash, or cash plus barter deal. In radio, syndication is a simpler business. Most syndicators produce their own material that they then sell to commercial radio stations.

The production companies, which used to produce primarily for networks and the ensuing syndication, now have many more places to pitch their products. Both basic and pay-cable networks are looking for theatrical movies, original series, and made-for movies, and production companies can produce first-run-syndication materials. Production companies can also produce for public broadcasting and the videocassette market. A number of them produce material for the business world.

Networks are very important suppliers of programming in both the broadcast and cable worlds. Radio stations often receive feeds from several networks, while TV stations affiliate with only one network. These networks are very competitive and use such tactics as block programming, tentpoling, hammocking, counterprogramming, and bridging to try to be number one. Cable systems affiliate with many networks so that they can fill all their channels.

Public broadcasting is somewhat of a closed circuit, with stations providing many of the programs that go on networks and networks providing most of the programming that goes on stations. They deal more with foreign programming than commercial stations do, although some independent stations program only foreign-language programming.

Most of the electronic media produce material for themselves. Radio networks produce news and music programs. Commercial radio stations have disc jockey shows that follow a format, as well as local news and special features. NPR produces most of its own material from Washington. Public radio stations air local music and features. Commercial networks produce their own news and sports and sometimes do entertainment programming such as soap operas and children's programs. Local news is a big element for broadcast stations, as are sports and public affairs. Public TV stations produce local educational and public affairs programs. Some cable TV networks, such as ESPN and CNN, produce almost all their own programming. Cable systems have public access and local origination channels for locally produced material. Corporate video consists primarily of material of interest only to one company.

Programming is a difficult job—one with a high burnout factor.

Thought Questions

1. What type of local programming could radio and TV stations in your area offer that they are not offering now?
2. Should TV networks be allowed to program as much of an affiliate's day as they do? Why or why not?
3. Should there be a requirement that all network series run at least thirteen weeks before they can be canceled? Give reasons for your answer.
4. Should commercial networks be allowed to produce more of their own programming? Why or why not?

Physical Characteristics

Telecommunications is totally dependent upon electronics and mechanics. Eliminate the electron and the total of telecommunications would disappear. For better or for worse, complicated equipment is needed in order to produce and distribute programming. When this equipment functions properly, it can be a great aid to the creative process. When it becomes cantankerous, it can damage or even destroy concepts, careers, and temperaments. Ingenuity and curiosity keep the electronic concepts in a constant state of flux. This leads to a spiral of improvements spiced with problems.

PART 5

Production Equipment

Introduction

Whether it be the coverage of the Olympics or a public access show about mind reading, production involves people using equipment in creative ways to produce programs. Throughout the years this equipment has changed, mainly in ways that make it smaller and more flexible. However, the basic elements of production have remained essentially the same: A program must have an interesting, producible idea that interests at least a small segment of the public; the images and/or sounds must be captured in a way that is discernable so that they can be distributed and exhibited. Even though this chapter is not intended to teach production techniques, it will provide the reader with a general understanding of much of the equipment needed for the production process.

Television is a triumph of equipment over people.

Fred Allen
comedian

Cameras

The camera is at the heart of television production. Sometimes cameras are used in TV studios, and sometimes they are used for **remotes**—productions away from the studio. At some facilities, such as industrial companies, the camera that is used in the studio one day may be taken into the field for a remote the next. At others, such as major networks, heavy-duty cameras remain

in the studio and portable ones are sent out for remotes. Cameras can be tied together so that three or four of them are shooting different angles of the same scene at the same time, and the director can select, on the fly, which shot to use. This is usually the way talk shows, game shows, and newscasts are shot in the studio and the way sports is shot in the field. Cameras can also be used by themselves in single-camera shooting style. In these cases, the footage shot is usually edited at a later time. This is common for on-the-scene news stories, documentaries, and dramas. (Often dramas are shot with film cameras rather than video cameras, but the single-camera style is used.)

Cameras consist of a viewfinder, a lens system, and an electronic system.

The **viewfinder** is simply a small TV monitor mounted on the camera to allow the camera operator to see the picture.

The **lens** system is similar to that of a photographic camera. With it, the camera operator selects and focuses the image and varies the light that strikes the electronics within the television camera. Selecting can be done by pointing the lens and zooming in or out until the proper amount of picture appears in the frame. Focusing is accomplished by adjusting a focus ring until the picture is sharp and clear. Light adjustment is accomplished by opening and closing an **iris** in the lens. If there is scant light in the room, the iris must be opened wide in order to take advantage of all available light; if there is profuse light, the iris can be partially closed.

The camera itself contains the electronic components that convert the image focused by the lens into electrical impulses. Most cameras manufac-

tured today do this by utilizing **charge-coupled devices (CCDs).** The CCD is a solid-state, light-sensitive chip that consists of a large number of individual silicon-sensing devices called **pixels.** The image from the lens is focused on the CCD, and each pixel builds up a charge containing information about the particular part of the image that hits it. This information relates to the color and brightness characteristics of the scene being photographed. Each line of pixel information is then read by a synchronized clocking signal, and this information is transferred momentarily to a storage area that is similar to the imaging area. The pixels in the original imaging area then pick up new information from the scene. The information that is being stored is sent from the camera in the form of electrons, representing the color and brightness characteristics. The more pixels a CCD camera has, the better the quality of the image it can produce. Cameras made for broadcast programming have more pixels than the less expensive consumer cameras, but even consumer cameras brag of more than 300,000 effective pixels.

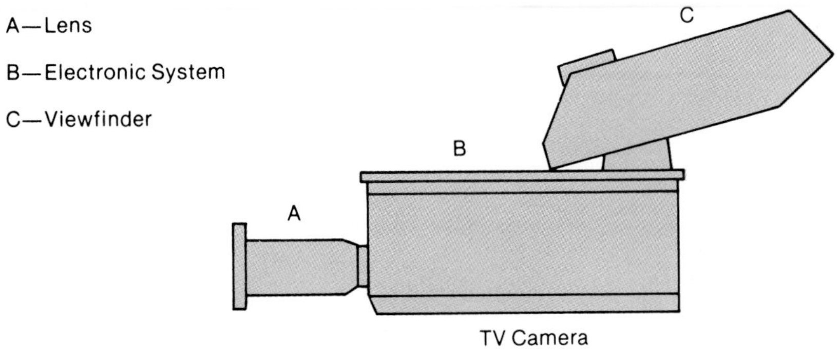

A—Lens

B—Electronic System

C—Viewfinder

TV Camera

FIGURE 14.1

Parts of a television camera.

FIGURE 14.2

A CCD. *(Courtesy of Sony Corporation)*

The other type of camera still in use today is the tube camera, which consists of an **electron gun,** a **target,** and a **photosensitive surface.** In this type of camera, the picture to be televised goes through the lens to the photosensitive layer of chemicals arranged in microscopic, but individual, dots. Each dot emits an electrical charge that varies in intensity depending on the amount of light hitting it. For example, the dots of the photosensitive layer reacting to a white shirt would give off a stronger electrical charge than the dots reacting to dark hair.

Electrical charges emitted from the back of this photosensitive plate hit a target and reproduce an electronic equivalent of the image. Each little speck of information on the target is then picked up individually by an electron gun mounted at the back of the tube. This gun emits a steady stream of electrons that scan the target from top to bottom.

In both the CCD camera and the tube camera, a new configuration of electrons is generated each one-thirtieth of a second. In the CCD camera, these electrons wind up in the memory unit; in the tube camera, they wind up in the electron stream generated by the electron gun. From there they are transmitted by wires to videotape recorders or other electronic components.

tube cameras

Production Equipment   359

FIGURE 14.3
Television camera tube.

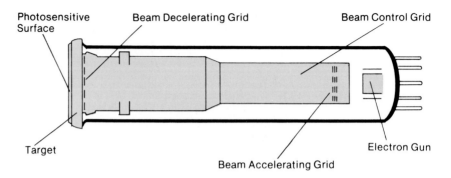

If the camera is black-and-white, it has only one chip or tube. If the camera is color, it usually has three chips or tubes working simultaneously; one for greens, one for blues, and one for reds.

Because CCD cameras use solid-state devices rather than bulky tubes, they can be smaller and lighter. They also operate better in low-light and bright-light conditions. Eventually, CCD cameras will probably totally replace tube cameras.[1]

In the United States all cameras, be they chip or tube, generate 525 lines of picture information each one-thirtieth of a second, referred to as a **frame.** So a new frame consisting of 525 lines of information from the top of the frame to the bottom is generated each one-thirtieth of a second. This information is never in the form of a total picture; however, it is a constantly moving dot. Because the human eye can retain an image, we perceive total images rather than the moving dot.

Camera technology improves almost daily, making for cameras that produce sharper, clearer pictures while becoming more lightweight and portable. One improvement that has been developed and tried but is not generally available involves **high-definition TV (HDTV).** A number of different systems have been developed, but they all essentially double the number of lines seen on the TV set. The main advantage of higher definition is that the picture has much better resolution, which makes it look very similar to 35mm film. The systems that have been developed also change the **aspect ratio** of the TV picture from four units wide by three units long to a five-by-three aspect ratio.[2] This more closely approximates the wide-screen aspect ratio used for theatrical films, so it would improve the transfer possibilities between film and tape. Several movies and documentaries have been produced with HDTV equipment and then transferred to film or to regular 525-line TV, but the technology is not widely used for production as yet.[3]

The camera unit (consisting of viewfinder, lens, and electronics) must be mounted on something so that it can be moved about. When cameras are taken to remote locations, this "something" is often a person's shoulder. Sometimes the person wears a **Steadi-Cam,** a harness device that attaches around the waist and keeps the camera from jiggling. Some of the newest consumer cameras are so small they can be held with one hand. However, for a more

frames

HDTV

mounts

FIGURE 14.4
A Cinema Products
Steadicam EFP® used
to minimize camera
movement. *(© Joel Lipton,
Courtesy of Cinema
Products)*

FIGURE 14.5
A camera mounted on
a tripod and fluid head.
*(Courtesy of O'Conner
Engineering Labs.)*

steady picture, the camera can be mounted on a **tripod,** a three-legged stand that usually has a movable head on top of it so that the camera can easily be moved from side to side and up and down.

When a camera is in a studio, it is also likely to be mounted on a tripod, but the tripod is usually placed on wheels, called a **dolly,** so that it can be moved around the studio. Another type of studio mounting device is a **pedestal,** a large tube used to move the camera up and down by a counterweight, hydraulic, or air-pressure system. When shots need to be taken from overhead,

tripods

pedestals

FIGURE 14.6

A pedestal. *(Courtesy of Quick-Set, Inc.)*

FIGURE 14.7

A crane. *(Courtesy of Chapman Studio Equipment)*

cranes

a **crane** is used. This is a large machine that can raise the camera to a high level. Usually one or more people are needed to operate a camera and its mounting device, but sometimes, for studio shoots, the cameras are **robotic**— they rove, unmanned, in the studio, responding to commands from the director in the control room.[4]

Occasionally unique devices are invented for holding cameras. One such device is the Skycam, which enables a camera to be suspended above a site such as a football stadium. It includes cables attached to rooftops and winches to move the camera holder about.[5]

Video Recorders

Videotape recorders are used for a myriad of purposes: taping studio productions, rolling pretaped on-the-scene news stories into newscasts, recording a sports event so part of it can be excerpted later, editing program material, airing programs and commercials, accepting a network feed for delayed broadcast, recording material at a remote location.

Sometimes the recorders are stand-alone units, and sometimes they are one part of a **camcorder**—a unit that incorporates both the camera and the recorder. The camcorders are most likely to be used in the field, and the stand-alone units are most likely to be used in the studio.

camcorders

The first videotape recorders marketed during the 1950s used two-inch tape. These recorders were modeled after audiotape recorders in that they placed information on the tape by rearranging iron particles to correspond to the various patterns of light and dark given off by the electrons coming from the camera. However, much more needed to be recorded for a video signal than for an audio signal—nearly one hundred times more.

two-inch tape recorders

In the past, audio recorded information was placed on the tape in a straight-line horizontal fashion by a stationary head. If video had been recorded the same way, the tape would have had to move at fifty-five feet per second, or nearly thirty-six miles per hour in order to place the information on properly. A one-hour videotaped program would have required approximately 198,000 feet of tape. Obviously, sending tape past stationary heads was not the way to go.

The first method engineers designed for videotape recording involved rotating heads that put the information on the two-inch tape vertically instead of horizontally. This was referred to as **transverse quadraplex.** These recorders were **open reel,** so that tape had to be threaded from one reel to another.

Later, engineers developed a method of putting information on diagonally; this is the method used for modern-day tape recorders and is referred to as **helical** recording. Now most tape recorders also use tape **cassettes** instead of open reels.

helical

A wide variety of formats exist for helical videocassette recorders. Some, primarily **Type C,** use one-inch tape while **U-matic** uses three-fourths-inch tape. Two half-inch formats have been in competition for the professional market; they are **Betacam** from Sony and **M-format** from Matsushita. These two are not compatible with each other—that is, a tape recorded with a Betacam cannot be played back on an M-format recorder even though they are both half-inch. Both of the companies have come out with improved versions of their professional half-inch formats, called **Betacam SP** and **M-II.** Two half-inch formats were also developed for the consumer market, **VHS** from Matsushita and **Betamax** from Sony. These also were not compatible with each other. VHS won over Betamax, and now an improved version of VHS, **Super-VHS** exists. Its proponents say it is good enough for broadcast use as well as consumer use. Sony has introduced two 8mm (approximately one-fourth-inch) formats, **Video-8,** and an improved **Hi-Band 8** that it says is good enough for all users. All of this has led to a dizzying array of choices for broadcasters, cablecasters, corporate production facilities, and the home consumer.

Although different machines have different features, all of them have the standard stop, record, fast forward, and rewind buttons. In addition, most of them can pause on a particular frame, record audio and video together or

functions

FIGURE 14.8
A Sony videotape
recorder using three-
quarter-inch tape.
*(Copyrighted by Sony
Corporation)*

FIGURE 14.9

A VHS videocassette
recorder. *(Courtesy of
Panasonic)*

FIGURE 14.10

A video-8 videocassette
recorder. *(Courtesy of
Sony Corporation)*

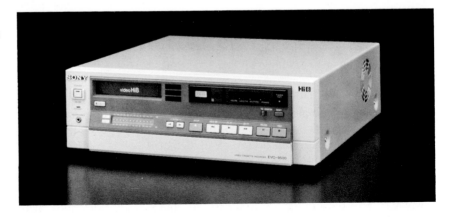

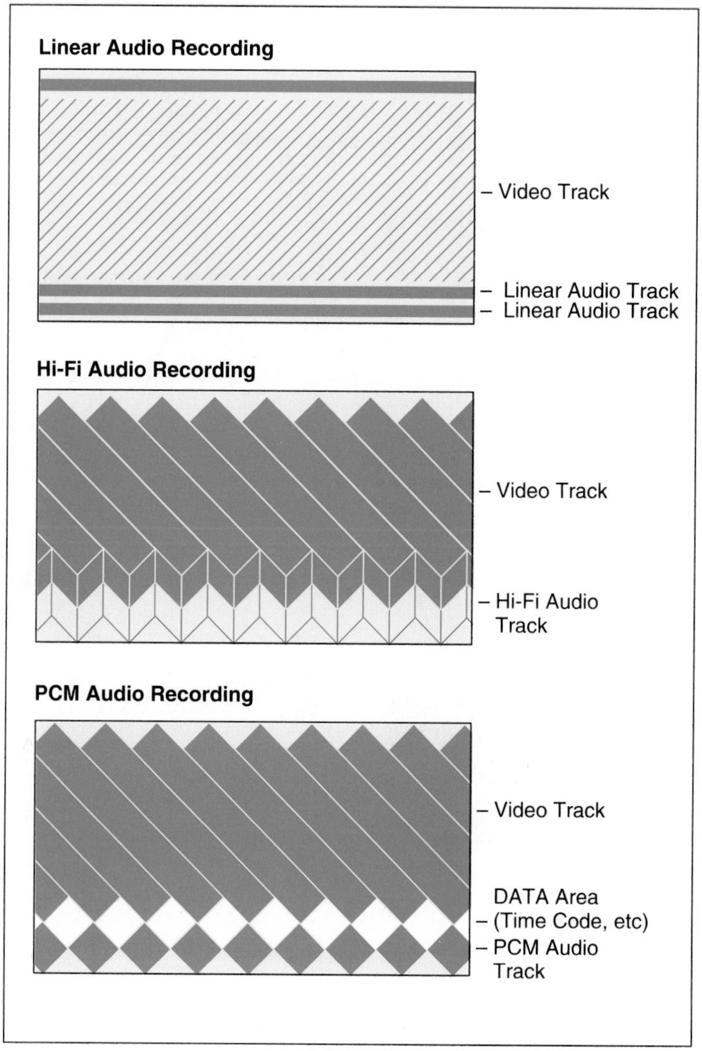

FIGURE 14.11
Linear hi-fi, and PCM
audio recording.

Linear Audio Recording

– Video Track

– Linear Audio Track
– Linear Audio Track

Hi-Fi Audio Recording

– Video Track

– Hi-Fi Audio
  Track

PCM Audio Recording

– Video Track

DATA Area
– (Time Code, etc)
– PCM Audio
  Track

separately, and play back viewable images in a fast mode. Most also have meters on them so that the operator can make sure all the signals are being recorded properly.[6]

The way that audio is recorded on videotape has also changed over the years. At first audio was recorded horizontally, even though the video was being recorded by four rotating heads or by one or more heads in the helical configuration. This form of audio recording was usually referred to as **linear** and was **monaural.** Some of the newer recorders place the sound as well as the picture on the tape helically and in the linear fashion. This helical recording is well-suited to **stereo.** Two forms of helical audio recording are available: **hi-fi,** which records the audio in the same area as the video, and **pulse code modulation (PCM),** which records the sound on its own section of the tape. A disadvantage of hi-fi audio is that it cannot be separated from the video for editing purposes.[7]

audio placement

Production Equipment

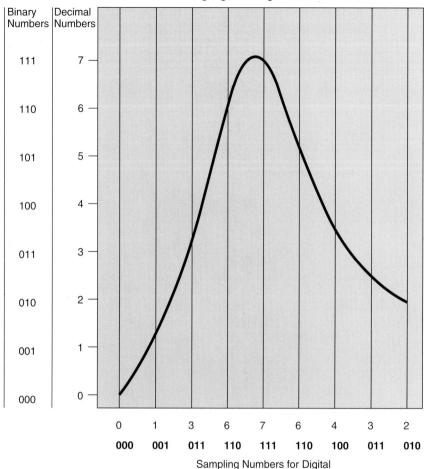

FIGURE 14.12

Analog and digital
representations.

Analog Signal Using Decimal Numbers

Binary Numbers	Decimal Numbers
111	7
110	6
101	5
100	4
011	3
010	2
001	1
000	0

0	1	3	6	7	6	4	3	2
000	**001**	**011**	**110**	**111**	**110**	**100**	**011**	**010**

Sampling Numbers for Digital

VCRs, like cameras, continue to improve almost daily. They, too, become smaller and lighter and record more information with the introduction of each new model.

digital

The newest advancement in videotape recording is **digital.** Until recently all video recording was **analog.** Analog recording involves a continuous method of signal recognition, whereas digital recording responds to discreet on and off impulses. Analog is similar to creating a line graph for a statistical analysis. All measurements are on a continuum, and the line curves up and down. Digital sound is similar to a person looking at individual numbers and writing them down in a set order. It is, in essence, many samples of the analog signal.

When the analog method is used to take an electronic signal from one place to another (e.g., one tape recorder to another), the signal changes shape and degrades the quality slightly. This is similar to someone trying to trace a curve on a graph. The reproduction will not be exactly as the original. When the digital method is used to transport the signal, this signal does not waiver

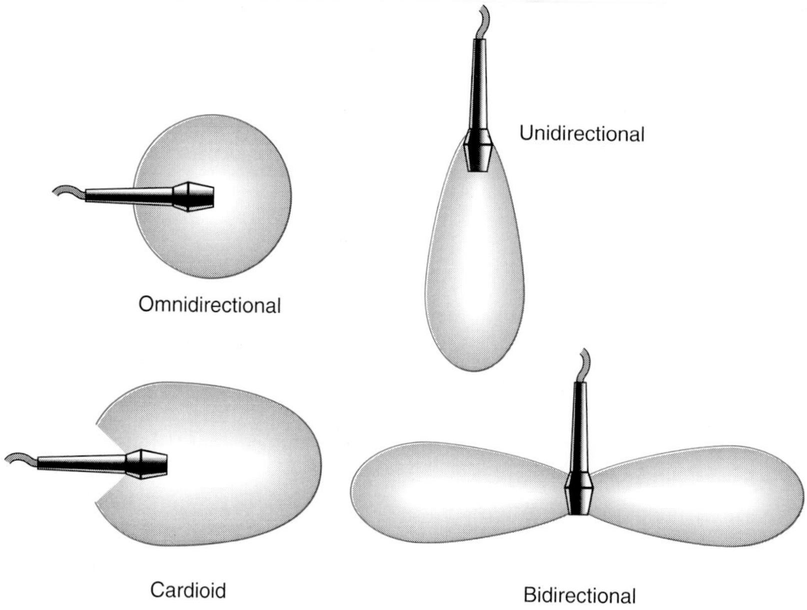

FIGURE 14.13
Microphone pickup
patterns.

Unidirectional

Omnidirectional

Cardioid

Bidirectional

or change. As it travels from one source to another, separate bits of information transmit with little loss of quality. This would compare to someone copying the numbers of the statistical study. The numbers would be copied accurately, so the results would not be distorted.

Digital technology is just coming to video, but it has been used for audio for quite awhile. In fact, the early name for digital recording was pulse code modulation, and the PCM method of recording audio on videotape is digital.[8]

Microphones

Microphones come in a vast array of configurations, but all are devices for converting sound into electrical impulses so that it can be sent through wires as variations in voltage. Similar ones are used for radio and TV and for studio, field, professional, industrial, and consumer situations.

Different mikes are constructed to convert sound in different ways. Some methods maximize fidelity at the expense of microphone ruggedness; others sacrifice fidelity for low cost; and still others emphasize the shape and ease of movement, but sacrifice frequency response.

construction

Microphones are also designed to pick up sound in varying ways. The four most common are: **unidirectional,** which picks up sound mainly from one side; **bidirectional,** which picks up sound mainly from two sides; **cardioid,** which picks up sound in a heart-shaped pattern; and **omnidirectional,** which picks up sound from all directions.

directionality

Each directional aspect has its own special uses. When only one person's voice is to be heard, such as that of a newscaster or sportscaster, a unidirectional mike is the best choice. It will pick up the person's voice sitting in front of it and minimize distracting noises in the background. Bidirectional mikes

FIGURE 14.14
A lavaliere
microphone. *(Courtesy of Beyer Dynamics)*

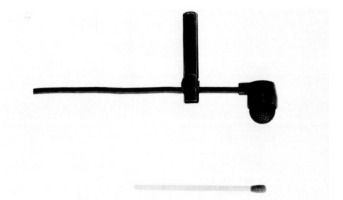

had their heyday during the era of radio drama—they enabled actors to face each other to deliver their lines. Cardioid mikes are very satisfactory for two people seated next to each other and are frequently used on TV talk shows. Omnidirectional mikes pick up overall crowd noises well and are often used for plays with large casts. When sound is to be recorded in stereo, two microphones that have several elements within one housing can be used.

positioning

Microphones can also be positioned in a number of different ways. When a mike is used for a news interview, it is usually hand held by the interviewer. Small mikes called **lavalieres** are attached to a person's clothing in such a way that they do not show. On men they look like a tie clasp and on women they can look like a pin. These mikes are very common for talk shows. Often a mike is suspended from a **boom,** which is a pole attached to a movable stand. A boom mike can follow the talent as they move about a set and are often used for dramas to comedies. Mikes on **table stands** can be positioned directly in front of a sitting person such as a disc jockey or sportscaster. If a person needs to move around through a complicated area, a **wireless mike** is often used. This mike has a small transmitter in it that sends the signal to a receiver that can be connected to other equipment such as a tape recorder. Wireless mikes are often used for singer/dancers or for narrators moving through a large area such as a museum. When sounds must be picked up from a great distance, **shotgun** mikes are used. These mikes have very narrow directionality (usually supercardioid) and are sometimes used for documentaries where it is impossible to get close to the subject. For example, someone trying to record the roar of a lion in the wilds of Africa would be best advised to use a shotgun mike.

The type of microphone to use, in terms of construction, directionality, and positioning, depends on the purpose and need for which it is intended.[9]

Lights

Lights are not needed for every television production. In some cases, such as news standuppers, the light provided by the sun is sufficient. However, anything shot in a studio needs lights, and outdoor scenes often need lights, too,

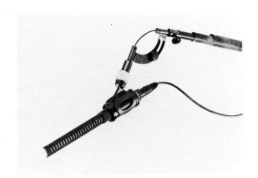

FIGURE 14.15

A microphone mounted on the end of a boom. *(Courtesy of Beyer Dynamics)*

FIGURE 14.16

A desk-top microphone on a stand. *(Courtesy of Shure)*

FIGURE 14.17

A wireless microphone. *(Courtesy of Shure)*

FIGURE 14.18

The case, windscreen, microphone, and shock mount of a shotgun mike. *(Courtesy of Shure)*

FIGURE 14.19
A quartz light. *(Courtesy of Mole Richardson)*

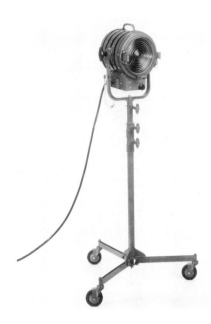

because the sun casts too harsh a light that needs to be softened by artificial lights.

tungsten lights

The most common type of lights used indoors are **tungsten-halogen lamps.** These are small but powerful bulbs made of quartz glass that is filled with halogen gas and a tungsten filament. They give off a somewhat orangish light that is different from sunlight, which has more of a bluish tinge. As a result, something that is shot indoors with a TV camera will have more of an orangish tinge than something shot outside. For this reason, cameras have orange filters built into them that need to be in place whenever footage is shot outdoors.

HMI lights

If lights are needed outdoors, a special type of light called an **HMI light** (hydrogen medium-arc-length iodide) is used. These give off the same kind of bluish light as sunlight. Usually they are used when the sun is so bright that it is creating harsh shadows. The HMI lights are positioned so that they hit the subject from the opposite side as the sun, making the shadows less harsh.

key, fill, back

This is the basis behind most lighting, whether it be in a studio, in someone's office, or outdoors. One bright tungsten-halogen light (or the sun) serves as a **key light.** It is usually placed to the front of the subject at a forty-five degree angle and provides primary illumination. Another light (HMI for outdoor or tungsten-halogen for indoor) is placed in front of the subject forty-five degrees in the other direction. This **fill light** softens the shadows created by the key light. A third light, called a **back light,** is placed behind the subject to separate it from the background, giving a three-dimensional effect. Although key, fill, and back lights are considered the basic lights for most sets, there are many variations of lighting that alter this pattern in order to create special dramatic effects. And there are other lights, such as lights that illuminate the background, that are added to the basic pattern.

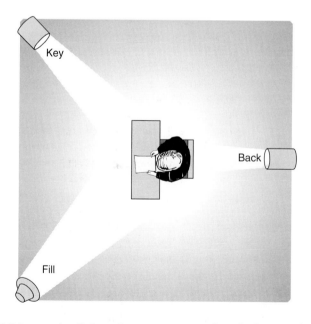

Key

Back

Fill

FIGURE 14.21
Basic lighting setup.

In addition to the lights, there are many other devices used to achieve appropriate lighting. Sometimes a lamp will be directed into a reflective **umbrella** that is used to bounce the light back onto a scene. This bounced light is much softer than the light that comes directly from the lamp. Metal flaps called **barndoors** are used to direct light or keep it off of some surface. **Reflectors,** which at their simplest can be large sheets of white cardboard, are often used outdoors instead of HMI lights to capture rays from the sun and bounce them back toward the subject.[10]

other devices

FIGURE 14.22

A barndoor. *(Courtesy of Mole Richardson)*

FIGURE 14.23

Reflectors being used for a scene on the beach.

In a studio, lights are usually suspended from a **grid,** high and out of the way of moving cameras. In the field, lights are usually mounted on stands, although special equipment can be used to hang them from doorways, bookshelves, or other places. Sometimes for news coverage, one light mounted on the camera is all that is used for illumination. This casts a harsh unflattering light, but it is quick to set up and easy to operate, both prerequisites for covering fast-breaking stories.

Newer television cameras do not need as many lights as older cameras in order to produce a discernible picture, so often people working at the consumer level do not use any lights at all. At the professional and corporate levels, however, the art of lighting is being given more emphasis than ever

FIGURE 14.24
A Sony editing control
unit. *(Courtesy of Sony
Corporation)*

before because cameras are more capable of responding to subtle lighting variations than they have been in the past.

Editing Equipment

Editing equipment is almost always needed when material is shot in the field with only one camera. Unless the event being taped is very straightforward, the camera operator will shoot a lot of unnecessary material and will not be able to tape shots in the order that they are to appear in the final product. This is true for everything ranging from the family vacation to a network movie of the week. However, the sophistication of the editing equipment used to assemble program material varies greatly depending on the intended final use.

The basics needed for editing include two VCRs, an edit controller, and monitors. One VCR, called the **source deck,** is used to play back material and the other, called the **edit deck,** is used to record the material from the source deck that is desired in the final product. The **controller** is used to mark edit points so that the right material is transferred from the source deck to the edit deck. The **monitors** are used to view the outputs of both decks.

basic equipment

At the consumer level, an editing system can consist of a camcorder as a source machine, a regular home VCR or another camcorder as the edit deck, and the family TV set as the monitor. The editing controller is often part of the VCR or its remote control. Pictures from both the source and edit decks appear on the TV set at the same time. Editing can be a simple process that involves setting up the desired pictures on the source deck so that they will record at the desired spot on the edit deck. Once something is transferred from the source to edit deck, it cannot be easily changed.[11]

consumer editing

At the professional level, greater speed and adaptability is needed, so more elaborate editing systems exist. Often these are divided into **off-line** and **on-line** systems. After the original material is shot, it is dubbed onto another tape called a **workprint.** (Sometimes the original material is shot on film rather than tape. In these instances the film is transferred to tape using a **flying spot scanner,** a machine that can handle film very gently. This tape then becomes the master and a workprint is made from it.) The workprint is used to make the editing decisions, saving the original tape from all the wear and tear involved with shuttling back and forth trying to decide where the proper edit

off-line

FIGURE 14.25

An offline editing
system that uses
multiple videotapes.
(Courtesy of Ediflex)

points should be. Sometimes multiple workprints are made so that they can be placed in off-line equipment that has multiple source decks. The person editing decides which shots he or she wants to preview, and the edit controller finds the source machine that is cued up closest to that material and displays it. In this way, the editor can see a whole edited production without actually recording anything. The outputs of the different cassette machines are seen one after another and the person editing can easily change shots as desired. A computer keeps track of all the editing decisions on a floppy disc.

on-line

Once all the decisions have been made on the off-line system, the computer disc is transferred to an on-line system. This uses the original tape material to make the final product according to the decisions made on the off-line system. In this way the original material needs to be shuttled very little.

time code

The computer can be used to keep track of edit points only if all the footage (both the original and the workprint) has **time code** on it. Time code gives a specific address to each frame. For example, a time code number of 01:23:15:02 would indicate that the material is one hour, twenty-three minutes, fifteen seconds, and two frames into a tape. Each frame has a time code number that always stays with it. This is how the decisions made with the off-line system can be transferred to the on-line system. The time code numbers of all the edit points decided upon while using the workprints are stored on the computer disc. When this disc is transferred to the on-line system, the time code numbers can be used to find the edit points on the master tape.

discs

Some of the professional off-line systems use optical discs rather than videocassettes. The discs can access the information quickly because a laser beam moves quickly to the proper place; nothing needs to rewind as in the case of videocassettes. This means less discs are needed for the decision-making process, but transferring the original to discs is more expensive than transferring to tapes.[12]

other editing systems

A wide variety of editing systems exist that are more complex than the consumer products and less complex than the professional off-line, on-line systems. These are often used for news, industrial, and educational productions. Some of these have controllers capable of storing five or six edit points that can be viewed before they are actually executed. In this way, they serve as both off-line and on-line systems. Others edit without time code, like the consumer systems, but they have controls that give greater flexibility for setting and changing edit points and previewing edits to make sure the points are properly set.

Although editing is usually associated with material taped in the field, programs shot in a studio are often edited also. Segments of a variety show can be shot out of order so that all the scenes that involve one set can be shot together. Then everything can be edited together in its final program order. Editing can also be used to correct mistakes or to shorten programs.

Processing Equipment

Often when material is edited on-line, equipment other than two decks, a controller, and monitors is utilized. The main purpose of some of this equipment is to process the electronic signals in such a way that the quality of the picture or sound is improved. Sometimes this equipment is also employed during studio productions, whether they are taped or live. When such equipment is used in conjunction with a studio shoot, it is usually located in a control room that is adjacent to the studio. The director and technicians sit in the control room undertaking their duties while the camera operators and talent perform in the studio.

FIGURE 14.27

A combination
waveform monitor
(*left*) and vectorscope.
*(Courtesy of Leader
Instruments)*

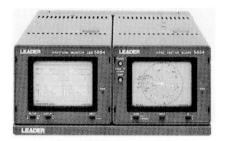

drop-out compensator

One piece of processing equipment is a **drop-out compensator.** It fills in gaps in video information where the head loses contact with the information on the tape, often because some of the oxide has fallen off the tape. Many drop-out compensators do this by filling in missing information in one scan line with information from the scan line above or below it.

time base corrector

A **time-base corrector** stores frames or parts of frames coming from various source videotape machines so that their timing will be in sync, and the picture will not roll when a switch is made from one input to another. They are essential anytime two shots from two different VCRs need to be seen on the screen at the same time, such as when one picture gradually merges into another.[13]

An **image enhancer** corrects for loss of sharpness and smearing of color

color correction

that results from recording a picture, and a **color corrector** can be used to make subtle changes in the red, blue, or green elements of a picture.

waveform monitor

Two other pieces of equipment, the **waveform monitor** and the **vectorscope** measure the video signal and are used to help an operator set camera or VCR inputs in an optimum way. The waveform monitor measures brightness. Generally the brightest point in a scene should be adjusted so that it

vectorscope

peaks at 100 percent on the waveform monitor. The vectorscope measures color, and it is the reason why color bars are recorded at the beginning of a tape. Each color—red, blue, green, magenta, cyan, and yellow—has an area on the vectorscope where it should register. If the colors do not register accurately, an engineer needs to make adjustments in the signal.[14]

audio

In the realm of audio, **filters** can be used to alter frequencies. For example, if a high-pitched whine occurs in a sound, it can sometimes be eliminated by filtering out the particular range of frequencies where the whine occurs. Of course, any other sounds that are also in that frequency range will be eliminated, too, so filtering is not always the answer to unwanted sounds. **Dolby** and **dbx** eliminate noise in audio systems, and **compressors** control characteristics related to volume.[15]

Many facilities, including universities, operate with very little processing equipment because the quality of the signal is not of uppermost importance, and processing equipment is expensive. Facilities that broadcast definitely need processing equipment, however, because signals further deteriorate when they go through the airwaves.

FIGURE 14.28
A character generator.
(Courtesy of Chyron Corporation)

Character Generators and Computer Graphics

During the early 1970s, **character generators** (**c.g.**s) were introduced as a method of creating graphics digitally. Prior to this, graphics, such as opening and closing credits, charts, graphs, and diagrams, were created by graphic artists. To create words, such as titles or credits, the artists would print white letters on black paper. Then one camera would focus on the letters and another would focus on a scene from the TV program being produced. The two pictures would be combined, the black of the paper would disappear, and what would appear on screen would be white letters over a scene from the TV program. previous graphics

With the advent of c.g.s, these cards were no longer needed. A character generator consists of a typewriter keyboard that electronically types words onto the TV screen. Usually c.g.s contain a variety of fonts so that different letter styles can be typed. They also have provisions for changing letter size, position, color, background, and method of display. In addition, some c.g.s have limited ability to draw pictures or charts. c.g.s

The ability to create complicated pictures digitally, however, came with the advent of **computer graphics,** which experienced large-scale development during the 1980s. Computer-generated imagery sometimes encompasses the ability to type words and numbers and, therefore, includes a keyboard. However, its main function is to create, display, and manipulate pictorial information. computer graphics

For this reason, the computer graphics devices usually include something to draw with, such as a light pen that creates lines wherever it touches the TV screen, or a mouse, which can be moved on a separate pad that translates the movements into lines on a TV screen. Some of these computer graphics systems are stand-alone units, and other are programs that can be used with personal computers such as the Macintosh and Amiga.

With most computer graphics systems, artists can select from a "paint box" the colors they wish to use and then mix these colors together, or they can select the size of "brush" they want and lines will appear in different thicknesses. Systems have also been designed that can utilize pictures or TV images that have already been created. For example, a photograph of a car

FIGURE 14.29
An example of
computer graphics.
*(Courtesy of Aurora
Systems)*

can be fed into the computer and can then be outlined, recolored, or manipulated in other ways. Similarly, one frame of a TV picture can be frozen and operated on graphically. Some computer graphics systems can also create animation, as well as figures that look three-dimensional.[16]

Computer graphics have become very popular, particularly in the creation of commercials. They represent one step in the video process that is fully digital. Both c.g.s and computer graphics are usually located in a control room if they are used in conjunction with a studio shoot, or in an editing suite if they are needed during the editing process.

Switchers and Digital Effects Generators

Most TV programs produced in a studio or edited in a postproduction configuration incorporate inputs from a number of sources. For example, a studio program might start with a computer generated graphic over which c.g. program titles are shown, then change to a long shot of the host on camera 1, followed by a close-up of a guest on camera 2. A commercial being edited might start with computerized graphics in the top half of a screen and a pretaped moving automobile in the bottom half of the screen. In order to accomplish effects like these, a **switcher** and/or **digital effects generator** must be utilized. These, like other pieces of equipment, are found in control rooms or editing suites.

switchers

A switcher, in its simplest form, consists of buttons and levers that place the proper picture on the air or on tape. For example, if the director wants the picture on camera 1, the person operating the switcher would punch the button labeled "camera 1." When the director wants camera 2, the "camera 2" button would be pushed. This is called a **take**—a quick change from one picture to another.

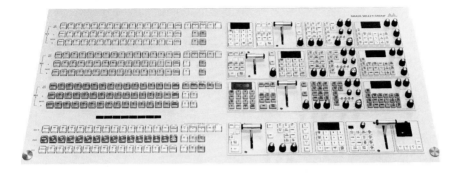

FIGURE 14.30

A switcher. *(Courtesy of the Grass Valley Group)*

Theoretically, a switcher could consist of two buttons if the only inputs available were two cameras and the only switching desired were takes. A third button for "black" could be added if it is desirable to have a blank screen before the program or at the end of it.

If a **dissolve** (a slow change in which one picture gradually replaces another) is desired, then three more buttons and a lever can be added to enable a slow transition from one picture to another. This also enables a **fade,** which is a transition between a picture and black.

Many switchers are used in conjunction with digital effects generators that allow complicated effects such as wipes, turns, inserts, squeezing, stretching, and flipping. These are all possible because a digital effects generator can grab, at any time, any video frame from any video source, change it into digital information, manipulate it in a variety of ways, store it, and retrieve it on command. The digitized signals are inputted to the switcher where they can be mixed with other inputs or sent on to some other piece of production equipment.

Some of what a digital effects generator does is very similar to what can be accomplished with computer graphics. In fact, sometimes it is impossible to tell from what source, in the overall TV system, a particular effect was created.[17]

digital effects generators

Audio Boards and Outboard Equipment

For TV studio shoots, for most editing, and occasionally for taping in the field, audio mixing is a must. It is also a major element of radio. The various audio inputs (music to open a show, the chatter of a disc jockey, sound effects to accompany the right answer on a game show) are sent through an **audio board** where they are combined so that each is at the proper volume. They are then

audio boards

Production Equipment   379

FIGURE 14.31

This audio board can accept up to twelve inputs at once. *(Courtesy of Opamp Labs, Inc.)*

sent to other pieces of equipment such as a videotape recorder, audiotape recorder, or radio station transmitter.

Various types of equipment are used to input audio. Microphones, of course, are one. Equipment that sits near the audio board so that it can be operated during a production is referred to as **outboard equipment.** Most radio and TV stations have **turntables** as one form of outboard equipment. These are quickly being replaced by **compact disc players (CDs).**[18] CDs play back sound digitally, which is superior in quality to the analog sound of records. Tape recorders are also used to record and play back sound. Some of these recorders use audiotape cassettes that operate very much like home cassette decks. Others use **cartridges,** a continuous loop of tape wound onto a single spool. Radio commercials and sound effects are usually recorded on cartridges because the amount of tape on cartridges can be very short. Also, such material can profit from a cartridge recorder's cue function that automatically starts and stops the machine.

In radio stations, disc jockeys sit in the control room in front of an audio board and introduce songs through a microphone. They then play the records or CDs and bring in taped commercials or other announcements as required. At some large stations, the DJs sit in a studio talking into a mike, and an engineer works in a nearby control room cuing up all the outboard equipment and operating the board. Other radio stations, particularly public radio stations that tape dramas, concerts, and interview programs, have spacious audio studios where multiple microphones can be set up to record large groups.

Audio equipment is also used in the field by both radio and television personnel. If several microphones are needed to tape a television segment, they are often routed through a small portable audio mixer before being inputted to the videotape recorder. Radio station news crews often use microphones and cassette recorders in the field to conduct interviews or to record on-the-spot reactions.

outboard equipment

radio studios

field uses

Physical Characteristics

FIGURE 14.32

A radio studio that includes cassette machines, cart machines, CDs, turntables, reel-to-reel recorders, and other related equipment. *(Courtesy of KLOS-FM, Los Angeles)*

When sound is mixed for an editing-intensive production such as a drama, a movie, or a documentary, **multi-track tape recorders** are often used. These usually use two-inch audiotapes capable of holding twenty-four different sounds. Dialogue, music, and sound effects are placed on separate tracks without regard to how loud each should be. Then at some point all the different sounds are played back in sync with the picture, and people sitting at a large audio board raise and lower the volume for the various sounds so that they work well in conjunction with each other. The mixed sound is then recorded on another track of the audiotape and is eventually transferred to the master videotape.

multi-track recording

Digital audiotape (**DAT**) is starting to be used, especially in the cassette format. This offers the same high quality as CDs and is quite likely to replace the analog recording methods presently being used for cassettes and cartridges.[19]

DAT

Digital technology is also becoming common as part of editing and mixing. Computer-driven pieces of equipment called **digital audio workstations (DAWs)** can be used to store, edit, mix, process, and manipulate sound all within one unit. No tape is needed. The sounds are stored on discs and manipulated by mouse or keyboard. Sounds can be inputted from microphones, CDs, tapes, or any other outboard equipment, but once the sound is in a DAW, it is converted to digital and kept in the digital form throughout editing and mixing. This keeps the sound at a very high quality.[20] Programs have also been written for personal computers that can be used to manipulate sound. These are more limited than what is possible with full-blown digital audio workstations, but, because they are relatively inexpensive, they do enable consumers and people working at educational institutions and in industry to manipulate sound digitally.[21]

DAW

FIGURE 14.33
A digital audio workstation. *(Courtesy oj Waveframe Corporation)*

automated equipment

Some radio stations operate with audio equipment that allows them to broadcast on an automated basis. These stations have multiple tape decks, some of which contain reel-to-reel music and some of which contain cartridge commercials. Subaudio tones, which cannot be heard over the air, signal one tape recorder to stop playing music and another to start playing a commercial, or vice versa. Taped station break announcements can also be inserted with specialized clock mechanisms. In this way, a station can program for a whole day without having either a disc jockey or an engineer present; the sound is sent directly from the tape deck to the transmitter.

satellites

Now that programming can be transmitted by **satellite,** some stations pick up their programming from the dish and retransmit it. In this way, they do not need automated equipment, but they can be fully or partially automated. The satellite can send the bulk of the programming and then, through a special signal, turn on local cartridges that play station breaks and commercials.

Most equipment used for radio and television is in a constant state of flux. Technical improvements lead to new techniques, and new techniques lead to the need for technical improvements.

Conclusion

Radio and television are both equipment intensive. The major equipment used for radio consists of microphones, audio boards, outboard equipment, and processing equipment. Microphones are constructed differently because they are needed for a variety of purposes. For the same reason, they have differing directionality and a large number of ways of being positioned.

Audio boards are used primarily for mixing sounds. Inputs come from various types of equipment and are combined and then sent to a transmitter or other piece of equipment.

Outboard equipment consists of turntables, CDs, and tape recorders—both cassette and cartridge. Processing equipment is used to improve the quality of the sound.

The most common way this equipment is used in radio is for a DJ to sit by the audio board cuing and mixing the various audio inputs as needed. Some of the audio equipment is also used remotely to gather news. Some radio stations are automated so they have little need for disc jockeys. Others receive much of their programming from satellite services.

All the equipment used for radio is also used for television. In addition, television production needs cameras, videocassette recorders, lights, editing equipment, video processing equipment, character generators, computer graphics generators, switchers, digital effects generators, and some additional audio equipment.

Most new cameras are CCD, but there are still many tube cameras in existence. American cameras create a 525-line image each one-thirtieth of a second, but at some point in the future this may change to a higher number of lines with HDTV. Cameras can be mounted on a person's shoulder, a tripod, a pedestal, a crane, or some more exotic device.

Recorders are often part of the camera and as such are called camcorders. Cameras used with studios and control rooms are usually stand-alone units. The first recorders used two-inch tape, but since then many helical formats of tape have been developed that use three-fourths-inch, half-inch, and 8mm tape. Audio laid on VCRs is linear, hi-fi, and or PCM. Digital video recording is possible and will probably become more widespread.

Lights used indoors are usually tungsten, and lights used outdoors are usually HMIs. The basic lighting pattern involves key, fill, and back lights. Many auxiliary pieces of equipment help control and enhance light.

Editing systems vary greatly depending on the sophistication of the needs. Consumer editing can involve no more than two recorders and a TV set. At the professional level, editing is often first undertaken off-line with time-coded workprints. A computer disc of decisions from the off-line system is then taken to an on-line system where the master tape is used for editing.

Often other pieces of equipment are used for editing and for studio shoots. These include processing equipment such as drop-out compensators, time-base correctors, and color correctors. Waveform monitors and vectorscopes are used to set brightness and color properly.

Equipment often found in control rooms includes character generators and computer graphics generators, both of which can be used to create words and graphics.

Switchers and digital effects generators enable several pictures to be seen at once. Dissolves, fades, and special effects are undertaken with these pieces of equipment.

In the audio area, television production often employs multi-track recorders and digital audio workstations in addition to mikes, audio boards, and outboard equipment.

Thought Questions

1. What do you think will be the main characteristics of cameras in the near future? Will home video cameras differ from those used by broadcasters? If so, how?
2. What improvements do you foresee in editing equipment?
3. In what different ways will digital technology be used in the production process?

Distribution

Introduction

Once a program is produced, it must be distributed in some manner to the viewer. Within the telecommunications field there are a variety of distribution methods, some of which are used in conjunction with one another. These methods range from the very simple pick-up-and-carry bicycle method, whereby a person literally carries a tape to and from locations, to more elaborate methods that involve wires, satellites, microwave dishes, and broadcast transmitters.

CHAPTER 15

The messages wirelessed ten years ago have not yet reached some of the nearest stars.

Guglielmo Marconi
inventor

The Spectrum

frequencies

Most systems of telecommunications distribution involve the use of the electromagnetic spectrum. This is a continuing spectrum of energies at different **frequencies** that encompasses much more than telecommunications distribution. The frequencies can be compared to the different frequencies involved with sound. The human ear is capable of hearing sounds between about sixteen cycles per second for low bass noises and sixteen thousand cycles per second for high treble noises. This means that a very low bass noise makes a vibration that cycles at a rate of sixteen times per second. These rates are usually measured in **hertz** in honor of the early radio pioneer Heinrich Hertz. One hertz (Hz) is one cycle per second. A low bass note would therefore be at the frequency or rate of 16 Hz, a higher note would be at 100 Hz, and a very high note would be at 16,000 Hz.[1]

hertz

As the numbers become larger, prefixes are added to the term hertz so that the zeros do not become unmanageable. One thousand hertz is referred to as one kilohertz (kHz), one million hertz is one megahertz (MHz), and a billion hertz is one gigahertz (GHz). Therefore, the 16,000 hertz high note could also be said to have a frequency of 16 kHz.

radio waves

Radio waves, which are part of the electromagnetic spectrum, also have frequencies and are measured in hertz. However, their frequencies are high, and in many ways they act more like light than sound. They are capable of carrying sound, and it is with these radio waves that most of the telecommunications distribution takes place.

Above radio waves on the electromagnetic spectrum are infrared rays and then light waves, with each color occupying a different frequency range. After visible light come ultraviolet rays, X rays, gamma rays, and cosmic rays.

radio wave divisions

The radio wave portion of the spectrum is divided into eight sections, each composed of a band of frequencies measured in Hz. Because most of the telecommunications applications are in the megahertz range, these frequencies can best be discussed in terms of MHz.

.003 to .03 MHz are very low frequencies

.03 to .3 MHz are low frequencies

.3 to 3 MHz are medium frequencies

3 to 30 MHz are high frequencies

30 to 300 MHz are very high frequencies or VHF

300 to 3,000 MHz are ultrahigh frequencies or UHF

3,000 to 30,000 MHz are super high frequencies

30,000 to 300,000 MHz are extremely high frequencies

wave length

Sometimes the low frequencies will be referred to as **long waves,** the medium frequencies as **medium waves,** and the high frequencies as **shortwaves.** These terms refer to the length of the wave rather than to its frequency, although the two are related. When a wave cycles slowly, it can stretch out farther than when it cycles quickly because it has more time between cycles. Therefore,

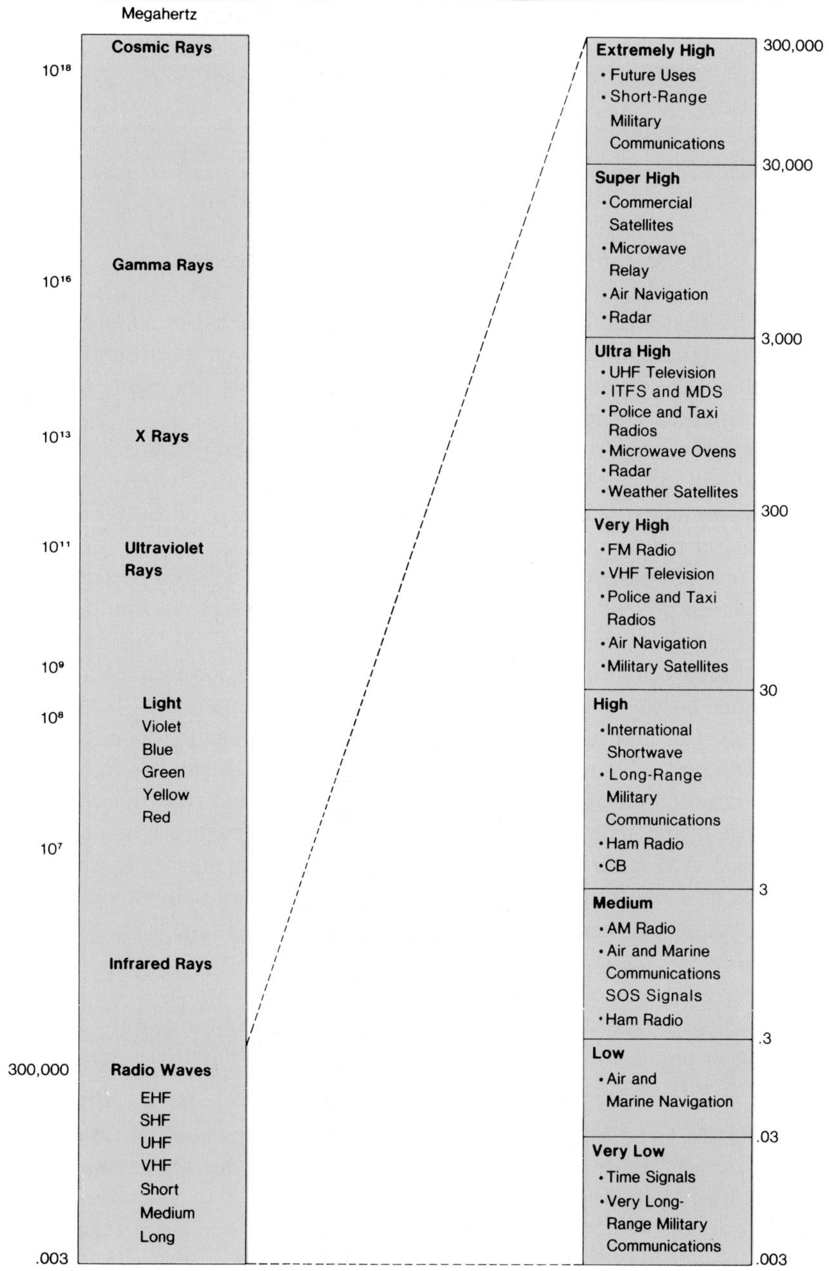

Megahertz

10^{18}	Cosmic Rays
10^{16}	Gamma Rays
10^{13}	X Rays
10^{11}	Ultraviolet Rays
10^{9}	
10^{8}	Light
	Violet
	Blue
	Green
	Yellow
	Red
10^{7}	
	Infrared Rays
300,000	Radio Waves
	EHF
	SHF
	UHF
	VHF
	Short
	Medium
	Long
.003	

Extremely High	300,000
• Future Uses	
• Short-Range Military Communications	
Super High	30,000
• Commercial Satellites	
• Microwave Relay	
• Air Navigation	
• Radar	
Ultra High	3,000
• UHF Television	
• ITFS and MDS	
• Police and Taxi Radios	
• Microwave Ovens	
• Radar	
• Weather Satellites	
Very High	300
• FM Radio	
• VHF Television	
• Police and Taxi Radios	
• Air Navigation	
• Military Satellites	
High	30
• International Shortwave	
• Long-Range Military Communications	
• Ham Radio	
• CB	
Medium	3
• AM Radio	
• Air and Marine Communications	
SOS Signals	
• Ham Radio	
Low	.3
• Air and Marine Navigation	
Very Low	.03
• Time Signals	
• Very Long-Range Military Communications	
	.003

FIGURE 15.1
The electromagnetic spectrum.

the lower the frequency, the longer the wave and the higher the frequency, the shorter the wave.

Although all radio frequencies in the electromagnetic spectrum are capable of carrying sound, they are not all the same. The frequencies that are toward the lower end of the spectrum behave more like sound than do the

characteristics of radio waves

FIGURE 15.2

In the same distance, the short wave has a greater frequency than the long wave.

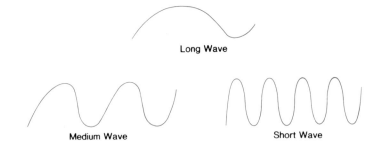

Long Wave

Medium Wave

Short Wave

frequencies that are at the higher end. These higher frequencies, in turn, behave more like light. For example, the lower frequencies can go around corners better than the higher frequencies in the same way that you can hear people talking around a corner but cannot see them.

uses of radio waves

Radio waves also have many uses other than the distribution of entertainment and information programming usually associated with the telecommunications industry. Such uses range from the opening of garage doors to highly secret reconnaissance functions.

placement of services

The spot on the radio portion of the spectrum at which a particular service is placed depends somewhat on the needs of the service. The lower frequencies have longer ranges in the earth's atmosphere. For example, long-range military communication is transmitted on the very low frequencies, while short-range communication is transmitted on the extremely high frequencies.

However, many placements are an accident of history. The lower frequencies were understood and developed earlier than the higher frequencies. In fact, not too long ago people were not even aware that the higher frequencies existed, and today the extremely high frequencies are still not totally understood. AM radio was developed earlier than FM, so the former was placed on the part of the frequency that people then knew and understood. The ultrahigh frequencies were discovered during World War II and led to the FCC's reallocating television frequencies after the war into two categories, very high frequency (VHF) and ultrahigh frequency (UHF).

The portions of the electromagnetic spectrum that are of greatest importance to the distribution of entertainment and information programming connected with electronic media are the following:

.535 MHz to 1.705 MHz: 117 AM radio channels (this band of frequencies is usually referred to as 535 to 1,705 kHz; prior to 1990 it only went to 1,605 kHz)

54 MHz to 72 MHz: TV channels 2, 3, and 4

76 MHz to 88 MHz: TV channels 5 and 6

88 MHz to 108 MHz: 100 FM radio channels

174 MHz to 216 MHz: TV channels 7 to 13

470 MHz to 890 MHz: TV channels 14 to 83

Consider this to be an electrical wave representing the original sound.

FIGURE 15.3

Diagram of AM wave.

Consider this to be the carrier wave of a particular radio station. Notice it is of much higher frequency than the electrical wave.

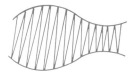

This would be the modulated carrier wave taking the sound signal. Note that the sound signal makes an image of itself and that the amplitude, or height, of the carrier wave is changed—hence, amplitude modulation.

Other portions of the spectrum are obviously important to broadcasting-related operations, too. Portions of the spectrum between 800 MHz and 14,000 MHz (.8 to 14 GHz) are used for microwave and satellite communication. The area around 2,500 MHz is used for instructional television fixed service (ITFS) and multichannel multipoint distribution service (MMDS). A large portion of the high-frequency band is used for shortwave broadcasting, primarily to foreign countries.[2]

Radio Broadcast

AM radio broadcasting emanates from the medium frequencies and FM comes from VHF. The position in the spectrum, however, has nothing to do with AM (amplitude modulation) and FM (frequency modulation).

Sound, when it leaves the radio station studio in the form of variations of electrical energy, travels to the station's **transmitter** and then to the **antenna.** At the transmitter it is **modulated,** which means that the electrical energy is superimposed onto the **carrier wave** that represents that particular radio station's frequency.

modulation

The transmitter generates this carrier wave and places the sound wave on it using the process of modulation. Modulation can occur as either **amplitude modulation (AM)** or **frequency modulation (FM),** regardless of where the carrier wave is located on the spectrum. In amplitude modulation, the amplitude (or height) of the carrier wave is varied to fit the characteristics of the sound wave. In frequency modulation, the frequency of the carrier wave is changed instead.[3]

amplitude modulation

frequency modulation

FIGURE 15.4
Diagram of FM wave.

Consider this to be the electrical wave representing the original sound wave.

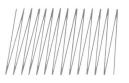

Consider this to be the carrier wave.

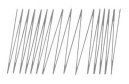

This would be the modulated carrier wave. The frequency is increased where the sound wave is highest (positive) and the frequency is decreased where the sound wave is lowest (negative). The amplitude does not change.

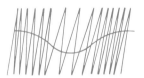

Perhaps this can be better seen by superimposing the sound wave over the carrier wave.

modulation differences

AM and FM have different characteristics that are caused by the differing modulation methods. For example, AM is much more subject to static. This is because static appears at the top and bottom of the wave cycle. Because FM is dependent on varying the frequency of the wave, the top and bottom can be eliminated without distorting the signal. AM, however, is dependent upon height, so the static regions must remain with the wave.

spectrum differences

AM and FM also have differences because of their placement on the spectrum. AM signals can travel great distances around the earth, while FM signals are more or less line of sight and sometimes cannot be heard if an obstruction such as a large building comes between the transmitter and the radio that is attempting to receive the station. This is because FM is higher up on the spectrum and closer to light than is AM. Light waves do not travel through buildings or hills, and FM signals are similarly affected.

AM, at the lower end of the spectrum, is not so affected, but these lower frequencies are affected by a nighttime condition of the ionosphere that, when hit by radio waves, bounces the wave back to earth. AM waves can be bounced great distances around the earth's surface—from New York to London, for example. Because of this phenomenon, some AM stations are authorized to broadcast only during daylight hours so that they will not interfere with other radio station signals that are traveling long distances because they are bouncing. FM is not affected by this layer because of its position in the spectrum, not

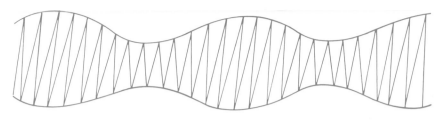

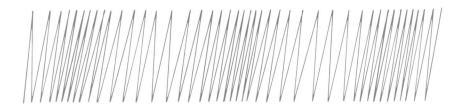

FIGURE 15.5

Comparison of AM and FM waves.

because of its manner of modulation. Theoretically, if FM waves were transmitted on the lower frequencies, such as 535 kHz, they could bounce in a manner similar to AM.

Another difference between AM and FM is that the **bandwidth** is greater for FM stations. Although each station is given a specific frequency, such as 550 kHz or 88.5 MHz, the spectrum actually used covers a wider area. For an AM radio station this width (bandwidth) is 10 kHz and for FM the bandwidth per station is 200 kHz. An AM station at 550 kHz actually operates at from 545 to 555 so that it can have room to modulate the necessary signals and prevent interference from adjacent channels. Because FM stations have a broader bandwidth, they can produce higher fidelity than AM stations.

bandwidth

This makes FM more adaptable to **stereo** broadcasting. An FM wave has sufficient space to carry more than one program at a time. Therefore, part of the wave can carry sound as it would be heard in the left-hand side of a concert hall, and another part of the wave can carry sound as it would be heard in the right-hand side; the result is a stereo transmission. Similarly, quadraphonic sound can be transmitted if four signals are carried on one channel.

FM stereo

AM stereo is also in operation now. Although there were initially several different methods for producing AM stereo, the one that is being adopted most widely is the **Motorola** system called C-Quam (Compatible Quadrature Amplitude Modulation). For this system, two separate carrier waves of the same frequency are modulated with separate left and right channels. These channels are then combined for transmission and are later separated in the radio receiver.[4]

AM stereo

The broader bandwidth of FM enables independent signals other than those needed for stereo and quadraphonic sound to be multiplexed on an FM radio station signal. This is referred to as **subcarrier authorization (SCA).** The information is carried along on the FM signal, but can only be received by those with special receivers. Some of the SCA services include piped music to doctors' offices, reading for the blind, and paging and dispatch services.[5]

SCA

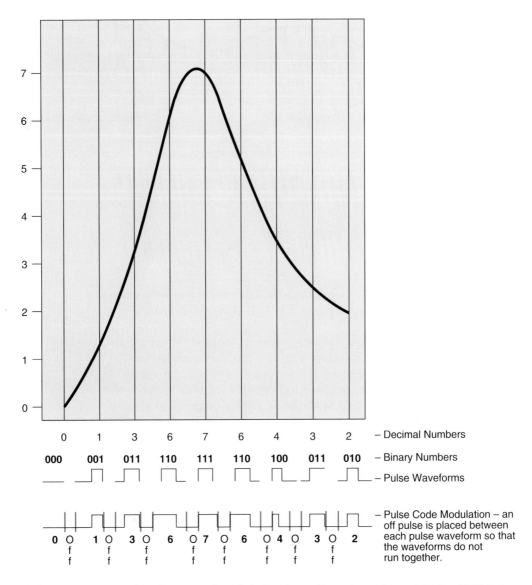

FIGURE 15.6
Pulse code modulation.

Another type of modulation for radio, **pulse code modulation (PCM),** has been proposed. This type of modulation would encode sound in digital form rather than the analog form used by AM and FM. The result would be even clearer sound than that generated by FM. Present plans call for this type of modulation to be used for a new type of radio service called **digital audio broadcasting.** This would use a part of the spectrum not presently used for radio and would beam audio signals from satellites directly to home and car radios. This form of broadcasting has not been authorized yet; it is just in the discussion and experimentation stage.[6]

For the broadcasting that presently exists, after the sound waves have been modulated (either AM or FM) onto the proper frequency carrier wave at the transmitter, the carrier wave is radiated into the air. It is sent into the airwaves at the assigned frequency and power from the radio station antenna. As noted earlier, because FM frequencies are line of sight, FM antennas are usually positioned on high places overlooking a large area. AM antennas can operate effectively at relatively low heights.

The FCC has established a complicated chart of frequency and power allocations to allow a maximum number of stations around the country to broadcast without interference. As of 1989, there was room for 107 stations of 10 kHz each between 535 and 1,605 kHz when in actuality there were about 4,900 AM stations nationwide. This was possible because many stations throughout the country broadcast from their antennas on the same frequency, but in a controlled way that considers geographic location and power. As of 1990, an additional 100 kHz was added to the AM band taking it to 1,705 kHz. This will make room for many more stations. However, new receivers will need to be manufactured and purchased before these stations can be heard by the public.

The allocation chart that exists for AM stations divides them into four categories: class I (clear), class II (secondary), class III (regional), and class IV (local). Clear channels usually operate with 50,000 watts and serve large centers of population and remote areas. There are only one or two class I stations per frequency, and there are forty-five frequencies on which these channels transmit.

Class II stations used to be clear channels, each having its own frequency. But as pressure for more stations increased, the FCC gradually allowed greater use of these frequencies. Now these stations operate from 250 to 50,000 watts serving population centers and adjoining rural areas. There are twenty-nine secondary frequencies that operate in such a way that they do not interfere with the clears.

Class III stations use 500 to 5,000 watts and serve cities and rural areas. There are forty-one regional frequencies and more than two thousand stations.

Class IV stations serve local areas and have a maximum power of 1,000 watts during the day and 250 watts at night. There are six frequencies for these local channels, each occupied by 150 or more stations.[7]

In order to avoid interference, many AM stations are required to have **directional antennas.** When a signal is sent out from an antenna, it will travel equally in all directions, causing a circular coverage pattern. In order to prevent the signal from going in a direction where it might interfere with another station, directional antenna systems are set up. Such systems require at least two towers—one normal tower and one or more to cancel out the signal coming from the normal tower. By varying the signal sent to each tower, the direction of the station signal can be limited.

Some hope the expansion of the AM band will also reduce interference. If the stations that must broadcast only during the day are given preference in applying for the new frequencies, they could leave their present positions

FIGURE 15.7

A directional antenna
setup for an AM
station. *(Courtesy of KHJ,
Los Angeles)*

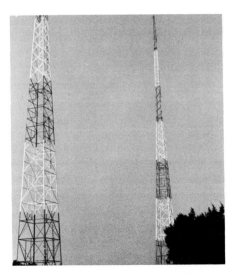

in the AM band, move to new positions above 1,605 kHz and broadcast twenty-four hours a day. This would mean both the old daytimer licensees and the licensees remaining on the old AM band would have less interference.[8]

FM stations are divided into class A, class B, and class C stations. Class A stations use power from 100 to 6,000 watts and generally serve a very limited area. Class B stations serve a larger area and use power from 5,000 to 50,000 watts. Class C stations, the most powerful, can use up to 100,000 watts.[9]

The waves sent out from a radio station antenna are always available, but they do not succeed in distributing program material unless someone turns on a radio receiver. The purpose of the receiver is to pick up and amplify radio waves, separate the information from the carrier wave, and reproduce this information in the form of sound.

The first stage in radio reception involves the antenna, which picks up the signal being sent out by the radio station antenna. From here the signal goes to the **radio frequency amplifier,** which has two purposes: (1) to amplify the signal (carrier wave plus information); and (2) to select the frequency. This is accomplished by a device within the amplifier called a tuner, which makes it possible to tune in the signal of one particular radio station and exclude all other stations. If the tuner were not used, all stations broadcasting in the area of the receiver would be heard simultaneously.

Only the station tuned in gets to the end of the radio frequency amplifier. From here it goes to the **detector,** where the carrier wave is separated from the information it is carrying. This detector then discards the carrier wave, which has already done its job, and feeds the information in the form of electrical energy to the next piece of equipment, the **audio frequency amplifier.** Here the impulses are strengthened and sent to the speaker, where the electrical energy is converted back into sound waves that can be heard.[10]

FM categories

receivers

When radio was first invented, a huge console, which usually sat in the living room, was used to house all the required equipment. Now, of course, it can all be housed in the smallest of receivers that can be worn in the ear, on the wrist, or in a pocket.

A home radio receiver need not rely on its own antenna to receive transmitted radio signals. They can be picked up and distributed to subscribers by a cable TV system.

cable radio

AM radio stations are the oldest form of telecommunications program distribution, but the methods devised many years ago both for modulation and transmission are still quite effective. If the bandwidth were wider, AM radio would be more flexible, but it is still a useful medium. FM radio, with its higher fidelity, has made great strides in recent years, but it is hampered somewhat by its shorter range and line-of-sight characteristic.

Over-the-Air Television Broadcast

As with radio, the television transmitter is the device for superimposing or modulating information onto a carrier wave. With television, audio and video are sent to the transmitter separately—the video signal is amplitude modulated and the audio signal is frequency modulated in either monaural or stereo. The two are then joined and broadcast from the antenna.

Again, as with radio, each TV station has a frequency band on which it broadcasts. However, a television station uses a great deal more bandwidth than a radio station. While AM stations span 10 kHz and FM stations 200 kHz, the bandwidth for a TV station is 6,000 kHz or 6 MHz. This means that a TV station takes 600 times as much room as an AM station and that all the AM and FM stations together occupy less spectrum space than four TV stations. The reason for this, of course, is that video information is much more complex than audio information and, therefore, needs more room to modulate and to protect the modulated information from interference.

bandwidth

TV channels are placed at various points in the spectrum. There is a small break between **VHF** channel 4 and channel 5 and a larger break between channel 6 and channel 7 that encompasses, among other services, FM radio. In general, in any particular area, two adjacent channels cannot be used because they would interfere with each other. For example, Detroit could not have operating stations on both channel 2 and channel 3. However, because of the break in spectrum between 4 and 5 and between 6 and 7, communities can use those adjacent channels. For example, New York has stations on both channel 4 and channel 5. In addition, **low-power TV** (**LPTV**) stations can sometimes operate on channels adjacent to regular high-power channels because the low power does not cause interference.

placement of channels

Channels 14 through 83 are well above channels 2 through 13 in the **UHF** range. Because these channels have not been highly utilized, the FCC has given some of the space set aside for channels 71 to 83 to land mobile radio. Even higher yet in the UHF range are the channels used by **MMDS**

(multichannel multipoint distribution service), and **ITFS (instructional television fixed service).** These cannot be received on a regular TV set even though they are broadcast in a manner similar to VHF and UHF stations.

UHF and VHF

Both UHF and VHF waves follow a direct line-of-sight path, so both are most effective if located at a high point. UHF signals, because they are closer to light, are more easily cut off by buildings and hills and are more rapidly absorbed by the atmosphere. Therefore, they require higher power at the transmitter to make up for these losses.

wattage

Although wattage varies considerably, generally VHF stations operate between 10 and 400 kilowatts and UHF between 50 and 4,000 kilowatts. Low-power TV, of course, transmits with lower power—10 watts for VHF and 1,000 watts for UHF.

multiplexing

Multiplexing on a TV station signal is not as easy as multiplexing on an FM station signal because of the greater complications and requirements of TV, but this process is still undertaken. For example, **teletext** and **closed captioning** for the hearing impaired are now broadcast along with the regular station signal. The words, numbers, and figures of these are much simpler than moving video pictures and can be carried in a much smaller space, namely the retrace of scanning. After the electron gun in a camera scans a frame from top to bottom, it turns off momentarily to return to the top of the picture and begins scanning again. The teletext information or captioning is placed in this space where the gun is turned off.

distribution

reception

Once a TV signal is sent from an antenna, it must be received in order for the distribution process to take effect. Signals that are transmitted are intended primarily for individual homes, but like radio, they can be picked up by cable TV systems for retransmission. Sometimes they are also picked up by low-power unattended receiver/transmitters called **translators.** These translators then rebroadcast the signal into an area that has poor reception.

Regardless of whether the signals are retransmitted or not, however, they wind up on a home TV set. The process of displaying the picture on the face of the set is essentially the reverse of its creation. The modulated impulses are received by the home antenna, demodulated, and sent to the TV picture tube and speaker. TV audio is, of course, produced very similarly to audio in a radio receiver. Video is produced by a scanning method similar to that used in tube cameras.

This method includes **electron guns** located in the rear of the **cathode-ray tube,** or picture tube. If the set is black-and-white, it holds a single electron gun. If the set is color, it contains three guns—one for reds, one for blues, and one for greens. The incoming video signal causes this gun (or guns) to scan the **phosphor screen** in the same manner as the camera gun—left to right, top to bottom, odd lines, then even lines—so that it creates what the eye perceives as a moving picture. When the beam strikes the phosphor layer, it causes the layer to glow according to the intensity of the signal, thus creating blacks, grays, whites, and various shades and combinations of color. A television picture is actually an array of small glowing dots blinking very rapidly in various degrees of brilliance.

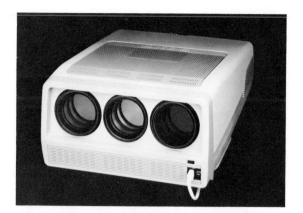

FIGURE 15.8
A video projector.
(Courtesy of Sony Corporation)

TV sets have improved greatly throughout the years. One of the most significant developments was the remote control because it allowed viewers to change channels from the comfort of their armchairs. Because they did not need to get up and go to the set, they changed channels much more frequently. Throughout the years, the remote control has added many features such as volume control and brightness control. Some remote controls can be used to allow interaction among over-the-air signals, cable TV, and a VCR. They allow for temporary muting of a program so that the viewer can answer the phone or talk to someone, and many have provision for displaying the time of day on the screen temporarily. With some remote controls, by pushing a button, a picture being telecast can be freeze-framed until the button is pushed again, returning the viewer to the original program. Several programs can be viewed at once by sets that can be made to insert different signals in different corners of the tube. Most of these special features are possible because of the same **digital** technology that allows for special effects in the production process.[11]

TV pictures also have the potential to be much larger than they were during the early years of television. Picture tubes as large as thirty-five inches in diameter exist, but most really large video displays are the result of **projection TV.** Various methods of projection TV exist, but the most common consists of a unit that takes the signals from a conventional cathode-ray tube, magnifies it, and projects it on to a large screen. Usually these projectors have three lenses—one to project reds, one to project blues, and one to project greens. Others have only one lens because they converge the colors inside the projector. The three lens projectors convey a better picture, but they tend to get out of alignment, so one lens systems are often more satisfactory if the projector is going to be moved about frequently. Many of the newer projectors are able to interface with computers as well as video signals. This is a real plus for businesses with a need to project the output of a computer screen so that a large group of people can see it. Not all projectors use the conventional tube-lens configuration. Some use CCD (charged-coupled device) technology, some use lasers for projection, and still others convert the signal by using liquid crystals.[12]

remote control

digital

large screen

FIGURE 15.9
A flat-screen TV.
*(Courtesy of Sony
Corporation of America)*

small screen

Liquid crystal displays (LCDs) have also played a major role in developing very small, portable TV sets. Small sets are difficult to manufacture because the cathode-ray tube requires depth so that the electrons from the gun can scan the phosphor screen. Liquid crystals, however, can be held between two glass plates. When voltages are sent to the crystals, they turn black. This characteristic can be used to create a black-and-white TV picture. Color can be created through the use of a color filter that creates a spectrum of hues. Many of the paperback-size TV sets use this liquid crystal display technology. Other small TV sets modify the cathode-ray tubes, primarily by placing the electron gun off to the side or below the screen and then bending the electron beam so that it hits the screen. A newer technology, called microtip, employs a large number of electron guns, each pointed at a very narrow area. These guns can be closer to the screen and still scan the area that they need to scan. This makes for a flatter TV set. As the technology for small sets continues to improve, it will be able to be used for very large sets, too. Eventually, **flat-screen TVs** that can hang on the wall like a picture will be developed.[13]

flat screen

3-D TV

Occasionally **three-dimensional television (3-D TV)** is transmitted. Generally, when a program is produced for 3-D, two different cameras are placed side by side, each taking a slightly different picture. This is to approximate the eyes, which are able to perceive depth because they are placed slightly apart. When the program material is played back, two images appear. In order for viewers to see only one image, they must wear special glasses, each lens of which filters out one of the images. Sometimes these glasses are different colors, and sometimes they are different shades of grey. Watching a 3-D program without special glasses is very annoying because of seeing the two images. Therefore, some methods that utilize only one image have been tried. With one, the person viewing 3-D wears special glasses that delay the picture in one eye, letting it get to the brain later in order to give a 3-D effect, but a person without special glasses sees a normal two-dimensional picture.[14]

Wire Transmission

Broadcasting that utilizes a transmitter and antenna is fine for disseminating programming in a local area; however, if the programming is to go a great distance, some other form of distribution must be used.

One such form of distribution is copper wires. Telephones are the main users of wires, and now that data services are available, both voice and data are large customers for telephone wires. Special telephone wires that give higher fidelity than regular telephone wires can also be used for audio transmission. For many years, radio network programming was sent across the country on such wires, but now much of it has gone to satellite. A wire system is augmented by amplifiers so that the signal remains strong. AT&T owns most of the telephone wires that cross the country, so networks or other entities that wish to use the wires lease time on them from AT&T.

special telephone wires

Wires can also be used in the production process to distribute audio signals. When a radio station wishes to broadcast a sports remote, it rents phone lines so that the game announcing can travel from the stadium back to the station. On an even simpler level, microphones and tape recorders are connected to audio boards by wires.

If a station is not located at the same place as its transmitter, the program can be sent over wire from the station to the transmitter. Videocassette recorders and videodisc players that are in homes and businesses are connected to TV sets by wires.

Phone wires are not adequate for television signals because they cannot carry the amount of information needed for video. A special type of wire, **coaxial cable,** has been developed for television. Coaxial cable, amplified periodically, can transport television programs great distances and can also carry programs from a station to a transmitter or from a remote location to the station. Coaxial cable also carries the TV signal to various apartments in SMATV setups.

coaxial cable

One of the most common uses of cable, however, is in cable TV systems. A cable system consists of a headend where all of the inputs, such as local TV stations, local radio stations, pay-cable services, and basic cable services, are received. At this headend, all the services are put on wire that is either buried underground or hung on telephone poles. This wire can then be taken into a subscriber's home and connected to the TV set.

Because the wire is physically in the home, signals can be sent back up the wire to the headend, enabling cable to be two-way or interactive. Although coaxial cable must be shielded, signals on the cable are not as subject to interference as signals traveling through the airwaves. Therefore, it is possible

FIGURE 15.10

A fiber optic. *(Courtesy of Corning Glass Works)*

fiber optics

to put a large number of signals in a fairly small wire and use all the channel numbers. In other words, both channels 2 and 3 can be used for cable TV, though not for broadcast TV.[15]

A newer form of wire transmission, **fiber optics,** now seems to be replacing telephone wires and coaxial cable. With this technology, audio, video, or data information can be sent through an optic strand that is less than a hundredth of an inch in diameter. Because this strand is made of glass, it is relatively inexpensive, lightweight, strong, and flexible. A fiber optic cable that is less than an inch in diameter can carry 400,000 phone calls simultaneously—ten times the amount that can be carried on conventional wire. For most fiber optic systems, the information is first converted from analog to digital form. Then it is carried on light produced by laser diodes through the glass fiber at a rate of 44.7 megabits per second. When the information gets to its destination, it is translated back into analog form. Using light makes transmission less susceptible to electrical interference than coaxial cable or telephone lines.[16]

Fiber optics are being employed at an increasingly rapid rate. Phone companies routinely replace old wiring systems with fiber, and some of the cable TV systems are starting to do likewise. Fibers have been used since 1980 to handle intercommunication and broadcast needs for the Olympics, and in 1988 they were used in a similar way for the Democratic National Convention in Atlanta. The Home Shopping Club uses them to handle their phone calls, to link computers for inventory updates, to handle videoconferencing with telemarketing subsidiaries, and to monitor warehouses for security purposes. Fiber optics are used to link the various pavilions at the Walt Disney EPCOT Center

FIGURE 15.11

A microwave tower.
(Courtesy of AT&T Long Lines)

in Florida; to interconnect CNN's Atlanta, Washington, and New York facilities; and to bring baseball feeds from the Astrodome to station KTXH in Houston.[17]

Although phone lines for audio and data, and coaxial cable for video, are still the primary devices for distribution, they could eventually be replaced by fiber optics. In addition, satellites are losing some of their business to fiber, in part because fiber optics are secure; it is much more difficult to pirate signals from enclosed glass strands than it is from signals being sent through the air.[18]

Microwave

Stringing cable is not always the most effective way to accomplish long distance distribution of programming material. Installing cable in mountainous, uninhabited terrain is not efficient in terms of initial installation or maintenance.

One alternative to cable is **microwave.** Microwaves are very shortwaves higher up in the spectrum than are radio or TV station allocations. They are line of sight, so relay stations must be in sight of each other and not more than about thirty miles apart. Each microwave station across the nation is mounted on a tower or tall building that occupies a relatively high place. The first station in the chain sends the signal to the next station, which receives the signal, amplifies it, and sends it on to the next station. The last station need only receive the signal.[19]

Along with the telephone lines, AT&T owns the microwave links, and the users of the microwave links lease time from AT&T. For this reason, no user relays material indefinitely on any particular microwave frequency. AT&T decides which frequencies are available for its various customers. In general, microwave relays are common in rural areas and coaxial cable distribution is common in urban areas.

Microwave can be used for a variety of purposes. Stations often place a microwave dish on the top of a remote truck to send the signal back to the station. Sometimes network signals are microwaved across the country. Cable systems often import distant signals by way of microwave.

Cable systems have yet another use for microwave. If a cable system covers a large area, it will often microwave relay its signals to various points within its franchise area called hubs and then wire from these hubs. In other words, the cable system will receive the various signals from satellites, transmitters, microwave stations, wires, and the like at its headend and then retransmit those signals in two different ways. It will send them by wire to the homes located near the headend and will send them by microwave to hubs in areas distant from the headend. At the hub, a microwave receiver will pick up the signals that will then be sent by wire to homes in that location.

Satellites

Satellite distribution is essentially glorified microwave distribution. Instead of sending signals from one point on the earth's surface to another by microwave, signals are sent by microwave from the earth to a satellite in outer space and back to the earth again. The satellite can be powered, at least in part, by energy that it gathers from the sun.

Satellite distribution has many advantages over microwave relay, one of which is that the cost of satellite distribution is less than the cost of conventional microwave distribution. Also, when waves travel through outer space, they do not encounter all the interference and inhospitable weather conditions that waves traveling through the earth's atmosphere encounter. Once a wave leaves the earth's atmosphere, it is virtually home free for the rest of its trip to the satellite and back again to the earth's atmosphere. Furthermore, with terrestrial microwave, a signal can travel only from one microwave relay station to another. Once a signal has reached a satellite, it can be received by any earth facility that is equipped to receive that signal. In other words, microwave is point to point and satellite is point to multipoint.

Being equipped to receive the signal means owning a **ground station** (also known as an earth station) satellite dish positioned in such a way that it lines up with the signal being sent from the satellite. Signals are sent to the satellite by large ground station dishes and are received on the satellite by individual **transponders.** Each satellite contains about twenty-four transponders, which are roughly analogous to channels. A satellite with twenty-four transponders can receive twenty-four different program signals at one time.

relays

remotes

hubs

compared to microwave

ground station

transponders

Physical Characteristics

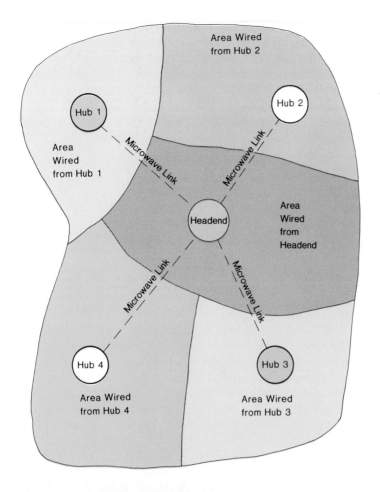

FIGURE 15.12
This hypothetical
franchise area shows
how the headend and
hubs relate to each
other.

FIGURE 15.13
ESPN's satellite
ground station dishes.
(Courtesy of © ESPN)

These transponders then transmit the program signals back to earth, where they are picked up by the ground station satellite dishes. The dishes that send information to a satellite are more expensive than the dishes that only receive, so reception points, such as cable TV headends, usually buy receive-only dishes.

The satellites used for telecommunications transmit in two main frequency bands, **C-band** and **Ku-band.** Those in the C-band operate between 4 and 6 GHz, and those in the Ku-band operate between 11 and 14 GHz. The C-band satellites were put up first and carry most of the cable TV programming. Some cable programming is carried on Ku-band satellites, and direct broadcast satellite (DBS) plans to transmit from this frequency range.

orbits

The satellites are positioned 22,300 miles above the equator. These **synchronous satellites** travel in an orbit that is synchronized with the speed of the rotation of the earth, thus appearing to hang motionless in space. This way they can continually receive and send signals to the same points on earth.

footprints

The satellites that transmit to the United States are positioned along the equator between 55 and 140 degrees longitude. Generally, they can be placed two degrees apart without interfering with each other. From there they can cast a signal, called a **footprint,** over the entire United States. Satellites positioned at other longitudes have footprints over other sections of the world. Most of the earth's surface can be covered by three strategically placed satellites. In this way, instant worldwide communication is possible through a worldwide satellite network and, of course, pictures can be beamed from anywhere in the world to the United States.[20]

Satellites do wear out, however, and need to be replaced. During the 1990s, many satellites will end their useful life and new ones will need to be launched in their place.[21]

The first communication satellite, Telstar I, was launched in 1962 by AT&T and the National Aeronautics and Space Administration. Several years after that three other organizations, Comsat, Western Union, and RCA, launched satellites naming them Comstars, Westars, and Satcoms respectively. Later other companies, such as Hughes, GTE Spacenet, GE Americom, Alascom, and National Exchange Satellite entered the satellite launch business.[22]

The early communication uses of satellites included the transmission of one-time international events such as Pope Paul's visit to the United States, splashdowns of U.S. space missions, and Olympic games. The broadcast networks used this material either as special programs or as inserts for news broadcasts.

uses

network remotes

Gradually, other uses were found for satellite interconnects. They are, of course, used for many purposes other than communications, including military and weather-gathering applications. But within the communications field, networks began using satellites to send remote programming, such as a football game, from its location to network headquarters in New York, where the football game could then be sent to affiliate stations through the network's microwave link leased from AT&T.[23]

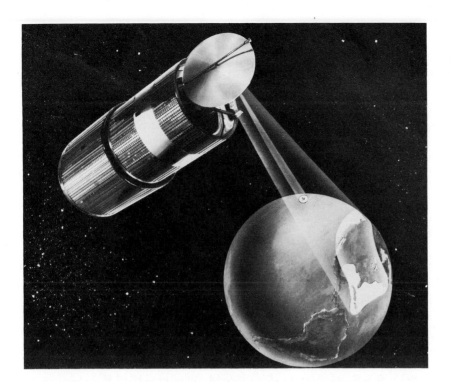

FIGURE 15.14
A satellite and its footprint. *(Courtesy of Hughes Aircraft Company)*

During the 1970s the Public Broadcasting Service and National Public Radio became the first networks to distribute all their programs to their affiliates by satellite.[24] A number of corporations and professional associations began experimenting with **teleconferencing** by having people in various parts of the country interact through satellite hookups.[25]

public broadcasting

However, HBO's placement of its pay service on RCA's Satcom I in 1976 opened the floodgates. Once the cable industry took to satellite, the demand for transponders well outstripped the supply.[26] When Satcom III, launched by RCA in December 1979, was lost, the cable industry companies that were planning to place their services on that satellite complained so bitterly that RCA leased time on other companies' satellites and re-leased it to the cable companies.[27]

cable TV transmission

Syndicators of such programs as "The Merv Griffin Show" and "Hour Magazine," who previously had mailed or hand carried tapes to stations, began simultaneously distributing by satellite and convincing the stations who carried the program to buy satellite receiving dishes. New radio networks sprang up and delivered their programming exclusively by satellite, and established networks, both radio and TV, converted from phone lines and microwave to satellite in order to deliver to their affiliates.[28]

syndicators

Some individuals bought satellite receiving dishes for their backyards or rooftops and were then able to receive most of the signals that were being transmitted by cable services, syndicators, networks, and others. This started what became a rather healthy backyard satellite dish business.

backyard dishes

Distribution   405

Local TV stations began **satellite news gathering** and, as a result could access national and international events and give them a local twist. New services such as **direct broadcast satellite** and digital audio broadcasting put in their bids to use satellite time and space.[29]

During their short history, satellites have become an important method of program distribution. They have been placed in synchronous orbits with specific footprints and their transponders can both receive and transmit to earth stations. Uses for satellites have gone from occasional special pickups to twenty-four-hour program services. They will no doubt remain an important element in the distribution process.

Pick Up and Carry

From the ethereal to the mundane, many forms of program distribution involve the physical transporting of a program from one place to another. Usually this means that one or more human beings on foot or in some sort of vehicle actually pick up a tape and carry it. One term used to describe this type of distribution is **bicycling** because, at one time, messengers on bicycles did carry material from one point to another.

educational TV

The old educational television network that mailed copies of tapes from one station to another used bicycling. Some cable TV access programming currently involves bicycling in that the community people who produce the programs hand carry the tapes to several different cable TV systems, where they are placed on videotape recorders and cablecast on access channels.

cable TV

videocassettes

One of the best examples of pick up and carry is the distribution pattern by which videocassettes and videodiscs reach the consumer. The program material is duplicated onto many cassettes or discs, and these are then shipped to wholesale houses or retail stores. Consumers, on foot, in a car, or perhaps even on a bicycle, go to the store, buy or rent the program, and take it home and put it in their machine.

Although the pick-up-and-carry method of distribution is not as glamorous as satellites, microwave, coaxial cable, optical fibers, or broadcast transmission, it is often the most effective and least expensive means of distribution.

High-Definition Television

High definition television (HDTV) is a major improvement on the horizon that could affect all methods of television distribution. Its main advantage is that it places many more horizontal lines of information on the screen than any of the present systems do, making for a much sharper picture.

Japanese development

Development of HDTV was begun by the Japanese during the early 1970s. The main players were Sony, Panasonic, Ikegami, and NHK (the Japanese broadcasting system). Most of the basic engineering was undertaken by Sony, and NHK tested the concepts over the air. Panasonic and Ikegami (along with Sony) developed cameras, videotape recorders, and other equipment needed for an entire HDTV package.

The HDTV that the Japanese developed scanned at 1,125 lines as opposed to the 525 lines of the American **NTSC** system and the 625 lines of the **PAL** and **SECAM** systems used in other parts of the world. Its aspect ratio was 5.33 to 3, (also referred to as 16:9) which was more wide-screen than the 4:3 aspect ratio of all the other systems. And HDTV scanned at thirty frames per second, which was like NTSC but unlike the fifty frame-per-second PAL and SECAM systems.[30]

This HDTV system was demonstrated in various places around the world, generally receiving positive reviews.[31] It was shown a number of times at conventions of the **Society of Motion Picture and Television Engineers (SMPTE),** whose members were very impressed with its quality. In 1983 SMPTE formed a committee to look into the adoption of a worldwide HDTV system. It looked at several other HDTV systems that had been proposed, most notably a 750-line system developed by RCA and a 655-line system developed by Compact Video, but these did not have the clarity of the 1,125-line system.[32] In 1985 the SMPTE committee recommended that the Japanese system (now called **MUSE**) be adopted worldwide.[33]

SMPTE involvement

SMPTE and many other organizations who had been frustrated by the variety of television systems that were in use felt that a worldwide system would be ideal. An HDTV tape produced anywhere in the world could be shown anywhere in the world. The constant dubbing of programs from NTSC to PAL and SECAM and vice versa would not need to take place.

world standard possibilities

SMPTE's recommendation was sent to a United Nations international committee of engineers, the International Radio Consultative Committee. Although it looked like MUSE would have clear sailing, it was opposed by a number of European countries. Their main complaint was the scanning rate of sixty frames per second. The European countries use either PAL or SECAM fifty frames-per-second systems. This is tied to their electrical current, which operates at fifty cycles per second, and they did not feel that a sixty frames-per-second system would work effectively for them. In 1986, the European countries refused to approve the Japanese system, and, as a result, the whole idea of one international system broke down.[34]

IRCC rejection

A consortium of European broadcasters and telecommunications companies, including the BBC, RAI (the Italian government network) and Philips (the Dutch equipment manufacturer), developed a new HDTV system called **Eureka.** It scans 1,250 horizontal lines at fifty frames per second, with the same 16:9 aspect ratio as the Japanese system. Because of its 1,250 lines, it is sometimes referred to as Vision 1,250.[35]

European developments

The United States, by now far behind both Japan and Europe in the development of HDTV, began to play catch up. The Defense Department pledged to spend $30 million on the technology so that the United States wouldn't be left behind, in part because the superior picture quality has many applications for military reconnaissance and pilot training.[36] The House Telecommunications Subcommittee held a hearing, the intent of which was to insure that this new technology would flourish in the United States.[37]

American developments

In 1988 the FCC set up some guidelines regarding the type of system that would be adopted in the United States. The most important of these stated that any system used in the United States would have to be compatible with current broadcast TV. In other words, anything transmitted in high definition would also need to be capable of being seen on a regular NTSC TV set.[38]

The term "high definition TV" began to be used less because it was so closely tied to the Japanese system. The FCC set up an advisory committee, calling it the committee on "advanced television" service. The words "improved definition" and "enhanced definition" also started being used.

A number of American companies began developing advanced TV systems, including Zenith, the only American company still manufacturing TV sets; Faroudja, a consortium of American companies that includes ABC, Comcast, Scientific Atlanta, TCI, and Westinghouse; the Sarnoff Labs, which are supported by GE-RCA-NBC; Massachusetts Institute of Technology, which has backing from such organizations as PBS, Ampex, Kodak, ABC, and Time; and General Instruments, the company that makes most of the scrambling and descrambling boxes for the cable TV network feeds.[39]

All the companies are to demonstrate their systems to the FCC's advanced television service committee. This committee will then select a standard. The Japanese have developed a version of MUSE that is compatible with the current NTSC system, so they hope to be under consideration, too.

bandwidth problems

One of the main problems inherent with advanced television that engineers have been working on for years involves spectrum space. HDTV cannot be transmitted on the same 6 MHz bandwidth that conventional broadcasting uses. Finding additional spectrum space while making the signal compatible with NTSC is difficult. One solution that is developing quickly involves digital **band compression.** If the analog signal is converted to digital, or if the original signal is digital, then all of the information for each frame does not need to be transmitted. Just the information that changes for each frame needs to be sent because the other information can be held. This means that far less needs to be transmitted, and the additional scan lines can fit into a 6 MHz bandwidth or less.[40]

Bandwidth is not a problem, of course, if the information is transmitted over wire. For this reason, videocassettes, videodiscs, and cable TV lend themselves well to experimentation with high definition. However, worldwide, broadcasters have dug in their stakes in terms of making sure they are in the

politics

forefront when the technology is developed, so political turf battles may prevent these other media from becoming the main purveyors.

In fact, the whole high definition TV process has become more political than technological. The business is seen as one that will garner $40 million a year, so no country is likely to let another country take the whole pie.[41]

present status

At present, the only country actually broadcasting in HDTV is Japan, where signals are sent several hours a day by satellite to public viewing places. The Europeans feel their system will be ready for broadcast shortly. American broadcasting of advanced TV is no doubt years away because a standard must first be chosen.[42]

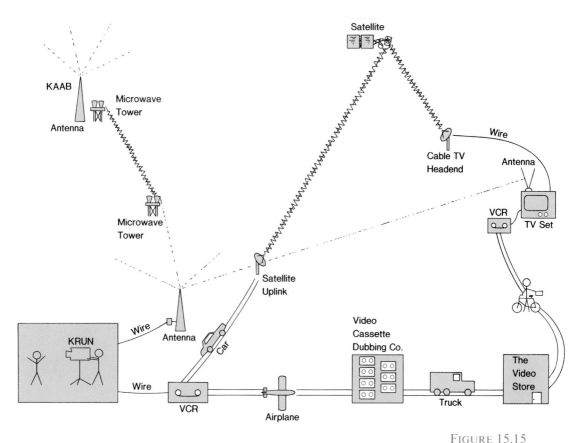

FIGURE 15.15
A distribution pattern.

Conclusion

Many methods of distribution of program material are currently in use. Most of them make use of the electromagnetic spectrum that encompasses a continuum of frequencies. These frequencies have varying characteristics and uses that involve terms such as hertz, VHF, UHF, shortwaves, AM, FM, modulate, line of sight, bandwidth, stereo, multiplexing, clear channels, and carrier waves.

Often several methods of distribution will be used in conjunction with one another before the programming gets from the producing agency to the consumer. For example, a radio station may send a program from the station to the transmitter by wire and then send it out the antenna through the airwaves where it can be received by radio sets in homes and cars. This same program may be sent to a satellite and received by other stations with satellite receivers; these other stations then send it out their transmitters to radio receivers in their areas. The program might even be duplicated onto an audio cassette and copies sold in stores by means of bicycle distribution.

An over-the-air broadcast TV program can be sent by wire, microwave, or satellite, or it can be bicycled. Although TV set reception has improved greatly over the years, it will improve more if HDTV is implemented.

Cable TV reception and distribution probably illustrate the ultimate in combining distribution systems. At the cable headend, local AM and FM radio station signals are received by an antenna as they are transmitted through the air. Sometimes a cable system will also pick up a multiplexed signal of an FM station, particularly if this signal is in the public interest, such as readings for the blind. Local TV stations will also be received by an antenna.

UPI cable news and various other video data information can be received through wires that also come to the headend. One or two distant stations may be received by a microwave link involving two or three microwave relay stations that bring the signal from the distant station to the cable system headend.

The headend will also have at least one satellite dish to receive pay-cable, superstations, and other cable network services. The headend may also include videotape recorders to which members of the public carry their access programs for cablecasting.

All of these inputs come to the headend where they are converted in such a way that they can be placed on the various cable system channels.

Now the inputs become outputs and they are sent through coaxial cable to subscribers' homes. If the cable system covers a large area, the signals may be microwaved to a hub and then sent through cable to homes.

With all the distribution processes in mind, try to describe what is happening to the exercise program being produced in the drawing in figure 15.15.

Thought Questions

1. If frequency modulated radio were placed in the medium frequencies rather than VHF, would the technical quality of radio be better than the present AM or present FM? Explain.
2. What will be the future for satellites? For fiber optics?
3. What do you think will be the status of HDTV in ten years?

Regulatory Controls

There are few countries of the world in which the government has such a hands-off policy toward telecommunications as in the United States. In some countries, the government totally controls both the finances and the programming of telecommunications entities. In other countries, the government oversees and directs overall philosophy and content.

In the United States, government bodies affect and regulate telecommunications but not to any great degree. In addition, there are groups and individuals who exert informal regulatory control through the democratic process. These groups and individuals can bring pressure to bear upon broadcasters that is sometimes more potent than formal regulation.

PART 6

411

Regulatory Bodies

Introduction

Many entities are involved in regulation of the telecommunications industry, both formally and informally. They range from Congress to individual citizens, from the Federal Communications Commission to broadcast producers. The regulatory groups intermingle, in part because of the manner in which the nation's forefathers established the basic government. The legislative branch writes the laws, the executive branch administers them, and the courts adjudicate them.

This basic process carries over to telecommunications regulation. A broadcast station that feels it has been wronged by a decision of the Federal Communications Commission can appeal that decision to the courts and also lobby in Congress to have the offending law changed.

In addition, democracy accords many rights to individuals and to organizations. These, too, have their impact upon the regulatory process.

Various entities become involved in telecommunications regulation in varying degrees. The Federal Aviation Authority becomes involved only when improperly lit antenna towers may be a hazard to airplanes. On the other hand, the Federal Communications Commission's main function is the regulation of the airwaves.

CHAPTER 16

All legislative powers herein granted shall be vested in a Congress of the United States, which shall consist of a Senate and House of Representatives. . . . The Executive power shall be vested in a President of the United States of America. . . . The Judicial power of the United States shall be vested in one Supreme Court and such inferior courts as the Congress may from time to time ordain and establish.

The Constitution of the United States

Organization and Functions of the FCC

origins

The **Federal Communications Commission (FCC)** is an independent regulatory body that was created by Congress because of the mass confusion and interference that arose when early radio stations broadcast on unregulated frequencies and at unregulated power. First Congress passed the **Radio Act of 1927,** which created the **Federal Radio Commission (FRC)** to deal with the chaos. Then in 1934 Congress wrote a new law, the **Communications Act,** which formally established the FCC with powers similar to its predecessor, the FRC.

commissioners

The FCC is composed of five commissioners appointed for five-year terms by the president, with the advice and consent of the Senate. The president designates one commissioner to be chairperson, but generally no president has the opportunity to appoint many commissioners because their five-year terms are staggered. Each commissioner must be a U.S. citizen with no financial interest in any communications industry, and no more than four of the five commissioners are supposed to be from one political party. Usually commissioners have backgrounds in engineering or law.

offices and bureaus

The commission maintains central offices in Washington and field offices throughout the country. The commission staff is organized into eight staff offices (managing director, general counsel, engineering and technology, plans and policy, legislative affairs, public affairs, administrative law judges, and the review board) and four bureaus (mass media, field operations, common carrier, and private radio).

Policy determinations are made by all the commissioners, with the chairperson then being responsible for the general administration of the commission's affairs. Most of the day-to-day work, such as handling interference complaints, public inquiries, and station applications, is undertaken by the staff.[1]

nonbroadcast functions

The FCC has myriad functions, many of which are not related to electronic mass media. For example, it has jurisdiction over airplane communications, ship-to-shore radio, police and fire communications, telephone and telegraph common carrier services, ham radio operations, military communication, and citizens' band radio. During wartime it coordinates the use of radio and TV with the national security program and may set up a service to monitor enemy propaganda. The FCC is constantly encouraging new uses of radio waves, particularly those that will promote safety.

All of this makes for a heavy work load for the commission, particularly as new uses for radio waves surface. The sudden popularity of citizens' band radios during the mid-1970s swamped the commission with problems of interference and channel reallocation, and the interest in low-power TV, multichannel multipoint distribution service, and direct broadcast service during the 1980s did likewise.[2]

rules and regulations

The FCC has been given powers, under the Communications Act of 1934, to set up rules and regulations that relate to the general operation of the telecommunications industry. This is how policies regarding issues such as must-carry, financial interest-domestic syndication, and distant signal importation have come into being.

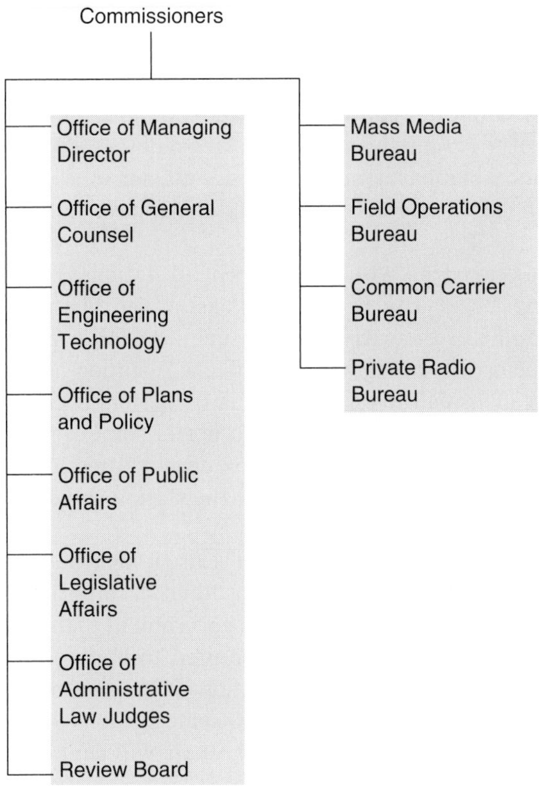

FIGURE 16.1
FCC organization
chart.

Commissioners

Office of Managing
Director

Office of General
Counsel

Office of
Engineering
Technology

Office of Plans
and Policy

Office of Public
Affairs

Office of
Legislative
Affairs

Office of
Administrative
Law Judges

Review Board

Mass Media
Bureau

Field Operations
Bureau

Common Carrier
Bureau

Private Radio
Bureau

Suggestions for these rules can come from telecommunications practitioners, the public, Congress, or other outside sources, or they can come from FCC bureaus or offices. In any case, the suggestion is referred to the appropriate office or bureau, and the commission then issues a notice asking interested parties to comment. These comments and replies are then evaluated and reported upon to the FCC Commissioners. Sometimes formal hearings are held before the Commissioners so that they may hear what specific individuals or groups have to say and ask pertinent questions. The Commissioners tnen decide whether to issue a new order, amend an old one, ask for more study, or do nothing. If some element in the telecommunications industry does not like the order, it can petition for reconsideration, and the process starts all over again.[3]

A great deal of what the FCC does in regulating radio and TV stations involves engineering. The FCC assigns **frequencies** to individual stations, determines the power each can use, and regulates the time of day each may operate. It then polices the broadcasters to make sure they stay within the frequency, power, and time regulations and to make sure unauthorized persons do not use the airwaves. In fact, about one-fourth of all FCC employees are employed in this fieldwork. The FCC also makes overall regulations to prevent interference among stations and regulates the location of station transmitters and the type of equipment used for transmission.

engineering functions

The FCC also controls the general allocation of frequencies, deciding which frequencies go to ship-to-shore communication, which to TV, which to FM radio, and so forth. Within the frequencies it allocates to radio and TV, the FCC creates a table designating the power and times that stations can exist in each section of the country and on each frequency. It also has the jurisdiction to set technical standards, such as those for color television or stereo TV. The commission can decide not to set a standard, however, as in the case of AM stereo.

The commission deals with **call letters** of all stations. Those stations west of the Mississippi begin with K and those east of the Mississippi begin with W (except for some of the early stations such as KDKA in Pittsburgh that had call letters before the ruling went into effect). A station can select or change the other letters of its call letters so long as the letters it chooses are not already in use by a competing station or do not in some way infringe on the rights of another station. Stations must broadcast station identifications that include the call letters and the community the station serves at least once an hour.[4]

The **Emergency Broadcasting System** is also under the jurisdiction of the FCC. This is a national hookup that ties together all radio and TV stations so that information can be broadcast from the government to the citizenry during a national emergency. All stations are required to maintain the equipment necessary for receiving emergency notification and to test this equipment regularly. If a state of emergency is declared, some stations will remain on the air broadcasting common information and others will shut down and remain off the air until the emergency is over.

Most of the technical responsibilities of the FCC are noncontroversial and broadcasters generally appreciate the FCC's role in this regard. The sensitive points arise in the area of licensing. The FCC has the power to grant, renew, transfer, and revoke licenses.

License Granting

Any citizen, firm, or group interested in a radio or TV license must file a written statement of qualifications with the FCC. Most of the regular radio and TV frequencies have been allocated, but new services such as LPTV or an expansion of the AM band become available occasionally.

One category of qualification for any license is character, which includes obvious matters such as felony convictions, participation in community organizations, and desire to be involved in the day-to-day operations of the stations. Aliens and foreign companies cannot own U.S. stations except in unusual circumstances.

Applicants for licenses must also describe their financial and technical qualifications. Financially, they must have access to enough capital to build and begin operation of the station.

Applicants applying for a new AM station or a low-power TV station must arrange for an engineering investigation to establish that the stations will not interfere with other stations because of its frequency, power, or hours of operation. An applicant for an FM or regular TV station can consult the allocation tables already set up by the FCC to find a place for a station.[5]

technical

The applicant must also set forth a full statement for its proposed program service that the FCC can take under cautious consideration. The FCC has no power to censor program materials. In fact, the Communications Act specifically prohibits censoring by stating that:

programming

> Nothing in this Act shall be understood or construed to give the Commission the power of censorship over the radio communications or signals transmitted by any radio station, and no regulation or condition shall be promulgated or fixed by the Commission which shall interfere with the right of free speech by means of radio communications.[6]

This indicates that the FCC cannot refuse a license to a financially, technically, and morally qualified candidate simply because it plans to broadcast some form of programming that the FCC finds objectionable. However, another section of the Communications Act states that:

> The Commission, if public convenience, interest, or necessity will be served thereby, subject to the limitations of this Act, shall grant to any applicant therefore a station license provided for by this Act.[7]

This **public convenience, interest, or necessity** statement has become a keystone of regulation.

Once an applicant has filed a written statement of character, financial, technical, and programming qualifications, the FCC must decide whether or not it should grant permission to build the station. If there is only one applicant, this process is fairly simple. If the paperwork indicates that the applicant will be successful and conscientious, then a **construction permit (CP)** is issued, and the applicant can begin to build the station.

CP

If there are multiple applicants, which is usually the case, then the FCC must decide which applicant will receive the CP. This used to be a very laborious process whereby the FCC staff members sorted through all the applications trying to determine which group or person was most qualified. Now, for the most part, the FCC decides station allocations by **lottery.** In essence, the names of all applicants are "put in a hat" and the winner is chosen by chance. That applicant is then carefully checked by the FCC to make sure it meets the qualifications.[8]

lottery

The chosen applicant then receives the CP and can begin building the station. After construction is completed, the license and permission to begin conducting program tests are granted. Once a station receives a license, it usually experiences little supervision from the FCC until license renewal time.

License Renewal

License renewal is the most controversial and publicized function of the FCC, mainly because the process has undergone radical changes throughout the years.[9] The power to renew licenses was given first to the Federal Radio Commission in 1927 and then to the Federal Communications Commission in 1934. Both of these agencies renewed licenses almost automatically in the early years. Two cases did arise, however, during the late 1920s and early 1930s where a license was not automatically renewed.

One concerned Dr. J. R. Brinkley who broadcast medical "advice" over his Milford, Kansas, station. He specialized in goat gland treatments to improve sexual powers and in instant diagnosis over the air for medical problems sent in by listeners. These problems could always be cured by prescriptions obtainable from druggists who belonged to an association Dr. Brinkley operated.

The other culprit was the Reverend Robert Shuler, who used his Los Angeles station to berate Catholics, Jews, judges, pimps, and others in his personal gallery of sinners. He professed to have derogatory information regarding unnamed persons who could pay penance by sending him money for his church.

The Federal Radio Commission decided not to renew either of these licenses and, although both defendants cried censorship, the Court of Appeals and U.S. Supreme Court sided with the FRC on the grounds that with only a limited number of frequencies available, the commission should consider the quality of service rendered. In ruling on the Shuler case, the court wrote that "if stations were possessed by people to obstruct the administration of justice, offend the religious sensibilities of thousands, inspire political distrust and civic discord, or offend youth and innocence by the free use of words suggestive of sexual immorality, and be answerable for slander only at the instance of the one offended, then this great science, instead of a boon, will become a scourge, and the nation a theater for the display of individual passions and the collusion of personal interests. This is neither censorship nor previous restraint."[10]

Aside from these two cases, the early FCC was quite lenient in license renewal, being more concerned with granting original licenses. But in 1945 the FCC decided to philosophize a bit about license renewal in order to clarify public convenience, interest, and necessity, while at the same time issuing temporary renewals to six stations so that it could have time to decide whether they were, indeed, serving the public interest.

In 1946 the commission issued an eighty-page document detailing its ideas on license renewal: "Public Service Responsibility of Broadcast Licensees." This document was almost instantly dubbed the **Blue Book** by broadcasters, in part because of its blue cover but more sarcastically because blue penciling denotes censorship. The document stated what became known as the **promise versus performance** doctrine. It stated that promises made when stations were licensed should be kept, and the performance on those promises should be a basis for license renewal.

Broadcasters probably would not have objected to that philosophy, but the document went on to detail proper broadcasting behavior. It was particularly adamant about avoiding the evils of overcommercialization, broadcasting public affairs programs and local programs, and maintaining well-balanced programming. The document was so lengthy that broadcasters looked at it as an affront to freedom of speech and a violation of the section of the Communications Act that stated the commission was not to have the power of censorship. On these grounds, broadcasters fought the Blue Book, and its provisions were never truly implemented.[11]

Licenses continued to be renewed almost perfunctorily. Stations did receive rebukes for such violations as failure to make proper entries in logs, failure to broadcast station identifications frequently enough, failure to have engineering instruments calibrated properly, failure to authenticate sponsorship of programs, failure to present controversial issues properly, and failure to give equal time to political candidates. For these violations, stations were fined, issued cease and desist orders, and issued short-term probationary licenses of less than three years, but these types of activities were few and far between.[12]

In 1960 the FCC softened its tone and issued a much briefer statement *1960 policy statement* of policy that listed fourteen elements usually necessary to meet the public interest:

1. Opportunity for local self-expression
2. Development and use of local talent
3. Programs for children
4. Religious programs
5. Educational programs
6. Public affairs programs
7. Editorializing by licensees
8. Political broadcasts
9. Agricultural programs
10. News programs
11. Weather and market reports
12. Sports programs
13. Service to minority groups
14. Entertainment programs

It also warned broadcasters to avoid abuses to the total amount of time devoted to advertising as well as the frequency with which programs are interrupted by commercials, but it did not specifically define what constituted abuses.[13] In addition, the FCC changed the license renewal forms so that broadcasters had to undertake **ascertainment**—a process by which they had to interview community leaders to obtain their opinions on the crucial issues facing the community so that broadcasters could design programming to deal with those issues.

The results of these interviews and the proposed program ideas were then submitted to the FCC. Along with this material, copious information concerning station operation for the previous three years, including a **composite week's** list of programming, was also sent. These seven days from the previous three years were selected at random by the FCC. The FCC's main concern was whether the programs listed adhered to what the station had proposed three years earlier.

For several years after the 1960 changes, licenses continued to be renewed rather perfunctorily. License renewal was primarily a private affair between the stations and the FCC. Over a thirty-five-year period dating from the 1920s, only forty-three licenses were not renewed out of approximately fifty thousand renewal applications.[14] But during the 1960s and early 1970s, this changed. Community people gradually became involved in the license renewal process, with the FCC scrutinizing license renewal much more carefully than in the past.

WLBT-TV

The first major community participation occurred in 1964 when a black group from Jackson, Mississippi, led by the United Church of Christ, asked the FCC if it could participate in the license renewal hearing of station **WLBT-TV.** The group felt the station was not serving its viewership properly and was presenting racial issues unfairly. The FCC refused this hearing because prior to this time, only people who would suffer technical or economic hardship from the granting of a license were permitted to testify at hearings, but it did issue WLBT a short-term license.

The UCC appealed to the courts stating that ordinary citizens should be able to be heard concerning license renewals. The courts agreed and ordered that a hearing be held. After this 1966 hearing, which some felt was a sham, the FCC decided WLBT should receive a full three-year license. The UCC once again appealed, and the courts decided WLBT should have its license withdrawn and also scolded the FCC for its administrative procedures during the hearing. WLBT was turned over to a nonprofit group, but the significance of this case was that it established the precedent for citizen participation in license renewal.[15]

WHDH

In 1969 a group of Boston businesspeople successfully challenged the ownership of TV station **WHDH,** which had been operated by the Boston Herald Traveler for twelve years. This was a very complicated case and involved rumors of improprieties on the part of all three companies that had originally applied for the station shortly after the freeze was lifted. The station had been awarded to the Herald Traveler in 1957, but various allegations had kept the ownership longevity in turmoil. In 1962 the Herald Traveler was given a four-month temporary license in the hope that the situation would be clarified within that time.

Four *years* later, an FCC hearing examiner recommended that the license be kept by the Herald Traveler, mainly because it had done a good job with programming during the four years that were supposed to have been only four months. Three years later (or seven years after giving WHDH a four-month license), the commission revoked the Herald Traveler license and awarded it to Boston Broadcasters, Inc., an organization that included some of the original unsuccessful applicants for the station back in the 1950s.[16]

This January 1969 decision sent shivers through the entire broadcasting community. Never before had a station license been transferred involuntarily unless the station licensee had first been found guilty of excessive violations. From all that was written about the case, most broadcasters concluded that the FCC did not base its decision on the fact that the Herald Traveler was an undesirable licensee, but rather on the opinion that the businesspeople were better.

The net result of the WHDH and WLBT cases was that a raft of renewal challenges were filed. In some instances, groups requested a **comparative license renewal** hearing because they wanted to operate the stations themselves.

In other cases, citizen groups filed **petitions to deny** license renewal mainly so that they could bargain for what they wanted. A 1969 case that became a model for bargaining between citizens and stations was that of **KTAL** in Texarkana, Arkansas. The citizens complained because KTAL, although licensed to Texarkana, was neglecting that city by moving its studios to Shreveport, Louisiana, a much more lucrative market. The groups met and worked out a thirteen-point plan that would assure that KTAL served Texarkana. The citizen group then withdrew its petition to deny the renewal. The FCC applauded the efforts of the citizens and station and said it would examine the record of KTAL when it came up for renewal again to make sure it was, indeed, serving Texarkana.[17]

KTAL

License renewal was also no longer perfunctory as evidenced by the fact that even public broadcasting stations had renewal troubles. In 1970 the **Alabama Educational Television Commission** applied for what it assumed would be routine renewals for its public television stations. However, the FCC refused renewal because of citizen petitions stating that the stations had systematically deleted all Public Broadcasting Service programs dealing with blacks or Vietnam. The Alabama commission was allowed to reapply for its stations with the understanding that it would mend its ways. Nevertheless, the case marked the first time such harsh action was taken against any public broadcasting stations.[18]

Alabama Educational Television Commission

One license renewal case that has dragged on for years involves **RKO** and the sixteen stations that it owned. Several of the RKO stations had been under attack by citizen groups as far back as the 1960s, and petitions to deny renewal had been issued against them. Then, in 1980, the FCC refused to renew the licenses of three of the RKO television stations including WWOR in New York and KHJ in Los Angeles. Its reason for doing this was that RKO's parent corporation, General Tire and Rubber, had admitted to the Securities Exchange Commission that it had bribed foreign officials. The FCC felt that RKO, therefore, did not meet the character requirements necessary to operate radio and TV stations. The license refusals were appealed and lengthy deliberations were begun.

RKO

The WWOR license went through an interesting evolution. The state of New Jersey did not have any VHF stations and had long objected to that fact. A senator from New Jersey introduced a bill into Congress that directed the FCC to grant a five-year license to the owner of any VHF station that would agree to relocate the station in a state having none. This bill, which became

APPLICATION FOR RENEWAL OF LICENSE FOR COMMERCIAL AND NONCOMMERCIAL AM, FM OR TV BROADCAST STATION

Federal Communications Commission
Washington, D.C. 20554

Approved by OMB
3060-0110
Expires 5/31/91

For Commission Fee Use Only	
FEE NO:	
FEE TYPE:	
FEE AMT:	
ID SEQ:	

For Commission Use Only: File No.

For Applicant Fee Use Only

Is a fee submitted with this application?   ☐ Yes ☐ No

If No, indicate reason therefor (check one box):
☐ Nonfeeable application

Fee Exempt (See 47 C.F.R. Section 1.1112)
☐ Noncommercial educational licensee
☐ Governmental entity

1. Name of Applicant

Mailing Address

City	State	ZIP Code

2. This application is for: ☐ AM ☐ FM ☐ TV

(a) Call Letters:

(b) Principal Community:
City   State

3. Attach as Exhibit No. _____ an identification of any FM booster or TV booster station for which renewal of license is also requested.

4. Have the following reports been filed with the Commission:

(a) The Broadcast Station Annual Employment Reports (FCC Form 395-B) as required by 47 C.F.R. Section 73.3612?   ☐ Yes ☐ No

If No, attach as Exhibit No. _____ an explanation.

(b) The applicant's Ownership Report (FCC Form 323 or 323-E) as required by 47 C.F.R. Section 73.3615?   ☐ Yes ☐ No

If No, give the following information:
Date last ownership report was filed _____
Call letters of station for which it was filed _____

FCC 303-S
May 1988

FIGURE 16.2

A license renewal form. *(Courtesy of the Federal Communication Commission.)*

law in 1982, was obviously aimed at RKO, giving it the opportunity to salvage its license by moving to New Jersey. WWOR did just that and received a five-year license renewal by the FCC. Before the station came up for license renewal in 1987, RKO sold it to MCA.

The courts sent the KHJ case back to the FCC for further deliberation. This was frustrating to both those for and against RKO because the KHJ case had actually begun in 1965 when a company called Fidelity challenged the license of the station. As part of the deliberation process, the FCC decided applications could be made for the other RKO licenses, just in case RKO was found an undesirable licensee in the KHJ hearings. The hearings dragged on and on and RKO gradually (and painfully) sold its stations to other companies. In 1988 Disney wound up buying KHJ and changed the call letters to KCAL. The FCC then stopped the hearing without deciding whether or not RKO was qualified to be a licensee. Thus, twenty some years of proceedings ended, not with a bang but with a whimper. The FCC called this "the most burdensome proceedings in the FCC history."[19]

Overall, however, the threat of losing a license abated during the late 1970s and early 1980s. The FCC issued several different statements, which, in sum, stated that the FCC would compare the station licensee and any challengers, but if the licensee had provided favorable service in the past, it would be given what is called a **renewal expectancy.** This would make it very hard

renewal expectancy

5. Is the applicant in compliance with the provisions of Section 310 of the Communications Act of 1934, as amended, relating to interests of aliens and foreign governments? ☐ Yes ☐ No

If No, attach as Exhibit No. ______ an explanation.

6. Since the filing of the applicant's last renewal application for this station or other major application, has an adverse finding been made or final action been taken by any court or administrative body with respect to the applicant or parties to the application in a civil or criminal proceeding, brought under the provisions of any law relating to the following: any felony; broadcast related antitrust or unfair competition; criminal fraud or fraud before another governmental unit; or discrimination? ☐ Yes ☐ No

If Yes, attach as Exhibit No. ______ a full description of the persons and matters involved, including an identification of the court or administrative body and the proceeding (by dates and file numbers) and the disposition of the litigation.

7. Would a Commission grant of this application come within 47 C.F.R. Section 1.1307, such that it may have a significant environmental impact? ☐ Yes ☐ No

If Yes, attach as Exhibit No. ______ an Environmental Assessment required by 47 C.F.R. Section 1.1311.

If No, explain briefly why not.

8. Has the applicant placed in its station's public inspection file at the appropriate times the documentation required by 47 C.F.R. Sections 73.3526 or 73.3527? ☐ Yes ☐ No

If No, attach as Exhibit No. ______ a complete statement of explanation.

The APPLICANT hereby waives any claim to the use of any particular frequency or of the electromagnetic spectrum as against the regulatory power of the United States because of the previous use of the same, whether by license or otherwise, and requests an authorization in accordance with this application. (See Section 304 of the Communications Act of 1934, as amended.)

The APPLICANT acknowledges that all the statements made in this application and attached exhibits are considered material representations and that all the exhibits are a material part hereof and are incorporated herein as set out in full in the application.

CERTIFICATION: I certify that the statements in this application are true, complete, and correct to the best of my knowledge and belief, and are made in good faith.

Name	Signature
Title	Date

WILLFUL FALSE STATEMENTS MADE ON THIS FORM ARE PUNISHABLE BY FINE AND IMPRISONMENT. U.S. CODE, TITLE 18, SECTION 1001.

FIGURE 16.2
(Continued.)

for a challenger to take away the license. Broadcasters would prefer legislation that would not allow a license to be given to a challenger unless the licensee was proven unfit. They have lobbied for such legislation, but it has not been passed.[20]

new licensing policies

New renewal license policies were initiated in 1981 that extended radio licenses to seven years and TV station licenses to five years. The license renewal forms were also changed to decrease the work undertaken by stations in order to renew licenses. Now stations only need to submit a postcard-size form indicating they have sent reports on employment practices and station ownership to the FCC, they comply with the rules related to interests in the station by aliens and foreign governments, they have not been subject to adverse legal actions, and they have placed a **public inspection file** in a place that is easily accessible to community members.[21]

This public inspection file must contain information about how the station has dealt with political candidates, as well as information about employment practices and station ownership. It must also contain a list of programs that the station has aired during the past three months dealing with community issues.[22]

License renewal obviously has had a checkered history, and future paths are as yet uncharted. The deregulation philosophy favored broadcasters, but a return to consumer advocacy could have the opposite effect.

Other Licensing Powers and Issues

revocation

If the FCC finds that a station is not fulfilling its obligations in serving in the public convenience, interest, and necessity, then the commission can revoke the license. The FCC has done this for such causes as unauthorized transfer of control, technical violations, fraudulent contests, overcommercialization, and indecent programs.[23]

other penalties

The FCC can also issue lesser punishments. It can impose a cease and desist order, which really is no punishment at all but a letter telling a station to stop a certain action. The FCC can fine stations up to $2,000 for every day that an offense occurs, up to a maximum of $20,000. The commission can issue a short-term renewal for six months or a year, which indicates to the licensee that it must mend its ways or the license will be revoked.[24]

license transfer

The FCC also becomes involved when a station is sold and the license is transferred from one party to another. It is the station management's prerogative to set the selling price and select the buyer, but the FCC does check on the character, financial, technical, and programming qualifications of the buyer. The FCC used to have a regulation that required new owners of a station to keep it for at least three years before selling it. This was to prevent people from buying stations just for the investment potential with no consideration for the community welfare. However, with deregulation, this regulation disappeared, and now stations are traded much more frequently than they used to be. Because of this, there is talk that the three-year rule should be reinstated.[25]

In the area of licensing, the FCC has no direct control over the networks, but it can control them indirectly through their owned and operated and affiliated stations. For example, when the FCC limited the number of hours of network programming that a station could air during prime time, a half-hour was chopped off network programming. When CBS was found to have rigged

networks

some tennis matches, the FCC slapped CBS's hand by giving a short-term license renewal to the CBS-owned Los Angeles station, KNXT.[26] But the networks themselves are not licensed. Consequently, they go through the license-renewal process only vicariously by way of the stations carrying their programs.

cross-ownership

An area of licensing that has been subject to fluctuations on the part of the FCC is what has come to be known as **cross-ownership.** Many of the early owners of radio and TV stations were companies that also owned newspapers. This was considered against the public interest, especially if a town had only one TV station that was owned by the newspaper. The fear was that the public would learn only one version of the news. On the other hand, the companies with news interests can operate stations more efficiently because their news functions can serve a double purpose. Similarly, controversies arose over whether one company should be allowed to own both radio and TV stations or even AM and FM stations in the same market. The FCC wrestled with this problem for many years, never finding a politically acceptable solution. Finally, the Justice Department became interested in the subject and urged the FCC to take action.

In 1975 the FCC issued guidelines on cross-ownership that prohibited companies from owning both a newspaper and a broadcast station in the same town. Later, it also prohibited such companies from owning cable TV systems. However, the FCC grandfathered any cross-ownership that already existed, so there still are companies that own combinations of newspapers, radio, TV, and cable TV in particular areas.

In addition, a great deal of cross-ownership exists as a result of mergers and acquisitions of the 1980s. When two companies merged, the new company often had multiple stations in one city because the two old companies had stations in the same city. The new companies petitioned the FCC to allow them to keep all stations, at least for a period of time. The FCC almost always granted the waiver because putting all the stations on the market would have artificially decreased their price.

In 1988, the FCC changed its cross-ownership rules and allowed common ownership of radio and TV stations in the top twenty-five markets. At present it appears the FCC is in a liberal mood and will probably reduce even further the restrictions on cross-ownership.[27]

One relatively new cross-ownership issue the FCC is considering involves the telephone companies. The phone companies are asking to own cable TV systems in areas they serve. At present this is not permissible, but the phone companies are constantly pleading their case.

The philosophical shifts that have occurred in licensing procedures have all taken place under one law—the Communications Act of 1934. This act, like much of the legislation of this country, allows for change and broad interpretation.

The Federal Trade Commission

Another independent regulatory body that becomes involved in broadcast regulation from time to time is the **Federal Trade Commission (FTC).** This commission was established by the Federal Trade Commission Act and the Clayton Act. Both acts were passed in 1914, basically to prevent unfair competition resulting from the giant corporate trusts. In 1938 the **Wheeler-Lea Amendment** to the FTC act was passed. It declared that "unfair or deceptive acts or practices in commerce are also illegal."[28] This opened the door for the FTC to investigate consumer deception, and it served as the takeoff point for the FTC's involvement in broadcasting.

This involvement concerns mainly fraudulent advertising that is deceptive to consumers. If the FTC feels that a company's ads are untruthful, it has

establishment

function

FIGURE 16.3
FTC organization
chart.

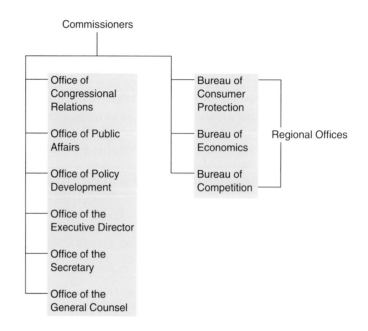

Commissioners

Office of
Congressional
Relations

Office of Public
Affairs

Office of Policy
Development

Office of the
Executive Director

Office of the
Secretary

Office of the
General Counsel

Bureau of
Consumer
Protection

Bureau of
Economics

Bureau of
Competition

Regional Offices

the power to order the company to stop broadcasting the ads. In some instances, the FTC can make the company broadcast a message publicly disavowing the false claims in an untruthful ad. Needless to say, this is distasteful to advertisers.

organization

Five FTC commissioners are appointed for a term of seven years by the president, with the consent of the Senate. The president appoints a chairperson, who in turn handles most of the management and personnel aspects of the commission. Working directly under the chairperson is the executive director, who as the commission's chief operating officer oversees day-to-day operations of the FTC offices and bureaus. The commission organization includes people to advise the commissioners on such issues as legal matters, policy development, competitive practices, and consumer education. Throughout the country there are regional FTC offices to handle local complaints and problems and to refer them to Washington, if necessary.

guidelines

The FTC attempts as much as possible to be preventive rather than punitive in its relationships with broadcasting, as well as other forms of business. In other words, it tries to disseminate its philosophies to businesses in the form of written statements known as Trade Regulation Rules, Industry Guides, and Advisory Opinions so that businesses know how to avoid the wrath of the FTC. If, for example, the commission decides to issue guidelines on toy advertising, it will inform all toy companies that it is planning to create rules in their area. Executives of toy companies, as well as members of the general public, then have a chance to express opinions about what these rules should include. After the members of the FTC have heard the various opinions, they write the guidelines and publicize them as widely as possible among toy companies, related trade associations, and consumers. The Trade Regulation Rules, Industry

Guides, and Advisory Opinions do not have the force of law; they are suggestions that the FTC hopes businesses will follow in order to avoid trouble.

If a company violates one of these guidelines or engages in other practices that the FTC believes will hurt competition or deceive consumers, then the FTC can bring action against the company.

Obviously, the FTC staff cannot act as a constant watchdog over every company in the country. Although the staff does occasionally bring charges, the FTC depends greatly on the public at large and businesspeople in general to bring complaints to it.

Often investigations are initiated by letters from consumers sent to the Washington FTC office or to one of the regional offices. These letters are first reviewed to make sure that the complaints expressed actually come under the jurisdiction of the FTC. If so, the investigation begins with FTC personnel questioning officials of the company involved in the alleged misdeed. Usually the FTC wants proof from the company that its ad is accurate. For example, if the complaint is against a car manufacturer who says its car is 50 percent quieter than any competitor's car, the FTC will require scientific proof of this claim.

At the end of the investigation, the FTC may close the case for failure to find a violation. If the FTC thinks a violation occurred, it can attempt to obtain voluntary compliance from the offending company by entering into a **consent order** with the company. In this case, the company does not admit to violations of any law, but it agrees to stop the disputed practice.

If a consent agreement can't be reached, the FTC can issue an administrative complaint or it can obtain an injunction to make the company stop its practice. These can then be taken through the court system, all the way to the Supreme Court, if need be, either by the FTC or the company.

Although the FTC is not as involved with the electronic media as the FCC, it occasionally handles issues of great concern to media advertisers.[29]

The Executive Branch

Obviously, the president has influence over the media, both formally and informally. Formally, the president can suspend broadcasting operations in time of war or threat of war and call into action the Emergency Broadcast System. The president also nominates commissioners and appoints both the FCC and FTC chairpersons. Most of the interaction between the president and the media, however, comes informally, with the president seeking positive coverage from radio and TV. This reached a peak during the Nixon administration when Nixon created an **Office of Telecommunication Policy** to advise him on the media and to make statements concerning media practices. This office's complaints about the type of coverage Nixon was receiving created an even wider gap between him and the press.

Partly as a result of this, President Jimmy Carter disbanded the OTP and formed the **National Telecommunications and Information Administration (NTIA),** which he placed under the Department of Commerce. Although this organization advises the president on media issues, its placement in the Department of Commerce keeps it somewhat distant from the president.[30]

Various executive departments also interact with telecommunications. The Justice Department oversees antitrust and as such was involved in the breakup of AT&T. The Department of State has a Bureau of International Communications and Information Policy that advises on media issues with possible international consequences. From time to time, other departments, such as defense and education, become involved with media-related issues.

other agencies

Other agencies that are either part of the executive branch or are independent occasionally affect broadcasting. The Equal Employment Opportunity Commission watches over telecommunications companies to assure that they are fulfilling commitments to affirmative action. The Federal Aviation Authority becomes involved with antenna towers and lighting so that planes will not crash into antennas. The Food and Drug Administration occasionally becomes involved with the misbranding or mislabeling of advertised products. The surgeon general's office became a party to the 1972 ban of cigarette advertising on radio and TV.

The Legislative Branch

Congress is heavily involved in telecommunications. It passed the Communications Act of 1934, which set up the basic broadcasting structure. Major changes in this structure must be approved by Congress as amendments to this act.

The whole basis for the government's intervention in radio frequencies was, and occasionally still is, debated in Congress. Ordinarily, the First Amendment would prohibit the government from infringing on the rights of citizens to communicate by whatever means they wish. However, Congress

scarcity theory

invaded this area because of the **scarcity theory.** Not everyone who wants to can broadcast through radio frequencies because this would cause uncontrollable interference. As a result, some body, namely Congress, needed to intervene and determine a mechanism for making decisions regarding who could and could not use the airwaves.[31]

Once given this right, Congress needed to find a provision in the Constitution that covered the subject. The provision chosen was the one empowering Congress to regulate commerce with foreign nations and among the states.

commerce

Although broadcasting was not commerce as such—goods did not change hands—the exchange of information by mail and wireless had already been accepted as a form of commerce, so broadcasting was added to this definition.

congressional laws

In addition to overall regulatory decision making that is tied to the Communications Act, Congress passes other laws, such as the copyright law and the cable TV law, that have important impact upon those dealing with telecommunications. Congress passed the law that set up the Corporation for Public Broadcasting, and it regularly decides on funding for public broadcasting. Congress approves FCC and FTC commissioners and budgets, and it monitors the FCC through both the House and Senate subcommittees on communications. These and other committees also often conduct special investigations into such aspects of telecommunications as the quiz scandals, rating practices, TV violence, or the effect of budget cuts on TV news.

State legislative bodies can also affect telecommunications. Under the Constitution, federal laws such as the Communications Act take precedence over state laws. Occasionally state laws dealing with libel, advertising, or state taxes have an effect upon elements of the industry. City councils, which have local legislative control, have jurisdiction over cable TV franchising. The legislative branch generally sets up broad policies that are handled on a day-to-day basis by other agencies. Needless to say, these broad policies are very influential in setting direction for the electronic media.

state and local laws

The Judicial Branch

The courts of the United States have had a significant impact on telecommunications, primarily through the appeal process. FCC decisions concerning licenses can be appealed through the U.S. Court of Appeals for the District of Columbia Circuit in Washington, D.C. This court can confirm or reverse the commission's decision or send it back to the FCC for further consideration. Appeals for FCC or FTC decisions that do not involve licenses can be taken to any of the eleven other U.S. Courts of Appeal, generally referred to as circuit courts. From any of these courts, final appeals can be taken to the Supreme Court, which can confirm or reject a lower court's decision. The Supreme Court can also refuse to hear a case, which means that the lower court's finding will stand.

appeals procedure

Courts are also significant in dealing with decisions other than those arising from FCC or FTC decisions. Rulings on freedom of speech, obscenity, censorship, copyright, monopoly, and many other subjects often have an impact that determines the direction of electronic media.

court decisions

Broadcasting Organizations

Organizations to which media practitioners belong often serve as regulators in an informal manner. They publish policy statements, lobby with Congress, and hold meetings at which pertinent issues in telecommunications are discussed.

The largest, most influential organization is the **National Association of Broadcasters (NAB),** formed in 1923 to counter demands from the American Society of Composers, Artists, and Publishers (ASCAP) regarding increases in the amount ASCAP was planning to charge radio stations for airing music.[32] When that issue was resolved, the NAB acquired many other purposes, such as lobbying in Washington for actions favorable to broadcasters, conducting broadcast-related research, acting as a public relations arm for broadcasters, and holding conventions and workshops to facilitate communication and professional growth among industry members.

NAB

At one time, the NAB also acted as a strong self-regulation force. It developed two codes, "The Television Code," and "The Radio Code," which were printed in two booklets that spelled out dos and don'ts for radio and television stations.[33] Stations were not required to follow these rules; they were merely advisory. But sometimes the inclusion of a provision in the NAB code prevented that provision from being legislated in a manner that would be compulsory.

The codes contained two major types of statements—those dealing with programming and those dealing with advertising. The advertising portion was fairly specific about the amount of time that could be devoted to commercials during different times of the day and was the part of the code that could be utilized most tangibly.

This portion of the code came under attack, however, in 1979. The Justice Department, acting on its own, started an antitrust suit against the advertising provisions stating that these guidelines artificially limited the supply of advertising time, increased the cost of commercial time, and violated the Sherman Antitrust Act. In 1982 the U.S. District Court sided with the Justice Department. The NAB stated that the decision made absolutely no sense because it would mean that the public would be subjected to more commercials. Nevertheless, the NAB announced that it would retire the time standards portions of the code, which, in effect, killed the self-regulatory function of the NAB.[34]

The NAB still exists as an effective voice for broadcasters, but it is no longer the self-regulatory body it once was.

Self-regulation is promoted by other organizations, too. Advertisers have an organization called the National Advertising Review Board, which handles complaints about ads from individuals, groups, or an advertiser's competition. The Radio-Television News Directors Association and the Society of Professional Journalists have codes dealing with news ethics. Although these are both fairly formal methods of self-regulation, many other organizations have more informal guidelines established at meetings.

Most of the organizations represent subgroups of the radio and TV business, the proliferation of which is often a matter for criticism. For example, broadcast executives can choose to join the National Association of Television Program Executives, the Association of Maximum Service Telecasters, the Broadcast Financial Management Association, the Broadcasters Promotion and Marketing Executives, the International Radio and Television Society, the Association of Independent Television Stations, and/or the National Association of Farm Broadcasters.

Engineers probably have more associations available to them than any other group. These include the Audio Engineering Society, the Institute of Electrical and Electronics Engineers, the Society of Broadcast Engineers, the Society of Motion Picture and Television Engineers, and the Society of Cable Television Engineers.

Advertising organizations also abound, including the American Advertising Federation, the American Association of Advertising Agencies, Broadcast Advertising Producers Society, the Association of National Advertisers, and the League of Advertising Agencies.

Cable TV's equivalent to the NAB is the National Cable Television Association (NCTA). A cable organization for access producers is the National Federation of Local Cable Programmers. People involved with multichannel multipoint distribution service have organized the Wireless Cable Association.

For educators there are organizations such as the Broadcast Education Association, the Association for Education in Journalism and Mass Communication, and the University Film and Video Association. Both educators and people involved with corporate video belong to the ITVA (International Television Association).

The Academy of Television Arts and Sciences is for those involved in the creative activities of programming; Alpha Epsilon Rho is a national honorary broadcasting society; and Broadcast Pioneers is open to those who were involved in early radio and TV.

Other representative organizations include American Women in Radio and Television, Country Radio Broadcasters, Television Critics Association, National Religious Broadcasters, and National Black Media Coalition.[35]

No one can belong to all relevant organizations, but membership in selected ones can help a person keep up to date and in touch with others who have similar interests and regulatory approaches.

Awards

Like the organizations, broadcasting awards have an informal regulatory function. Broadcasters covet awards and use them for personal and station promotion. Therefore, they will frequently air programs that they feel might qualify for awards.

Most famous of the TV awards are the **Emmys** that are bestowed by the Academy of Television Arts and Sciences for many categories of national prime-time programming. (Daytime awards are given by a different but similar organization, the National Academy of Television Arts and Sciences.) Each year the categories within which awards will be made are hotly disputed by committees of industry representatives. How many westerns should be on TV before an award is given for best western? Is sound mixing on videotape different enough from sound mixing on film to merit separate Emmys? Should writers of short series be placed in the series category or the specials category? These intramural disputes occur to some extent because of the changing nature of TV programs, the huge outpouring of material, and the importance placed on winning an Emmy.

Emmys

The NCTA has established **ACE** (Awards for Cablecasting Excellence) awards that are presented each year, the National Federation of Local Cable Programmers presents awards for access programs, and the ITVA gives awards for various categories of corporate video.

cable and corporate

FIGURE 16.4
The Emmy, awarded
for outstanding
television
programming.
*(Courtesy of the Academy of
Television Arts and Sciences)*

Clio and Peabody

For commercials, the top statuette is the **Clio,** given annually at the American TV and Radio Commercials Festival Awards for best U.S. radio and TV commercials. The prestigious George Foster **Peabody** Awards for Distinguished Broadcasting are bestowed annually in news, entertainment, education, youth programs, documentaries, and public service.

other awards

Other awards include the Armstrong Award for excellence in radio broadcasting; the Freedoms Foundation Awards for programs to bring about a better understanding of America; the National Press Photographers Association Awards for best news stories; and the Ohio State Awards for educational, informational, and public affairs broadcasting.[36]

Network and Station Policies

Many networks and stations have their own written or unwritten codes. For example, some stations will not accept beer and wine ads and others publish a list of words that they consider obscene or indecent.

broadcast standards

At the commercial broadcast networks, the business of making sure all programs and commercials adhere to good taste falls to the **broadcast standards** department, a group usually operating independently of programming or sales and reporting directly to top network management. This group reviews all program and commercial ideas when they are in outline or storyboard form and then screens them several times again as they progress through scripting and production. If any of the proposed ideas run counter to standards, the broadcast standards department requests changes before the idea can proceed to the next step. The department is also responsible for making sure all copyrighted material used within programs has been cleared and that all advertising claims can be substantiated.

Naturally, all is not roses and moonlight between broadcast standards and program or commercial producers. The latter can appeal decisions to top management or try to convince the broadcast standards department to compromise on certain issues. The power of the broadcast standards people to actuate change varies with the times. When the networks are under a great deal of pressure from consumer groups and/or the government regarding sex and violence, broadcast standards is listened to carefully. When the pressure is off, these people assume more of a rubber stamp capacity.

At local stations the broadcast standards function may not be as formal. Frequently the general manager performs the function on an "as needed" basis, and sometimes the function is delegated to whichever group or person handles station legal matters or public relations.

Within the cable TV industry, talk of self-regulation has arisen from time to time, particularly in regard to hard-R movies, but nothing has been implemented.

Network and station self-regulation occurs constantly and informally at all levels through the actions of individuals employed in broadcasting. A disc jockey who decides not to play a record he or she considers in poor taste is engaging in self-regulation, even though management provides no direction on the matter. Likewise, writers, producers, directors, actors and actresses, and editors are constantly basing decisions on their own ideas of propriety and appropriateness.[37]

Citizen Groups and Individuals

Community and national citizen groups can alter broadcasters' behavior and thereby execute informal regulation. Organizations such as the NAACP and NOW have been successful in changing both program fare and hiring practices as they relate to minorities. Action for Children's Television has had an important role in changing children's programming. Consumer groups have frequently organized boycotts against products advertised on certain programs, and broad consumer groups have established a power base to affect overall programming.

Professional critics occasionally serve as a catalyst for establishing self-regulation. Their individual influence is small since television programs live and die by ratings rather than by critical review, but their columns are sometimes used to substantiate points being made by other groups. Occasionally critics have been instrumental in saving a public affairs or documentary show that might have been dropped because of poor ratings.

Broadcasters are often the object of unfavorable comments made by national opinion leaders. It is hard to tell whether Newton Minow's "vast wasteland" speech or former Vice-President Agnew's blast at broadcast journalism led to any self-regulation, but both certainly caused a stir within the industry.

All of these forces—the branches of government, organizations, awards, media practitioners, and citizens—aid in shaping the regulation of telecommunications.

The Pros and Cons of Regulation

The relationship between the electronic media and the government is a forced marriage that has never enjoyed a honeymoon atmosphere. Broadcasters are ever fearful that their privileges will be regulated to oblivion, and government agencies are afraid that broadcasters will take advantage of their privileges to the detriment of the entire system. The upper hand is attached to a constantly swinging pendulum seeking a middle ground.

protection of broadcasters

Sometimes government regulation protects the broadcaster. Obviously the technical regulations fall into this category, preventing the interference and chaos that reigned during the 1920s. Also, the system generally ensures that stations are licensed only to people of proper character so that one or two bad apple stations cannot ruin the entire broadcasting barrel.

Licenses are also refused to station applicants who might promote unfair competition, and antitrust suits are filed against telecommunications organizations who tend to monopolize. This protects the free marketplace and helps all broadcasters stay in business once they have obtained a license.

Government regulations generally attempt to protect station owners from being overpowered by the networks. The ruling that limits the networks to three prime-time programming hours per night is an example of an attempt in this direction. The ruling that disallowed one company from owning two networks, thus separating NBC Red and Blue into NBC and ABC, was also an attempt to protect station owners.

protection against rapid change

Regulation also protects against rapid, irreversible change. Both the FCC and FTC are criticized for vagueness and slowness, but by setting down only vague guidelines and deciding most issues on a case-by-case basis, both organizations prevent cataclysmic changes that would jar the systems of individuals and companies. There is usually a great deal of advance warning and discussion before any sort of policy is set, and even what is set can easily be overturned if it does not work as intended. The net result is moderation and the kind of policy that survives a host of tests. This indirect method of regulation allows for compromises that can resolve conflicting points of view and for new interpretations of the Communications Act.

Criticism of government regulation abounds, particularly within broadcasting circles. Regulation is condemned because it drifts, stalls, and vacillates. Terms such as "convenience, interest, and necessity" on which major decisions are made are inadequately defined. Semantic problems plague the definition of "public service." Guidelines issued by the various agencies are often muddled to start with, altered frequently, and then reversed by the courts. The broadcaster trying to abide by the law becomes confused and befuddled and must wade through thick books that attempt to explain broadcast rules, regulations, and policies. FCC and FTC commissioners often have widely disparate views as to the extent to which their agencies should regulate, and many of their decisions are by a one-vote margin.

Even in areas where the FCC and FTC claim to regulate in a systematic manner, they are often accused of being ineffective. For example, the FCC is supposed to see that stations broadcast in the public interest, but educational, religious, and public service groups do not really benefit because the material they do get to air is low-budget and broadcast at undesirable times. The FCC does not regulate airing time or production cost for these groups, so, in effect, it is not aiding their cause significantly.

Federal Trade Commission rulings against false advertising can be particularly ineffective. By the time the FTC conducts an investigation and issues a cease and desist order for a particular commercial, the commercial has run its course anyhow and sold many products, so the company is quite willing to remove it from the airwaves. About the only real clout the FTC has regarding any particular ad is the bad publicity that might arise for the company involved. This effect can be unfortunate, too; if the company is innocent, it still suffers from the bad publicity.

Many people feel that the regulating agencies overstep the thin line that separates watchdog and dog watched. The primary criticisms are against the FCC's incursions into programming decisions. The Communications Act specifically prohibits censorship, but some of the actions taken by the FCC in the name of "public interest" are controversial enough to raise cries of censorship.

Regulation suffers from people problems that often seem insolvable. Commissioners who come to the FCC have little prior experience with broadcasting, and yet some of them use their FCC position as a stepping stone to high-paying executive positions within broadcasting. Senators and representatives depend on the support of broadcasters to become elected and, hence, are often heavily influenced by broadcasting lobbies. Commissioners, because they know their decisions can be appealed to the courts, often make conservative decisions that are unlikely to be overturned.

In the area of self-regulation, there are those who mourn the death of the NAB codes. One of the primary complaints about broadcasting is the proliferation of commercials. If broadcasters are prevented from keeping their own house in order on this issue, then how can public indignation be handled?

Self-regulation within cable and home video presents particularly interesting problems. Should X-rated movies be allowed on cable or, for that matter, on cassettes? Perhaps there should be self-regulation standards for those services that are on the basic tier of cable and that all subscribers must receive if they want to receive any cable TV, but no self-regulation for the upper-tiered pay services that subscribers are free to select or reject.

The extent to which citizens should be involved in station operation is also debated. On one hand, stations exist for the public, so the public should be allowed to have its say regarding policies and operations. On the other hand, many citizens do not understand what is involved in operating a station and, as a result, make unreasonable demands.

ineffectiveness

overstepping bounds

people problems

self-regulation

citizen groups

The informal vehicles of self-regulation—organizations and awards—can be criticized for over-proliferation, wheel-spinning, and inequality, and yet they can point with pride to their accomplishments.

Despite all the pros and cons of both formal and informal regulation, broadcasters have managed to exist with it through periods of growth and rapid change, and it is likely to remain with them in some form during the years to come.[38]

Conclusion

Broadcasters interrelate with all three branches of government—legislative, executive, and judicial—and with independent regulatory bodies. Congress gives the broad stroke to regulation by enacting and amending the Communications Act. It also approves the appointment of commissioners and holds hearings on communications-related issues.

Within the executive branch, the president appoints commissioners and can call into effect the Emergency Broadcast System. Most of the president's influence is informal, however.

The courts are the place of appeal for those who feel they have been dealt with unjustly by other branches of the government or by elements of society. The courts have heard cases on such subjects as freedom of press, obscenity, and copyright.

The main government agency interrelating with the electronic media, however, is the FCC, which issues specific guidelines and handles the main regulatory chores. Some of these, such as prevention of interference, are technical in nature and have not changed much over time. Others, such as licensing, are more philosophical and have been subject to changes as society's outlook alters. The areas of granting, renewing, transferring, and revoking licenses have caused broadcasters varying degrees of consternation throughout the years. Decisions regarding Brinkley, Shuler, WLBT, WHDH, KTAL, the Alabama Educational Television Commission, and RKO have had impact on the regulatory direction of the industry.

While the government can regulate in a formal manner, there are other institutions that can regulate informally. Organizations to which media practitioners belong are among these. Some of them have codes that serve as guidelines for behavior. The NAB is the largest organization for broadcasters, but its codes have been struck down by the Justice Department.

Some stations and networks have their own internal regulatory forces, often called broadcast standards departments. Citizen groups, critics, and awards can all strongly encourage telecommunications forces to produce programming to meet specific needs.

Although regulation is often criticized in our democratically run country, its values are evident and its influence is much less than in other countries.

Thought Questions

1. To what degree should citizens groups and individuals be involved in the operations of telecommunications organizations?
2. Should FCC commissioners be required to have broadcasting experience? Explain.
3. Was the Justice Department correct in bringing suit against the NAB code? Why or why not?

Laws and Regulations

Introduction

The government regulatory bodies that oversee telecommunications create and refine many laws and regulations affecting the conduct of the industry. Sometimes they are laws drafted by Congress; sometimes they are regulations or advisory statements issued by such agencies as the FCC or FTC; and sometimes they are policies or precedents set as the result of specific court cases.

Often all of these interrelate. Congress passes a law that the FCC executes, but a station may be unhappy with this execution and appeal to the courts. Sometimes the courts decide the FCC had improperly executed the law and request a new implementation. Other times the courts side with the FCC and the station's recourse is to go to Congress to try to have the basic law amended.[1]

This type of interrelationship can be seen in the actions that have arisen regarding the First Amendment, the Communications Act, the fairness doctrine, and many other statutes.

CHAPTER 17

A function of free speech under our system of government is to invite dispute. It may indeed best serve its highest purpose when it induces a condition of unrest, creates dissatisfaction with conditions as they are, or even stirs people to anger.

*William O. Douglas
former Supreme Court Justice*

The First Amendment

The **First Amendment** to the U.S. Constitution is basic to much of what occurs in the telecommunications industry. This amendment states:

> Congress shall make no law respecting an establishment of religion, or prohibiting the free exercise thereof; or abridging the freedom of speech, or of the press; or the right of the people peaceably to assemble, and to petition the Government for a redress of grievances.

The freedom of press and freedom of speech aspects of this amendment have a great effect upon broadcasting and have been subject to many court cases and formal and informal regulations.[2]

    Issues are raised under many banners, one of them being **censorship.** If newspeople feel the government is trying to censor or withhold information, they consider this a breach of freedom of the press. The government often replies that such withholding of information is a **clear and present danger** to the country. If the press were told everything, the government would not be able to compete with other governments of the world.

    Clear and present danger extends to individuals, also. In one of its decisions, the Supreme Court noted that freedom of speech does not give someone the right to falsely yell "Fire" in a crowded theater because this would create a clear and present danger to those in the theater. On the other hand, governmental agencies are reluctant to label words as dangerous if freedom of

speech is involved. During the 1960s, the NAACP wanted the FCC to censor speeches being made over Georgia TV by a candidate for the Senate on the grounds that the speeches contained racially inflammatory remarks that were a danger to the TV stations and to the people of Georgia. The FCC refused to issue prior restraint in that case, saying that the speeches did not literally endanger the nation.[3]

    News reporters often feel freedom of the press is being violated if they must disclose sources from which they have obtained news, if they must surrender outtakes from their stories, or if they must testify in court regarding information they promised would be kept confidential. For quite a number of years reporters were required to testify, but during the 1970s, most states passed **shield laws** that gave some protection to news reporters. In most states, they must testify only if the evidence is crucial to the case and can't be obtained any other way. Because news reporters must sometimes testify, they must be careful not to promise confidentiality because they might be required to break it.[4]

    The First Amendment has been used for a myriad of other subject areas. Producers, writers, directors, and actors of the Hollywood community sued under the First Amendment when **family hour** was instituted. They claimed that the FCC, NAB, and networks had conspired to censor prime-time programming by requiring that the content be suitable for children. They succeeded in having the family-hour concept repealed.[5]

Cable TV companies successfully used the First Amendment to win the right to build a cable system in a city, even though they had not been granted a franchise. The rationale was that cable TV operators are recognized First Amendment speakers and should not be denied the right to cablecast.[6]

cable franchising

Journalists used the First Amendment to win the right to continue to conduct **exit polls,** where they interviewed people right after they had voted, even though this had been strongly opposed on the grounds that it affected the outcome of the vote in the western time zones.[7]

polling interviews

Profanity, Indecency, and Obscenity

Profanity, indecency, and **obscenity** are all major areas where the First Amendment comes into play. They are outlawed by the U.S. Criminal Code, but there have been many instances where this criminal code and the First Amendment clash. Often it is difficult to determine when people's freedom of speech should be abridged because what they are saying is profane, indecent, or obscene. Part of the problem surrounding this is the changing definitions of these words. What was considered obscene in one decade may be perfectly acceptable in the next decade. Also, what is indecent in Nebraska may be fairly commonplace in Hollywood.

The three words, even when defined in their most concrete way, often overlap. Profanity is defined as irreverent use of the name of God. Indecency involves broadcasting something offensive in relation to the standards of a particular community. Obscenity is more extreme than indecency. Its present definition was determined in a 1973 court case, *Miller v California.* To be obscene, a program must contain the depiction of sexual acts in an offensive manner, must appeal to prurient interests of the average person, and must lack serious artistic, literary, political, or scientific value. Obscenity is the most serious of the three and the most likely to lead to station license revocation or heavy fines. However, the line between indecency and obscenity is fuzzy and, in fact, some profanity, combined with sexual references, can be part of indecent or obscene material.

definitions

In practice, the FCC rarely chastises stations for profanity. An occasional "damn" or "hell" is common in everyday life and is not something to raise the ire of the FCC.

profanity

Indecency is the area where broadcasters have had the most difficulty. One of the most famous indecency cases arose when a Pacifica Foundation public radio station in New York, **WBAI,** aired a program on attitudes toward language. It was aired at 2:00 in the afternoon and included a comical monologue segment that spoofed seven dirty words that could not be said on the public airwaves. A father driving in the car with his son happened to hear this monologue and complained to the FCC. The FCC placed a note in WBAI's license renewal file, which led the station to appeal through the courts all the way to the Supreme Court. WBAI claimed that the FCC was censoring—a violation of the Communications Act. The high court determined that censorship was not involved because the FCC did not stop WBAI ahead of time

"seven dirty words"

from airing the program. The Supreme Court did say, however, that the program should not have been aired, but it based this mainly on the fact that the program was aired during a time period when children were likely to be in the audience.

topless radio

In another instance, the FCC fined a station $2,000 for airing what became known as **topless radio.** A number of stations throughout the country aired talk shows to which listeners called and recounted explicit sexual experiences. After one station was fined for indecency, and this fine was upheld by the courts, this form of talk show disappeared from the airwaves.[8]

"seven dirty words" strengthened

In 1987 the FCC sent letters to three radio stations asking them to respond to allegations of airing indecent material. In doing this, the FCC strengthened its "seven dirty words" Pacifica stand by stating that language or material that depicts sexual or excretory activities in terms offensive to contemporary community standards would be in violation if broadcast at a time of day when there is reasonable risk that children are in the audience.[9] In 1988, KZKC-TV in Kansas City, Missouri, was fined $2,000 for airing "Private Lessons," a film about an older woman's attempts to seduce a young boy.[10]

indecency ban

Then in 1990, the FCC voted to ban indecent programming on radio and TV regardless of the time of day. Its reasoning was that children are likely to be in the audience at any time and should be protected from this type of programming.[11]

obscenity

Obscenity has been less of a problem for broadcasters than indecency. In their attempts to avoid indecent programming, they have usually managed to stay far away from what might be considered obscene. However, in 1990 a Chicago station, WSNS, was denied license renewal, in part because at one time the station aired adult movies.[12]

Cable TV is more likely to see action regarding obscenity than commercial broadcast radio and TV because some of cable is geared toward smaller audiences and includes provisions that make it difficult for the mass audience to be exposed to undesirable material. For example, Playboy Channel must be specifically subscribed to before it comes into the home. On that basis, the 1984 cable law did not prevent cable from showing indecent material, but it did prohibit obscene material. This gives cable more latitude than broadcasting, so it is more likely to cross the line between indecency and obscenity.

Utah case

In Utah a group tried to use a state obscenity statute to keep sexual acts and nudity off cable TV in the state. However, the courts ruled that this statute could not be used, clearing the way for Utah airing of such services as the Playboy Channel.[13]

videocassettes

The videocassette business is also ripe for obscenity problems. However, as with cable TV, someone must make a conscious effort to rent or buy a videocassette to show on a home recorder. Video stores have also been taken to court for stocking sexually-oriented cassettes, but store owners have been successful in winning the cases by using the First Amendment.[14]

Libel, Slander, and Invasion of Privacy

Most **libel** laws and principles that apply to electronic media have their roots in the written press.[15] In early days of radio and television, there was debate as to whether libel applied at all. Libel is defined as defamation of character by published word, whereas **slander** is defamation by spoken word. Slander carries less penalty than libel, ostensibly because it is not in a permanent form to be widely disseminated. Some people felt radio and television should be under slander laws because words were spoken rather than printed. But because these spoken words are spread far and wide, broadcast defamation has come under the libel category.

definitions

Libel was not a big issue with the broadcast media until the late 1970s. Part of this was due to the fact that radio and television engaged in very little investigative reporting, so libel suits were less likely to occur. Radio and television newscasts and documentaries are also limited in terms of time, so generally programs deal only with well-known issues and people. People who are **public figures** have great difficulty winning a libel suit because the rules and precedents applied to them are much stricter than for ordinary citizens. In order to win a libel suit, a public figure must prove that a journalist acted with **actual malice.** As a result of all this, very few public figures bothered to bring libel suits against the broadcast media.

public figures

By the late 1970s, however, TV was a major source of information and a dominant force in society that some felt had become overly arrogant. What was said about people on TV definitely affected their reputations and livelihoods. As a result, libel cases were on the upswing. During the early 1980s, almost 90 percent of the people who filed libel suits against broadcasters won. This was soon reduced to 54 percent as broadcasters began taking these cases more seriously.[16]

Cases brought to the courts included the following: a Chicago businessman who said ABC's "20/20" made him look like an arsonist because he had an interest in some buildings that burned; a Philadelphia mayor who claimed the CBS station there had broadcast, incorrectly, that he was the target of a federal investigation; a former army officer who felt he was depicted as a liar on a "60 Minutes" segment dealing with charges of army cover-ups of atrocities during the Vietnam War; a right-wing presidential candidate who sued NBC but wound up being told to pay NBC; and singer Wayne Newton, who claimed that an NBC newscast made it appear that he had strong ties with the Mafia. Some of these cases have been settled either in or out of court, and others are still winding their way through the appeals process.[17]

cases

In 1985 a very important case that had been called "the libel case of the century" came to a quiet close. The suit had been brought against CBS by General William **Westmoreland** and centered around a documentary aired in January 1982 called "The Uncounted Enemy: A Vietnam Deception." In this documentary CBS accused Westmoreland of purposely deceiving his military superiors and President Lyndon B. Johnson with estimates of Vietnam enemy troop strength that were much lower than was really the case. Westmoreland

Westmoreland

filed a libel suit asking for $120 million in damages, but because he was a public figure, he had to prove actual malice on the part of CBS. The trial began in 1984 with both Westmoreland's reputation and the credibility of network reporting at stake. The trial progressed for several months, then was settled out of court. CBS issued a statement saying Westmoreland had fulfilled his patriotic duties as he saw them, and Westmoreland withdrew from the suit. Because the trial did not go to jury and because it ended in a rather bewildering manner, the effect on the body of libel law was negligible. Most observers saw the case as a victory for CBS and predicted it would have a chilling effect on other public figures tempted to sue the media.[18]

Such seemed to be the case because succeeding libel cases, most of them against newspapers, were generally decided in favor of the media.[19]

invasion of privacy

Invasion of privacy is related to, but different from, libel. It often leads to information that is damaging to a person, but its main thrust involves how the information is gathered. Invasion of privacy laws vary from state to state; however, in general, invasion of privacy laws allow a person to be left alone.

state laws

Often the laws deal with giving a person physical solitude. For example, reporters cannot trespass on a person's property and take a photo through the bedroom window. The laws also take into account the publication of private facts. Generally, facts are private if the person involved does not want them known. However, in some states facts are not private if they can be obtained from public records or if they are part of an important news story. For example, the names of rape victims are public record, so they can be broadcast. Victims' names are often withheld, however, because stations know the victims do not want their names revealed.[20]

Invasion of privacy laws also prevent someone from being presented in a false light. This is very similar to libel except it is more likely to come from omission of material, and it does not have to be defamatory. Theoretically, people could claim invasion of privacy for published information that flattered them.

Invasion of privacy also keeps someone's name or image from being used for commercial gain by someone else. For example, a videofile could not tape a series of clandestine shots of a politician playing tennis and then sell copies of this tape to video stores.

Access to the Courts

A clash between the First and Sixth Amendments has led to difficulties for broadcast journalists in the courts. Although the First Amendment guarantees freedom of press, the Sixth Amendment guarantees **fair trial.** For many years lawyers and judges felt a fair trial was impossible if cameras were present in the courtroom.[21]

Canon 35

In 1937 the American Bar Association adopted a policy, known as **Canon 35,** that barred still cameras and radio from courtrooms. Later TV cameras were included under the same policy. This was not a law, but simply an ABA policy. Most judges abided by this policy, so it effectively kept cameras out of courtrooms for many years. The rationale was that cameras would detract from the proceedings, cause disruption, and lead defendants, lawyers, and jurors to act in ways that they would not act if no cameras were present.

In the early days of television there was reason for this rationale because of the bulky equipment and lighting needed for TV coverage. However, during the 1970s, when unobtrusive ENG equipment became available and when TV was no longer such a novelty, broadcasters began pressuring for access to the courtrooms.

In 1972 the American Bar Association liberalized Canon 35, redesignating it **Canon 3A.** It recommended that individual states and individual judges be given discretion as to whether or not to allow cameras into their courtrooms. One by one states began allowing cameras in courtrooms on an experimental basis. Nothing went awry, so by the mid-1980s all states were allowing cameras into the courtrooms under specific conditions. Each state and, in some instances, each judge set specific rules for media coverage of trials. For example, cameras had to be located in fixed positions; juries could not be photographed; and lights could not be used.

Canon 3A

The right of broadcast media to be present in the courts was further affirmed in 1981 when two Florida defendants claimed that televised coverage of their trial had denied them due process. The Supreme Court ruled against them, stating that the Constitution does not prevent states from allowing TV cameras in the courts.[22]

Florida case

The federal courts and the Supreme Court have been more and more receptive to the idea of cameras in their courtrooms and have allowed them in limited ways.[23]

Supreme Court

Controversies still arise regarding court trial coverage. Certain subjects, such as rape and child molestation, are sensitive and usually decided on a case-by-case basis. Overall, the broadcast media has covered trials in a tasteful manner that has not proven disruptive.

Editorializing

Editorializing was first ruled on in 1941 in a case involving WAAB, a Boston radio station that for several years had been expressing its views on political candidates and controversial issues. This case became known as the **Mayflower Decision** because the Mayflower Broadcasting Corporation filed a challenge to the license renewal of WAAB, stating that the station had not served in the public interest and that its license should be transferred to the Mayflower Broadcasting Corporation. Part of Mayflower's case stated that WAAB had been editorializing and that these one-sided presentations were not in the public interest. The FCC rejected Mayflower's bid, but on grounds having nothing to do with editorializing. The FCC stated in its report on the case that it disapproved of editorials and felt broadcasters should not be advocates.

Mayflower Decision

This decision, although resented, was not challenged by broadcasters because few of them were interested in editorializing. However, during the 1940s several top members of the National Association of Broadcasters became interested in the issue and the **Cornell University** radio station petitioned the FCC to reconsider the Mayflower Decision. As a result, after hearings held from 1947 to 1949, the FCC finally reversed itself, stating that stations should be encouraged to editorialize.

However, in order to limit one-sided points of view, the FCC ruled that stations had to make a positive effort to see that **opposing viewpoints** were also broadcast. Stations were also to make sure that the people chosen to refute editorials were qualified, so that the rebuttal had the same chance of convincing the public as the original editorial.

This decision did not have any immediate impact on commercial broadcasting. Stations did not immediately rush into editorializing, since they traditionally had not engaged in it. The stations were also unsure about what exactly the FCC meant when it said they had to seek opposing viewpoints. Over the years, editorializing guidelines have been defined and refined by court cases and FCC rulings. It is now generally accepted that all editorials do not need to be refuted, but if the character or integrity of a group or person is going to be attacked, that person or group must be notified and given reasonable opportunity to reply. Also, if the station plans to endorse a political candidate editorially, his or her opponents must be notified within twenty-four hours and provided opportunity for reply.[24]

For many years public broadcasting stations were forbidden to editorialize because they received federal funds. However, a public radio chain, Pacifica Foundation, took this restriction to court, and in 1984 the Supreme Court ruled that prohibiting public stations from editorializing was a violation of the First Amendment.[25]

Equal Time

Section 315 of the Communications Act has come to be known as the equal time provision. In reality, it deals with **equal opportunity** rather than equal time. Section 315, as written in the 1934 Communications Act, reads as follows:

> If any licensee shall permit any person who is a legally qualified candidate for any public office to use a broadcasting station, he shall afford equal opportunities to all other such candidates for that office in the use of such broadcasting station, and the Commission shall make rules and regulations to carry this provision into effect, provided that such licensee shall have no power of censorship over the material broadcast under the provisions of this section. No obligation is hereby imposed upon any licensee to allow the use of its station by any such candidate.[26]

This provision is in effect only during periods of election campaigns and deals only with candidates for political office. The people it covers must be legally qualified candidates who have publicly declared that they are running for political office. It does not apply to any particular person until he or she has officially declared to run for office. In other words, although it may be common knowledge that a senator is planning to run for reelection, opponents cannot demand equal time for the senator's appearances until he or she makes an official public declaration. Because there are deadlines for declaring one's candidacy for office, a person must become an official candidate well before the actual election.

Because Section 315 actually mentions equal opportunity, it guarantees much more than that all candidates for a particular office will be given the same number of minutes of airtime on a particular station. The airtime given must approximate the same time period so that one candidate is not seen during prime time and another in the wee hours of the morning. The cameras and other facilities that the first candidate is able to use must be available to opponents. All candidates must be able to purchase time at the same rate and if a station decides to give free time to one candidate, it must give equal free time to all opponents.

equal opportunity

However, if one candidate purchases time and other candidates do not have the money to purchase equal time, the station does not need to give free time to the poorer candidates. Because some candidates usually have larger campaign chests than others and because some candidates prefer to concentrate their campaigning on radio and TV, while others choose to emphasize other publicity methods such as direct mail or billboards, candidates rarely have equal time on any particular radio or TV station. The station is not required in any way to equal the actual time as long as equal opportunity for the time was afforded each candidate and the candidates were the ones who chose not to avail themselves of the time.

purchasing time

The provisions apply equally to all candidates, whether they be from the major Republican and Democratic parties or from smaller ones such as the Libertarian and Progressive parties. As long as candidates have met the requirements for the office for which they are running, they are covered under the provisions of Section 315.

all parties

Through the years, interesting cases have arisen in regard to Section 315. Shortly before the 1956 election campaign between Dwight Eisenhower and Adlai Stevenson, Eisenhower, who was the incumbent president, was given free time on all networks to talk to the American people about an urgent crisis in the Middle East. Stevenson's forces immediately demanded equal free time. However, the FCC decided that Eisenhower's speech was exempt from Section 315 because it did not deal with normal affairs, but with a crisis situation. This has become a precedent that has been followed in other election years with other presidential, crisis-oriented speeches.

Eisenhower's crisis speech

In 1959 Lar Daly, a candidate for mayor of Chicago, protested that the incumbent mayor, who was running for reelection, had been seen on a local news show and Daly demanded equal time on that station. Had he been an ordinary candidate, he might have been granted time and the issue dropped. However, he was an eccentric who dressed in an Uncle Sam outfit—complete with a red, white, and blue top hat. Daly ran, unsuccessfully, for some office in every election. Daly was granted his equal time, but Congress amended Section 315 so that the equal time restrictions would not apply to candidates

Lar Daly

Section 315 amendment

appearing on newscasts, news interviews, news documentaries, and on-the-spot coverage of news events.

The actual wording of this amendment was as follows:

Appearance by a legally qualified candidate on any—

1. bona fide newscast
2. bona fide news interview
3. bona fide news documentary (if the appearance of the candidate is incidental to the presentation of the subject or subjects covered by the news documentary), or
4. on-the-spot coverage of bona fide news events (including but not limited to political conventions and activities incidental thereto),

shall not be deemed to be use of a broadcasting station within the meaning of this subsection. Nothing in the foregoing sentence shall be construed as relieving broadcasters, in connection with the presentation of newscasts, news interviews, news documentaries, and on-the-spot coverage of news events, from the obligation imposed upon them under this Act to operate in the public interest and to afford reasonable opportunity for the discussion of conflicting views on issues of public importance.[27]

This amendment enabled broadcasters to continue to perform their usual news-oriented functions without being inundated with requests from minority candidates. However, the minority candidate issue still remains a thorny one. When large numbers of candidates are running for a particular office, radio and TV stations often find themselves devoting an inordinate amount of airtime to political commercials and programs. Stations that would like to give free airtime to major candidates so that the public could hear their views often refrain from doing so because of the huge amount of broadcast time that would then be taken up by splinter party candidates.

WDAY-TV case

In 1956 WDAY-TV in Fargo, North Dakota, granted time to a splinter candidate running for the Senate. In keeping with the regulation of Section 315, the station did not censor the candidate's speech, which later became the subject of a successful libel suit against both the candidate and the station. The state supreme court declared that the station should be immune from damages because it could not, by law, censor the speech. The U.S. Supreme Court agreed, stating that Section 315 grants a station immunity from liability in the case of such libelous material.

major candidate debates

The splinter candidate issue was raised again in 1960 when Democrat John F. Kennedy and Republican Richard M. Nixon were running for president. The networks wanted the two to debate on television and the two candidates agreed, but Section 315 would have necessitated equal debating opportunity to all the other candidates on the presidential ballot. Because Congress felt that this was a special situation, it suspended Section 315 temporarily for the 1960 presidential and vice-presidential offices only.

Broadcasters hoped that similar suspensions would be forthcoming, but the issue took a different turn. The FCC, under the 1959 Communications Act amendment that allowed on-the-spot coverage of bona fide news events, decided in 1975 that networks and stations could cover candidate debates if these were on-the-spot news events. The broadcasters themselves could not arrange the debates, nor could the candidates, but if some other group, such as the League of Women Voters, set up the debates, the media could cover the event in much the same way as they cover awards ceremonies or state fairs.

candidate supporters

In 1976 and 1980, presidential debates were shown in this way—sponsored by the League of Women Voters and telecast as news events. Then in 1983, the FCC ruled that debates could be sponsored by broadcasters. The League of Women Voters appealed, but the court sided with the FCC, stating that broadcasters should have the same right to hold debates and exclude candidates as a civic group.[28]

Another issue that arose was whether Section 315 applied only to candidates or to candidates and their supporters. At first it was only candidates, but in 1970 the FCC extended Section 315 to cover supporters as well. Now if a spokesperson comes on to tout a candidate, or if an announcement is broadcast that does not "star" the candidate, opposing candidates must be given equal opportunity.

Because of all the difficulties involved with Section 315, some stations decided they didn't want to offer time to anyone so that they would not need to worry about when they did or did not need to give time to opponents. But the FCC nixed that idea and, in 1971, stated that stations could have their licenses revoked if they did not allow access to candidates running for federal office. For state and local elections, stations were given the option of not dealing with any candidates if they so desired.

Also in 1971 Congress again amended Section 315 so that stations had to sell time to candidates at the lowest rate that they would sell that time to advertisers who regularly purchased large quantities of time. This significantly lowered the amount of money that candidates paid for announcement time.

lowest rate

In recent years, entertainers and other TV personalities have been entering the political arena at an increasing rate. This has created a dilemma for broadcasters in terms of Section 315. If, for example, a candidate has a legitimate profession as a station weathercaster, must the station give equal time to other candidates each time the weathercaster is seen or heard on radio or TV? The answer is yes, but the practical reality is no. In 1972 NBC offered free time to candidates because it inadvertently showed a movie in which actor Pat Paulsen, who was then a declared candidate for president, had appeared for thirty seconds. Overall, networks and stations stopped showing Pat Paulsen movies so that they would not be confronted with the Section 315 provision.

entertainers as candidates

length of spot

The issue, of course, became more acute when Ronald Reagan ran for governor of California and then president of the United States. Although Section 315 did apply to the movies in which he appeared, problems did not arise because the films, for the most part, were not aired during the election period and because his opponents made light of the situation, joking that showing the films might actually cause him to lose votes.

Waivers from opponents can also handle this particular situation. If a weathercaster is thinking of running for city clerk, he or she may obtain waivers from other candidates by promising not to mention the election on the air or use the weathercasting as a political platform. In such waivers, the opponents promise not to seek equal time or opportunity from the station or stations involved.

In general, people who appear regularly on radio or TV will not run for office if they cannot obtain waivers. However, if they feel strongly about entering the political arena, they can quit their jobs or ask for another job of an off-air nature. They can also try to convince their employer that the station should give the equal time to the opponents. In 1988 a Sacramento TV reporter took Section 315 through the courts trying to get a change so that equal time would not apply to people with on-air employment, but he was unsuccessful.[29]

During the 1976 presidential campaign, Gerald Ford's campaign committee tried to buy spots on WGN in Chicago. The management's position was that any candidate who wanted to appear on either the WGN radio or WGN-TV station had to buy time in units of at least five minutes. The philosophy behind this was that no candidate for something as complicated as a major political office should make announcements of a thirty-second or sixty-second nature, but should explore the issues for at least five minutes. The FCC, however, ruled that candidates should be allowed to buy whatever length of time they wanted and so ordered WGN to change its policy.

campaign beginning

During the 1980 campaign, Jimmy Carter's aides wanted to buy a half hour of network prime time early in December 1979 in order to launch Carter's campaign for reelection. The networks refused, saying that December was too early to start an election campaign. Carter forces appealed the decision and the Supreme Court sided with them, indicating that it is the candidates, not the broadcasters, who can decide when a campaign has begun.[30]

back-to-back statements

In 1988, King Broadcasting wanted to air, at its own expense, back-to-back programs of the Republican and Democratic presidential candidates. In other words, Michael Dukakis would speak for thirty minutes and then George Bush would speak for thirty minutes. The FCC said this would not be exempt from Section 315, and King would need to give time to all splinter candidates. King is now appealing this through the courts. Two half hours of the programs King proposed did occur on network TV, but the candidates paid for the time and splinter candidates could not afford to pay at the same rate.[31]

Each election year the courts are inundated with new cases regarding Section 315. Many of the ramifications from Section 315 are still largely uncharted, despite the thick tomes on political broadcasting that are issued by the FCC and the National Association of Broadcasters (NAB).[32]

The Fairness Doctrine

The **fairness doctrine** has a foggier history than either editorializing or equal time. It does not appear as a concrete doctrine, but rather as a series of actions and rulings dealing with presentation of controversial issues. In many respects its roots are located in both the editorializing decisions and Section 315.

At the end of its 1947 to 1949 editorial hearings, the FCC issued a statement, part of which read as follows:

editorial statement

> . . . the licensee must operate on a basis of overall fairness, making his facilities available for the expression of the contrasting views of all responsible elements in the community on the various issues which arise.[33]

This statement did not raise any ripples at the time, but in later years it was interpreted to mean that fairness applied to forms of programming other than editorials.

Likewise, when Congress amended Section 315 in 1959, it stated that nothing in the news exemptions should be construed as relieving broadcasters from the obligations imposed upon them under this chapter to operate in the public interest and to afford reasonable opportunity for the discussion of conflicting views on issues of public importance.[34] This phrase, too, seemed harmless at the time but was later used in conjunction with fairness issues.

Section 315 amendment

The 1960s was the decade in which fairness came to the fore. The political climate of unrest and distrust bred controversies that might have remained dormant in more settled times. The first cases to raise the issue revolved around station and network documentaries of the early 1960s. Several cities complained to the FCC that they had been depicted unfairly in the documentaries, but the FCC disagreed, saying that the producers had given opportunity for all points of view.

Since the issue of fairness was frequently being raised, the FCC issued a 1963 advisory statement to stations on the subject, followed by a 1964 fairness primer. Although both of these were weak in terms of actual guidelines, they did make reference to the 1949 statement that had appeared in the editorializing document. The FCC also established that fair presentation of material extended beyond editorializing and into other areas of programming.

fairness primer

Following this, the FCC seemed more inclined to favor entities that claimed fairness had been violated. It ruled that a Colorado station needed to give time to debt adjusters who claimed that they had been maligned in a documentary. It also ruled that two Alabama radio stations needed to present counterarguments to those presented in a syndicated program that dealt with a nuclear test ban treaty.

The case that has become the hallmark of the fairness doctrine is the **Red Lion** case. In 1964, station WGCB in Pennsylvania, operated by the Red Lion Broadcasting Company, broadcast a talk given by the Reverend Billy Hargis that charged Fred J. Cook, author of books critical of Barry Goldwater and J. Edgar Hoover, with Communist affiliations. Cook demanded that WGCB give him the opportunity to reply. The station said it would if Cook would pay for the time, but Cook objected and contacted the FCC, which ordered the station to grant Cook the time whether or not he was willing to pay for it. The decision was appealed through the courts and upheld by the Supreme Court on the grounds that free speech of a broadcaster does not embrace the right to snuff out the free speech of others. The significance of this

Red Lion case

decision went beyond the dispute between Cook and the Red Lion Broadcasting Company because it upheld the constitutionality of the fairness concept and of the FCC's right to implement it.

John F. Banzhaf III case

One creative application of the fairness doctrine was a case in 1967 brought up by **John F. Banzhaf III.** In 1964 the surgeon general's office had determined that there was a link between cigarette smoking and lung cancer, a point that was considered controversial. Banzhaf, several years later, petitioned WCBS-TV in New York saying that cigarette commercials showed cigarette smoking in a positive light, so free time should be given to antismoking groups to present the other side. WCBS replied that within its news and public affairs programs it had presented the negative side of smoking and did not feel that it needed to give equivalent commercial time to the issue. Banzhaf took the matter to the FCC, which sided with him—as did the appeals courts. Therefore, until 1972, any station that aired cigarette commercials had to provide time for anticigarette commercials. In 1972 Congress passed a law that banned cigarette ads on radio and TV and, therefore, the anticigarette commercials also stopped.

hate clubs

Another fairness decision involved WXUR, an AM/FM radio station in Media, Pennsylvania, owned by a right-wing fundamentalist preacher who only programed opinions that agreed with his point of view. He had what some people called "hate clubs of the air"—programs that attacked particular groups that he did not like. Citizens complained, and in 1965, the FCC warned the station it must abide by the fairness doctrine. The station did not mend its ways, and in 1970 its license was revoked because of fairness violations.

pendulum swing

All of these fairness decisions of the 1960s led broadcasters and the public to believe that future decisions by the FCC and the courts would favor those who felt fairness should be strictly interpreted, and complaints about unfair presentation of issues poured into the FCC. However, once again the pendulum began to swing. No doubt, part of the reason for this was that broadcasters were now being more careful about fairness issues and were voluntarily complying with some of the demands. At any rate, in general, the decisions of the 1970s began to make a turnaround.

Friends of the Earth

For example, in 1970 an ecology group, **Friends of the Earth,** tried to emulate Banzhaf's case by requesting that ads for gasoline be countered by antipollution announcements. However, this attempt was not successful. The FCC stated that the cigarette ruling was not to be used as a precedent because that case was unique.

Vietnam views

In another case, two different groups wanted to buy time to present their views against the war in Vietnam. One group said that it would be acting to counter an armed forces recruitment spot and the other wanted to respond to the president's views. Both were turned down, one by a Washington, D.C. station and the other by CBS, so both groups went to the FCC. The FCC also turned them down, stating that broadcasters, not public interest groups, should decide when an issue required that opposing viewpoints be expressed, and that broadcasters should select the person or persons to deliver those viewpoints.

CBS presented a documentary, "The Selling of the Pentagon," that showed how the military used public relations tactics to convince the public that the Vietnam War and other military projects were worthwhile. The military establishment, enraged by the program, went to the FCC to complain that the presentation had been one-sided and edited in a distorting manner. After much study, the FCC dismissed the charges by indicating that CBS had set aside an hour for opposing viewpoints and that there was no evidence of deliberate distortion intended to misinform.

"The Selling of the Pentagon"

In 1973, NBC broadcast a documentary about pensions provided by private industry. The FCC told NBC to make time available for the industry point of view. NBC, which thought it had given enough time to that point of view within the documentary, objected and took the issue to court. The court agreed with NBC, but did not directly attack the issue of whether or not NBC had fairly presented the points of view. Rather the court took the FCC to task because it had been inconsistent in applying the fairness doctrine.

pensions

Somewhat in response to this, the FCC, in 1974, issued a Fairness Report that came closer to a "doctrine" than anything that had been associated with the various rulings that came to be known as the fairness doctrine. The Fairness Report was an attempt to clarify for broadcasters what their duties and obligations were in this area. For example, it stated that a licensee had an obligation to devote a reasonable amount of program time to the discussion on controversial issues and that overall programming should be balanced concerning specific subjects. It said that commercials would be subject to fairness only if they directly addressed controversial subjects. The report included under fairness all aspects of political campaigns not included in Section 315.

Fairness Report

During the 1980s, the number of complaints brought forth under the fairness doctrine waned, and the doctrine itself became the focus of attention. In 1981 the FCC, in a deregulatory mood, asked Congress to repeal the doctrine, but Congress did nothing. In 1985 the FCC undertook a study that concluded that the fairness doctrine was not in the public interest. Also, a court decision stated that the doctrine was not law but merely an FCC regulation that the FCC could abolish.

abolishment

In 1987 Congress passed a bill that would have codified the fairness doctrine and made it a law rather than an FCC policy. However, President Reagan vetoed the bill. Shortly thereafter the FCC said it was going to stop enforcing fairness and, in effect, abolish the whole fairness concept. This meant stations could air programming about controversial issues without having to worry about complaints or contrasting points of view. Although this appeared to kill the fairness doctrine, the whole issue could rear its head again during the 1990s.[35]

If the fairness doctrine appears somewhat confusing, it's because it is. For better or worse, the future of this concept is likely to be as confusing as its past.[36]

Copyright

basis

protection vs availability

clearance

compulsory licenses

violations

cable industry

Most of what has been discussed so far regarding laws and regulations relates to over-the-air broadcasters and to some extent cablecasters. Other entities, such as the videocassette industry and corporate TV facilities, are not regulated to any degree by the FCC. However, the issue of **copyright** is one that affects all aspects of telecommunication.

The copyright principles stem from the U. S. Constitution, which has a clause in it that authorizes Congress to promote science and art by allowing authors and inventors to have exclusive rights to their works for a limited period of time.[37]

The current copyright law, which went into effect in 1978, states that copyright exists for as long as the creator of the work is alive, plus fifty years. After that, works are placed in **public domain** and can be used without obtaining permission. The law also provides for **fair use,** which allows for some use of the work without permission from or payment to the copyright holder. Copyright principles try to balance the need for the protection of works for the creator with the need to make those works available to the public. If artists cannot profit from their works, they are not likely to create them. However, totally protecting a work, such as a book, would mean it could not be quoted, even in everyday conversation, without the permission of the author. The copyright law and related court cases try to hit a happy medium by allowing such things as the videotape recording of TV programs so that they can be viewed at a different time, or the photocopying of a magazine article in order to use the material in it for research. Such uses are considered fair use.

However, for most telecommunications production, copyright must be cleared. TV stations, cable networks, corporate producers, and all others involved with audio or video production must obtain permission and often pay a fee before they can include copyrighted music, photographs, sketches, quotes, film clips, or other similar material.[38] Often, finding the copyright holders and making arrangements with each individually is quite a chore. This is why **blanket licensing,** such as that undertaken by music licensing organizations such as ASCAP and BMI, has arisen.

The copyright law also provides for **compulsory licenses,** such as those that cable TV systems must pay to the Copyright Tribunal to cover the material delivered to them by superstations such as WTBS.

One of the main problems facing various facets of the telecommunications industry is violation of copyright, often called **piracy.** This takes various forms and is very difficult to control. It can range from someone taping a song played by a radio station and making copies of the song for friends to an international ring that distributes illegally dubbed copies of movies to foreign countries even before the movies have been released.

The cable TV industry has been plagued with the problem of people receiving signals without paying for them. People who are somewhat technical have been able to tap off a neighbor's signal and bring the programming to their homes. Elaborate interactive antitheft equipment has now curtailed most of this. Another cable TV problem has involved the unauthorized reception of satellite programming intended for distribution by cable systems. The biggest problem has been home satellite dish owners who received the programming without paying for it. **Scrambling** has helped solve this problem, although many illegal scrambling **decoder boxes** are available. Home dish owners can now pay to receive the signals legally, but many still prefer illegal means.

The business most hurt by piracy is the videocassette business. Films and videocassettes are extremely easy to copy, and movie companies estimate that they lose one billion dollars a year because of illegal duplication. One of the thorniest problems is piracy in foreign countries. Many countries do not have copyright laws. They produce very little, themselves, in the way of creative works and do not understand the concept of protection for something that they assume has already been paid for. However, most international piracy is not innocent. It is conducted by mafia-like organizations that set up large duplicating facilities and have elaborate distribution networks. The FBI has made concerted efforts to stamp out piracy, but it still continues.[39]

videocassette industry

Other Regulations

Of course, there are many other laws and regulations that affect telecommunications practitioners. For example, the U.S. Criminal Code outlaws **lotteries** sponsored by radio or TV. Stations are allowed to have contests, though, and sometimes cases arise as to whether a station is conducting a contest or a lottery. Generally something is considered a lottery if a person pays to enter, if chance is involved, and if a prize is offered. Stations usually avoid lotteries by making sure people do not have to pay to enter contests.[40]

lotteries

Broadcasters and cablecasters must adhere to the regulations of the Equal Employment Opportunity Commission. Stations with five or more employees must file annual reports with the FCC listing employees by race, sex, and job category. Failure to hire and retain minorities can lead to license renewal problems. The FCC also has special policies aimed at increasing the number of minority people who own broadcast properties.[41]

equal opportunity

In addition, the FCC has numerous regulations that relate to such areas as **must-carry, syndicated exclusivity,** and **financial interest-domestic syndication.**

Regulation policies are in a constant state of flux—a situation having both advantages and disadvantages. A fluid state allows for changes to keep up with the times, but it also creates internal inconsistencies among the various regulatory bodies. It inhibits rigidity on the part of both elected and appointed officials, but it also causes confusion and uncertainty for those who must deal with regulation.

Conclusion

All branches of government are intrinsically involved in the laws and regulations that govern telecommunications.

The FCC has been the main overseer of station behavior. As such, it revokes licenses and issues warnings to stations for infractions. These have included fines against obscenity as in the case of "topless radio," disapproval and then approval of editorials as in the Mayflower and Cornell cases, and decisions of when equal time provisions were violated as in cases regarding length of political commercials and presidential crisis speeches. On occasions the FCC has issued guidelines or rulings such as fairness statements, the extension of Section 315 to supporters, and the indecency ban.

The executive branch is the one most likely to call clear and present danger if the president feels reporters are acting irresponsibly. The president can veto communication bills as in the case of the fairness bill.

The legislative branch occasionally amends the Communications Act. This has had a particular effect upon equal time. The amendment to Section 315 that removed equal time restrictions from bona fide newscasts had a great effect upon election campaigns, as did the amendment that required stations to sell time to candidates at their lowest rates. The fairness doctrine also grew out of a Section 315 amendment, and a bill was sent forth codifying fairness. Other criminal and civil laws passed by Congress, such as those dealing with lotteries and copyrights, affect broadcasting as do the state laws dealing with invasion of privacy.

The courts and telecommunications have a double-edged interrelationship. On one hand, a battle has been waged regarding the right of radio and TV to have access to the courts for news coverage. On the other hand, the courts have made many decisions that directly affect broadcasters. Obscenity rulings, such as WBAI's "seven dirty words" and Utah's cable TV channels, were eventually decided in the courts. Likewise, libel cases, including the Westmoreland case, often begin or end in the courts. Some Section 315 decisions have been appealed to the courts, such as the determination of when a campaign begins, who can sponsor presidential debates, and whether or not candidates who are entertainers can be exempt from Section 315. The fairness doctrine has kept the courts quite busy with cases that include Red Lion, cigarette advertisements, and Friends of the Earth.

Although the various branches of government disagree with each other from time to time, equilibrium is maintained by the checks and balances system.

Thought Questions

1. Give some examples of what you feel would constitute censorship and what you feel would be clear and present danger.
2. Give examples of what you think would constitute indecency and what would constitute obscenity in your particular community.
3. If you were going to make a list of rules for TV journalists to abide by regarding access to the courts, what would they be?
4. Should the fairness doctrine be reinstated? Why or why not?

1

1. Eric Barnouw, *A Tower in Babel: A History of Broadcasting in the United States to 1933,* (New York: Oxford University Press, 1966): 7.

2. "Summary of Broadcasting and Cable," *Broadcasting* (December 16, 1990): 115; "VCR Reach Rises, Per Arbitron Study," *Variety* (October 30, 1990): 16; and Bureau of the Census, *Statistical Abstracts of the United States, 1990,* (Washington, D.C.: U.S. Department of Commerce, 1990): 550.

3. *1990 Nielsen Report on TV* (New York: Nielsen Media Research, 1990); and *Radio Year-Round: The Medium for All Seasons* (New York: Arbitron Ratings Company, n.d.).

4. Neil Hickey, "Decade of Change, Decade of Choice," *TV Guide* (December 9, 1989): 29–34.

5. Ibid., 32.

6. Harry F. Waters, "The Future of Television," *Newsweek* (October 17, 1988): 86. This article reported that a study done by *Channels* magazine indicated that half the people in the 18–34 age group regularly use their remote controls to watch more than one program at a time, and 20 percent watch three or more shows.

7. Bert E. Bradley, *Fundamentals of Speech Communication* (Dubuque, Iowa: Wm. C. Brown Publishers, 1991): 5.

8. Beverly M. Payton, "Back from the Brink: Fans Rescue Endangered Series," *Emmy* (November/December 1987): 32–37.

9. Gerald S. Lesser, *Children and Television: Lessons from Sesame Street* (New York: Random House, 1974).

10. Aimee Dorr, Peter Kovaric, and Catherine Doubleday, "Parent-Child Coviewing of Television," *Journal of Broadcasting and Electronic Media* (Winter, 1989): 35–51; Peter Nikken and Allerd L. Peeters, "Children's Perceptions of Television Reality," *Journal of Broadcasting and Electronic Media* (Fall, 1988): 441–52; and Ali Reza Zohoori, "A Cross-Cultural Analysis of Children's Television Use," *Journal of Broadcasting and Electronic Media* (Winter, 1988): 105–13.

11. George Comstock et al., *Television and Social Behavior: A Technical Report to the Surgeon General's Scientific Advisory Committee on Television and Social Behavior* (Washington, D.C.: Government Printing Office, 1972).

12. Ronald Drabman and Margaret Thomas, "Does TV Violence Breed Indifference?" *Journal of Communication* (Autumn 1975): 86–89.

13. Barbara J. Wilson, "The Effects of Two Control Strategies on Children's Emotional Reactions to a Frightening Movie Scene," *Journal of Broadcasting and Electronic Media* (Fall 1989): 397–418.

14. Jinok Son, Stephen D. Reese, and William R. Davie, "Effects of Visual-Verbal Redundancy and Recaps on Television News Learning," *Journal of Broadcasting and Electronic Media* (Spring 1987): 207–16.

15. Dan Berkowitz, "Refining the Gatekeeping Metaphor for Local Television News," *Journal of Broadcasting and Electronic Media* (Winter, 1990): 55–68; David K. Perry and John T. McNelly, "News Orientations and Variability of Attitudes toward Developing Countries," *Journal of Broadcasting and Electronic Media* (Summer 1988): 323–34; and D. Charles Whitney et al., "Source and Geographic Bias in Network Television News 1982–1984," *Journal of Broadcasting and Electronic Media* (Spring 1989): 159–74.

NOTES

16. Tom Link, "Saving What We've Seen," *Emmy* (Winter 1980): 39–41; and Jon Krampher, "Making History," *Emmy* (April 1989): 14–19.

17. "Super Violence Comes Home," *Mother Jones* (January 1987): 15.

18. "TV's Toughest Year is Just a Preview," *Fortune* (November 19, 1990): 95–114.

19. "Is It Just TV or Most of Us?" *New York Times Magazine* (November 15, 1959): 15.

20. "How Blind is Hollywood to Ethics?" *Los Angeles Times/ Calendar* (April 22, 1990): 8–10.

21. David E. Tucker, "Teaching Broadcast Ethics," *Feedback* (Spring 1989): 17–22.

2

1. *1988 Products and Services Guide* (Long Beach, CA: A-VIDD Electronics, 1988): 200–201.

2. Phil Kurz, "Standards Conversion Demystified," *AV Video* (July 1990): 38–47.

3. "U.S. May Only Get to Watch TV's Leap into the Future," *Insight* (July 4, 1988): 40–41.

4. "The Path Once Taken," *Channels* (Field Guide, 1988): 20–21.

5. "CNN Reaching 3 More Nations," *Variety* (October 16, 1990): 10.

6. "Ted's Global Village," *Newsweek* (June 11, 1990): 49–50.

7. "Canada Wants to Raise CBC's Native Program Levels to 90 Percent," *Broadcasting* (March 9, 1987): 43.

8. Shawn Tully, "U.S.-Style TV Turns on Europe," *Fortune* (April 13, 1987): 95–98.

9. "April in Cannes: MIP Programing Blooms Profusely," *Broadcasting* (April 20, 1981): 77.

10. "Telso's Channel Crossings," *Channels* (November 5, 1990): 30–31.

11. "Going Hollywood: Foreign Companies Look for Part in U.S. Studios," *Broadcasting* (April 17, 1989): 40.

12. Sydney W. Head, *World Broadcasting Systems* (Belmont, Ca.: Wadsworth Publishing Company, 1985), 197.

13. Sydney W. Head, *Broadcasting in America* (Boston: Houghton Mifflin Company, 1972), 7–12.

14. "Fortress Europa," *Variety Anniversary Issue* (1990): 102–9.

15. "The Westernization of Eastern Europe: Opening Up the Airwaves," *Broadcasting* (June 25, 1990): 34–37.

16. Interview with Kate Harrison, Project and Research Officer for the Public Interest Advocacy Centre, Sydney, Australia, June 17, 1987.

17. "NHK's DBS Goes on Line in April, and With It HDTV," *Variety* (January 30, 1989): 14.

18. "Fortress Europa," 103–4.

19. "Challenged on TV, the Phone Official Dialed Home and Couldn't Get Through," *Variety* (August 1, 1987): 30–32.

20. "The New World of International TV Programming," *Broadcasting* (October 23, 1989): 75–77.

21. "Formation of the BBC," *BBC Fact Sheet 1* (May 1982): 1–2; and R. W. Burns, *British Television: The Formative Years* (London: Peter Peregrinos, 1986).

22. "The Constitution of the BBC," *BBC Fact Sheet 9* (May 1981): 1–2.

23. "License Fees," *BBC Fact Sheet 11* (May 1982): 1–2.

24. William E. McCavitt, *Broadcasting Around the World* (Blue Ridge Summit, Pa.: TAB Books, 1981): 236.

25. *It's Your BBC* (London: British Broadcasting Corporation, n.d.): 5–6.

26. "The Growth of Television," *BBC Fact Sheet 32* (May 1981): 1–2.

27. "Britain's BBC-TV," *Broadcasting* (November 3, 1986): 43–53.

28. *This is Independent Broadcasting* (London: Independent Broadcasting Authority, n.d.), 3–14.

29. *Advertising on Independent Broadcasting* (London: Independent Broadcasting Authority, n.d.), 3–10.

30. Alan Coren, "It's Sausages and Kidneys vs. Morning TV," *TV Guide* (June 4, 1983): 18–19.

31. Chris Auty, "The New English Channel," *American Film* (December 1982): 17.

32. *A Guide to the BBC's Programmes and Services for Education* (London: British Broadcasting Corporation, n.d.); and *Learning with Television and Radio* (London: Independent Broadcasting Authority, n.d.).

33. "House of Commons Defeats Bill Allowing Ads on BBC," *Variety* (January 24, 1985), 14.

34. Cheryl Rhodes, "The Growth of Videotext in Britain and France," *DataCast* (January/February 1982): 23–35.

35. "Technical Boom Around Corner for U. K. Video Biz," *Variety* (January 15, 1982): 24.

36. "Truce Ends Brit Sky War," *Variety* (November 5, 1990): 1.

37. "Thatcher Administration to Unveil B'casting Bill," *Variety* (December 5, 1989): 26.

38. William McCavitt, *Broadcasting Around the World*, 52–75; "TV: Medium With a Message," *U.S. News and World Report* (February 1, 1982): 34; and Ellen Propper Mickiewitz, *Media and the Russian Public* (New York: Preager, 1981).

39. "Gorbachev Lifts Party Monopoly on Radio, TV," *Los Angeles Times* (July 16, 1990): A–1; and "Soviets Shake Up B'cast Structure," *Variety* (November 15, 1990): 3.

40. *History and Development of the ABC* (Sydney: Australian Broadcasting Corporation, 1983); and Allan G. Brown, *Commercial Media in Australia: Economics, Ownership, Technology, and Regulation* (St. Lucia, Australia: University of Queensland Press, 1986).

41. *Tokyo Broadcasting System* (Tokyo: Tokyo Broadcasting System, n.d.).

42. Zhenzhi Guo, "A Chronicle of Private Radio in Shanghai," *Journal of Broadcasting and Electronic Media* (Fall 1986): 379–92; Max Wilk, " 'I Love Lucy' Would be a Great Leap Forward," *TV Guide* (October 7, 1978): 32–35; and "CBS TV Package Sold to China," *Variety* (July 20, 1984): 1.

43. Sydney Head, *World Broadcasting Systems,* 27–28.

44. Bruce Henstell, "Rede Globo," *Emmy* (January/February 1984): 40–42.

45. Alan Brender, "After a 25-Year Wait . . ." *TV Guide* (August 14, 1976): 24–26.

46. "Saudi TV Pace-Setter of Broadcasting in Arab World," *Variety* (April 16, 1982): 16.

47. Griffin Smith, Jr., "Guatemala: A Fragile Democracy," *National Geographic* (June 1988): 789.

48. Interview with Abdul Aziz Abbas, Controller of TV Programmes, Radio-TV Malaysia, January 6, 1987.

49. *Facts About VOA* (Washington, D.C.: United States Information Agency, 1987): 2; and *BBC Broadcasts to the World* (London: British Broadcasting Corporation, n.d.).

50. *The Voice of America: A Brief History and Current Operations* (Washington, D.C.: United States Information Agency, 1986).

51. Donald R. Browne, "The International Newsroom: A Study of Practices at the Voice of America, BBC, and Deutsche Welle," *Journal of Broadcasting* (Summer 1983): 205–31; and Philo C. Wasburn, "Voice of America and Radio Moscow Newscasts to the Third World," *Journal of Broadcasting and Electronic Media* (Spring 1988): 197–218.

52. *Facts About VOA,* 2; and *BBC Broadcasts to the World.*

53. "IFRB Backs U.S. Claims of Soviet Jamming," *Broadcasting* (September 22, 1986): 84.

54. "U.S. Broadcasts to Cuba Miss Targeted Start-Up," *Los Angeles Times* (January 30, 1985): VI, 1.

55. "GAO Report Slams USIA TV Marti Studies," *Broadcasting* (November 19, 1990): 46.

56. "USIA's High Tech Propaganda War," *Broadcasting* (February 6, 1984): 51.

57. *Armed Forces Radio and Television Service (AFRTS)* (Washington, D.C.: Department of Defense, n.d.); Tom Link, "The GI Joe Network," *Emmy* (August 1990): 70–74; and Craig R. Stephen, "The American Forces Network, Europe: A Case Study in Military Broadcasting," *Journal of Broadcasting and Electronic Media* (Winter 1986): 33–46.

58. Speech by David Leach, Managing Director, Orion Telecommunications, Ltd. at the Annenberg Summer Workshop, Washington, D.C., June 17, 1987.

59. "U.S. Prepared for ITU Conference on Wider AM Band," *Broadcasting* (February 10, 1986): 68.

60. "Space WARC Reaches Consensus," *Broadcasting* (September 16, 1985): 40.

61. *Twenty Years Via Satellite* (Washington, D.C.: COMSAT, 1986).

62. *INTELSAT Report 1985–1986* (Washington, D.C.: INTELSAT, 1986).

63. "International Communications Satellite Competition Examined," *Broadcasting* (June 29, 1987): 44; and "The Expanding International Horizon of Satellites," *Broadcasting* (July 25, 1988): 88.

3

1. For a more detailed account of Maxwell, see Orrin E. Dunlap, Jr., *Radio's 100 Men of Science* (New York: Harper and Brothers Publishers, 1944): 65–68.

2. For a more detailed account of Hertz, see ibid., 113–17.

3. For a more detailed account of Marconi, see Degna Marconi, *My Father Marconi* (New York: McGraw-Hill, 1962).

4. For a more detailed account of Fleming, see Dunlap, *Radio's 100 Men of Science,* 90–94.

5. For a more detailed account of Fessenden, see Helen M. Fessenden, *Fessenden: Builder of Tomorrows* (New York: Coward-McCann, 1940).

6. For a more detailed account of De Forest, see *Lee De Forest, Father of Radio: The Autobiography of Lee De Forest* (Chicago: Wilcox and Follett, 1950).

7. For a more detailed account of Sarnoff, see Eugene Lyon, *David Sarnoff: A Biography* (New York: Harper, 1966).

8. Erik Barnouw, *A Tower in Babel: A History of Broadcasting in the United States to 1933* (New York: Oxford University Press, 1966), 57–61; and Kenneth Bilby, *The General: David Sarnoff and the Rise of the Communications Industry* (New York: Harper and Row, 1986).

9. David Sarnoff, *Looking Ahead: The Papers of David Sarnoff* (New York: McGraw-Hill Book Company, 1968), 31–33.

10. For a more detailed account of Conrad, see Dunlap, *Radio's 100 Men of Science,* 180–83.

11. Robert E. Summers and Harrison B. Summers, *Broadcasting and the Public* (Belmont, Calif.: Wadsworth Publishing Company, 1966), 34.

12. Barnouw, *A Tower in Babel,* 85.

13. Ibid., 86.

14. Ibid., 100.

15. Ibid., 102.

16. William P. Banning, *Commercial Broadcasting Pioneer: WEAF Experiment, 1922–1926* (Cambridge, Mass.: Harvard University Press, 1946), 150.

17. Ibid., 155.

18. "The First 60 Years of NBC," *Broadcasting* (June 9, 1986): 49–64.

19. For a critique of CBS, see Robert Metz, *CBS: Reflections in a Bloodshot Eye* (Chicago: Playboy Press, 1975); and "The Winning Ways of William S. Paley," *Broadcasting* (May 31, 1976): 25–45.

20. For more information, see "The Silver Has Turned to Gold," *Broadcasting* (February 13, 1978): 34–46; and Sterling Quinlan, *Inside ABC: American Broadcasting Company's Rise to Power* (New York: Hastings House, 1979).

21. "After 50 Years the Feeling Is Still Mutual," *Broadcasting* (September 10, 1984): 43–50.

22. Joan Huff Wilson, *Herbert Hoover: Forgotten Progressive* (Boston: Little, Brown, and Company, 1975), 112–13.

23. Donald G. Godfrey, "Senator Dill and the 1927 Radio Act," *Journal of Broadcasting* (Fall 1979): 477–90.

24. Summers and Summers, *Broadcasting and the Public,* 45–50.

25. For a first-hand account of "Amos 'n' Andy," see Charles J. Correll and Freeman F. Gosden, *All About Amos 'n' Andy* (New York: Rand McNally, 1929); and Bert Andrews and Ahrgus Julliard, *Holy Mackerel: The Amos 'n' Andy Story* (New York: Dutton, 1986).

26. For more information on radio programming, see Irving Settel, *A Pictorial History of Radio* (New York: Grossett and Dunlap, 1967); Frank Buxton and Bill Owen, *The Big Broadcast, 1920–1950* (New York: Viking, 1972); J. Fred McDonald, *Don't Touch That Dial: Radio Programming in American Life from 1920 to 1960* (Chicago: Nelson Hall, 1979); Mary C. O'Connell, *Connections: Reflections on Sixty Years of Broadcasting* (New York: National Broadcasting Company, 1986); and Vincent Terrace, *Radio's Golden Years: The Encyclopedia of Radio Programs 1930–1960* (San Diego, Calif.: A. S. Barnes and Co., 1981).

27. Erik Barnouw, *The Golden Web: A History of Broadcasting in the United States 1933–1953* (New York: Oxford University Press, 1968), 8–18.

28. Giraud Chester, Garnet R. Garrison, and Edgar E. Willis, *Television and Radio* (New York: Appleton-Century-Crofts, 1971), 35.

29. James A. Brown, "Selling Airtime For Controversy: NAB Self Regulation and Father Coughlin," *Journal of Broadcasting* (Spring 1980): 199–224.

30. "Genesis of Radio News: The Press-Radio War," *Broadcasting* (January 5, 1976): 95.

31. For a more detailed account of radio during World War II, see Paul White, *News on the Air* (New York: Harcourt, Brace, 1947).

32. Barnouw, *The Golden Web,* 46.

33. Summers and Summers, *Broadcasting and the Public,* 46.

34. Ibid., 70.

35. Federal Communications Commission v. Sanders Brothers Radio Station, 309 U.S. 470, March 25, 1940.

36. *Broadcasting Yearbook, 1975* (Washington, D.C.: Broadcasting Publications, Inc., 1975), C–289.

37. Ward L. Quaal and James A. Brown, *Broadcast Management* (New York: Hastings House, 1976), 292.

38. Michael C. Keith, *Radio Programming* (Boston: Focal Press, 1987), 3.

39. For more insight into the rise of DJs, see Arnold Passman, *The DJ's* (New York: Macmillan, 1971).

40. "Gov't Gets 1st Payola Conviction," *Variety* (May 23, 1989): 1.

41. For a more detailed account of Armstrong, see Lawrence Lessing, *Man of High Fidelity: Edwin Howard Armstrong* (New York: Lippincott, 1956).

42. "AM Radio Plays Hard to Be Heard Again as Audience Dwindles," *Insight* (July 4, 1988): 44.

43. "FCC Ends Curb on Simultaneous AM, FM Programs," *Variety* (April 1, 1986): 1.

44. "AM Radio Plays Hard to Be Heard," 44.

45. Bruce Klopfenstein and David Sedman, "Technical Standards and the Marketplace: The Case of AM Stereo," *Journal of Broadcasting and Electronic Media* 34:2 (Spring 1990): 171–94.

46. "FCC Issues Prescription for AM's Ills," *Broadcasting* (April 16, 1990): 35.

47. "U.S. Firm Building Digital Audio System," *Broadcasting* (May 21, 1990): 34.

48. "Broadcasters Jubilant as FCC Votes to Lift Most of Regulations Covering Radio," *Variety* (January 15, 1981): 1.

49. "First with Eight," *Broadcasting* (September 17, 1984): 105.

50. *Broadcasting/Cablecasting Yearbook, 1987* (Washington, D.C.: Broadcasting Publications, Inc., 1987), H–72.

51. "ABC Opens New Network Radio Broadcast Center," *Broadcasting* (November 5, 1984): 62.

52. "Sounds of Success," *Newsweek* (December 3, 1979): 132–34.

53. "New Radio Networks Spring Up," *New York Times* (January 11, 1980): III,22: 6.

54. "Westwood One to Buy NBC Radio Networks for $50 Mil," *Variety* (July 21, 1987): 1.

55. "Capital Cities/ABC Realigns Radio Networks," *Broadcasting* (May 21, 1990): 32.

56. "The Expanding Universe of Radio Networks,," *Broadcasting* (February 19, 1990): 37–50.

4

1. The best overall chronicle of the history of television can be found in the three-volume history of broadcasting by Erik Barnouw published in New York by Oxford University Press. The titles and dates are: *A Tower in Babel: A History of Broadcasting in the United States to 1933* (1966); *The Golden Web: A History of Broadcasting in the United States 1933–1953* (1968); and *The Image Empire: A History of Broadcasting in the United States from 1953* (1970). This work will be cited as a major source of information because it gives the most back-up detail. It is also scholarly and thoroughly researched through innumerable interviews, unpublished papers, and oral history collections, as well as the more traditional books, articles, and documents. Barnouw has condensed and updated much of the television material into a one-volume book, *Tube of Plenty: The Development of American Television* (1975). Another excellent one-volume history is Christopher H. Sterling and John M. Kittross, *Stay Tuned: A Concise History of American Broadcasting* (Belmont, Calif.: Wadsworth Publishing Company, 1978). Of further interest is Michael Winship, *Television* (New York: Random House, 1988); this is a volume that was made to accompany a PBS series on the history of television.

2. Barnouw, *A Tower in Babel,* 210, 231. These early mechanical systems are best described in A. A. Dinsdale, *First Principles of Television* (New York: John Wiley, 1932).

3. Barnouw, *The Golden Web,* 42, 127, 145. See also "Portrait," *Business Week* (June 7, 1952): 106; and "Portrait," *Fortune* (February 1954): 47.

4. Barnouw, *A Tower in Babel,* 210, and *The Golden Web,* 39–40, 42, 283. See also George Everson, *The Story of Television: The Life of Philo T. Farnsworth* (New York: Norton, 1949); and Stephen F. Hofer, "Philo Farnsworth: Television Pioneer," *Journal of Broadcasting* (Spring 1979): 153–66.

5. Barnouw, *A Tower in Babel,* 66, 154, 210. For Zworykin's own view of the electronics of television, see Vladimir Zworykin, *The Electronics of Image Transmission in Color and Monochrome* (New York: John Wiley, 1954).

6. Barnouw, *The Golden Web,* 126.

7. Ibid., 126–30.

8. Ibid., 242–44.

9. Jon Krampner, "The Death of the Dumont Network," *Emmy* (August 1990): 96–103.

10. Barnouw, *The Golden Web,* 285–90.

11. Jack Slater, "The Amos 'n' Andy Show," *Emmy* (January-February 1985): 48–49.

12. A large number of excellent, highly pictorial works have been published that deal with TV programming through the years. These include Linda Beech, *TV Favorites* (New York: Scholastic Book Services, 1971); Arthur Schulman and Roger Youman, *The Television Years* (New York: Popular Library, 1973); Irving Settl, *A Pictorial History of Television* (New York: Frederick Ungar Publishing Company, 1983): Fred Goldstein and Stan Goldstein, *Prime-Time Television: A Pictorial History from Milton Berle to "Falcon Crest"* (New York: Crown Publisher, 1983); Alex McNeil, *Total Television: A Comprehensive Guide to Programming from 1948 to the Present* (New York: Viking Penguin, 1984); and "The Twenty Top Shows of the Decade," *TV Guide* (December 9, 1989): 20–27.

13. Barnouw, *The Golden Web,* 295.

14. Most of this information was gained from a 1976 interview with True Boardman, one of the writers blacklisted during the fifties. Three sources interesting to peruse are John Cogley, *Report on Blacklisting* (Washington, D.C.: The Fund for the Republic, 1956); *Red Channels: The Report on Communists in Radio and Television* (New York: Counterattack, 1950): and Larry Ceplair, "Hollywood Blacklist," *Emmy* (Summer 1981): 30–32.

15. Barnouw, *The Image Empire,* 22–23.

16. See note 12.

17. Tony Peyser, "Pat Weaver: Visionary or Dilettante?" *Emmy* (Fall 1979): 32–34.

18. See note 12.

19. Daniel E. Garvey, "Introducing Color Television: The Audience and Programming Problems," *Journal of Broadcasting* (Fall 1980): 515–26.

20. Larry Ceplair, "The End of the Trail," *Emmy* (September/October 1984): 50–54.

21. See note 12.

22. "Dress Rehearsals Complete with Answers?" *U.S. News* (October 19, 1959): 60–62; "Is It Just TV or Most of Us?" *New York Times Magazine* (November 15, 1959): 15+; "Out of the Backwash of the TV Scandals," *Newsweek* (November 16, 1959): 66–68; "Quiz Probe May Change TV," *Business Week* (November 7, 1959): 72–74+; and "Van Doren on Van Doren," *Newsweek* (November 9, 1959): 69–70.

23. "The First 50 Years of Broadcasting: 1962," *Broadcasting* (May 25, 1981): 94.

24. See note 12.

25. P. M. Stern, "Debates in Retrospect," *New Republic* (November 21, 1960): 18–19; and "TV Debate Backstage: Did the Cameras Lie?" *Newsweek* (October 10, 1960): 25.

26. "Covering the Tragedy: President Kennedy's Assassination," *Time* (November 29, 1963): 84; "Did Press Pressure Kill Oswald?" *U.S. News* (April 6, 1964): 78–79; "President's Rites Viewed Throughout the World," *Science Newsletter* (December 7, 1963):

355; and J. H. Winchester, "TV's Four Days of History," *Readers Digest* (April 1964): 204I–204J+.

27. "Satellites," *Broadcasting* (July 20, 1987): 33–44.

28. For more information on TV's relationship to Vietnam, see Michael J. Arlen, *Living Room War* (New York: The Viking Press, 1969); Oscar Patterson, "An Analysis of Television Coverage of the Vietnam War," *Journal of Broadcasting* (Fall 1984): 397–404; and Edward Fouhy, "Looking Back at 'The Living Room War,'" *RTNDA Communicator* (March 1987): 12–13.

29. Barnouw, *The Image Empire,* 197.

30. See note 12.

31. Les Brown, *Encyclopedia of Television* (New York: Times Books, 1977): 283.

32. Barnouw, *The Image Empire,* 251–52.

33. Elizabeth J. Heighton and Don R. Cunningham, *Advertising in the Broadcast Media* (Belmont, Calif.: Wadsworth Publishing Company, 1976), 268–71.

34. "The First 50 Years of Broadcasting: 1971" *Broadcasting* (July 27, 1981): 119.

35. "'Family Hour' OK but Not by Coercion," *Variety* (November 5, 1976): 1.

36. "FCC Opens Box on Financial Interest," *Broadcasting* (June 28, 1982): 30.

37. "Now an Open Season on All Newsmen," *Broadcasting* (November 24, 1969): 44.

38. "Ratings Down 10 Percent, Shares Off Four Points," *Broadcasting* (November 9, 1987): 35.

39. "Capcities + ABC," *Broadcasting* (March 25, 1985): 31.

40. "General Electric Will Buy RCA for $6.28 Billion," *Los Angeles Times* (December 12, 1985): I, 1.

41. "Welch Names Robert Wright to NBC Presidency," *Broadcasting* (September 1, 1986): 34.

42. "NBC's Farewell to Fred," *Newsweek* (July 13, 1981): 69.

43. L. J. Davis, "How Tinker Turned It Around," *Channels* (January/February 1986): 30–39.

44. "CBS Gleam in Ted Turner's Eye," *Broadcasting* (March 4, 1985): 35.

45. "Tisch Does What CBS Feared in Turner," *The Wall Street Journal* (November 20, 1987): 6.

46. "Metromedia Vid as Fox Subsid," *Variety* (May 7, 1985): 1.

47. "Murdoch Holds Full Fox Hand," *Variety* (September 24, 1985): 1.

48. "Fox Broadcasting Company: The Birth of a Network," *Broadcasting* (April 6, 1987): 88; and Alex Ben Block, *Outfoxed: Marvin Davis, Barry Diller, Rupert Murdoch, Joan Rivers, and the Inside Story of America's Fourth Television Network* (New York: St. Martin's Press, 1990).

49. "Turner Gets Financial OK to Buy MGM," *Electronic Media,* (August 19, 1985): 1.

50. "Japan Goes Hollywood," *Newsweek* (October 9, 1989): 62–65.

51. "Here's a Brief Look at 1985's Media Mergers," *Electronic Media* (January 13, 1986): 88.

52. Harvey Solomon, "The Meaning of the Time-Warner Merger," *Channels* (June, 1989): 62–63.

53 Stratford P. Sherman, "Are Media Mergers Smart Business," *Fortune* (June 24, 1985): 98–103.

54. "FCC Eliminates Number of Regs for TV Stations," *Variety* (June 28, 1984): 1.

55. "FCC Strikes the Flag on TV Ownership Rules," *Broadcasting* (August 13, 1984): 35.

56. "FCC Swamped with Applications for New Low-Power TV Stations," *The Wall Street Journal* (October 30, 1981): 29.

57. "LPTV Gets the Go Ahead," *Broadcasting* (March 8, 1982): 35.

58. "They Jes' Keep a-Growin'," *Channels* (Field Guide, 1987): 54.

59. "Hi-Fi Meets Television," *Newsweek* (June 24, 1985): 78; and "Momentum Builds for TV Stereo," *Broadcasting* (September 9, 1985): 117.

60. "FCC's FIN-SYN Surprize: Everybody Knocks It," *Broadcasting* (April 15, 1991): 37–38.

61. D. K. Knapp, "The New Golden Years," *Emmy* (August 1986): 50–55; and Michael Pollan, "The Contenders," *Channels* (October 1988): 50–51.

62. "Home Shopping Network's Shopping Spree," *Broadcasting* (March 9, 1987): 55.

63. "The Whys and Wherefores of Text Transmission," *Broadcasting* (July 5, 1982): 32–35.

64. Fred Bronson, "Live Aid," *Emmy* (August 1986): 76–83.

5

1. Erik Barnouw, *A Tower in Babel: A History of Broadcasting in the United States to 1933* (New York: Oxford University Press, 1966), 61.

2. Werner J. Severin, "Commercial vs. Non-Commercial Radio During Broadcasting's Early Days," *Journal of Broadcasting* (Summer 1981): 295–302.

3. Barnouw, *A Tower in Babel,* 215.

4. John Witherspoon and Roselle Kovitz, *The History of Public Broadcasting* (Washington, D.C.: Current, 1987), 8–9.

5. Donald N. Wood and Donald G. Wylie, *Educational Telecommunications* (Belmont, Calif.: Wadsworth Publishing Company, 1977), 32.

6. *Broadcasting/Cablecasting Yearbook, 1990* (Washington, D.C., Broadcasting Publications, Inc., 1990), A–3.

7. Frederic A. Leigh, "College Radio: Effects of FCC Class D Rules Changes," *Feedback* (Fall 1984): 18–21.

8. An excellent history of the NAEB can be found in Harold E. Hill, *The National Association of Educational Broadcasters: A History* (Urbana, Ill.: National Association of Educational Broadcasters, 1954).

9. *National Public Radio Fact Sheet* (Washington, D.C.: National Public Radio, n.d.), 1.

10. Susan Stamberg, *Every Night at Five: Susan Stamberg's "All Things Considered" Book* (New York: Pantheon, 1982).

11. *National Public Radio Fact Sheet* (Washington, D.C.: National Public Radio, 1987), 1.

12. For more information on NPR, see Robert K. Avery and Robert Pepper, "Balancing the Equation: Public Radio Comes of Age," *Public Telecommunications Review* (March/April 1979): 19–30; Joseph Brady Kirkish, "A Descriptive History of America's First Public Radio Network: National Public Radio, 1970–1974" (Ph.D. Dissertation, The University of Michigan, 1980); and John Kenneth Garry, "The History of National Public Radio, 1974–1977" (Ph.D. Dissertation, Southern Illinois University at Carbondale, 1982).

13. "National Public Radio to Alter Rate Structure," *Variety* (February 10, 1988): 13.

14. *American Public Radio* (St. Paul, Minnesota: American Public Radio, n.d.): 1–7.

15. "Public Radio's Programming Explosion," *Broadcasting* (May 29, 1989): 57.

16. "Public Radio Goes After Broader Programming Identity," *Broadcasting* (February 19, 1990): 47–49.

17. "NPR Programming Strategy: Pluralism," *Broadcasting* (March 19, 1990): 54.

18. *American Public Radio*, 2.

19. Frederic A. Leigh, "College Radio: Effects of FCC Class D Rule Changes," *Feedback* (Fall 1984): 18–21.

20. "Last Minute Salvation of NPR," *Broadcasting* (August 1, 1983): 21.

21. "Public Stations Approve NPR Reorganization," *Electronic Media* (May 30, 1985): 3.

22. "What Happened to Mankiewicz?" *Broadcasting* (April 25, 1983): 28.

23. Donald P. Mullally, "Radio: The Other Public Medium," *Journal of Communication* (Summer 1980): 189–97.

24. Lynne S. Gross, "A Study of College Credit Literature Courses Offered on Open Circuit Television," *NAEB Journal* (November/December 1962): 87–88.

25. Robert K. Avery and Robert Pepper, "An Institutional History of Public Broadcasting," *Journal of Communication* (Summer 1980): 126–38.

26. *Broadcasting Yearbook, 1975* (Washington, D.C.: Broadcasting Publications, Inc., 1975), A–7.

27. Clifford G. Erickson and Hyman M. Chausow, *Chicago's TV College: Final Report of a Three-Year Experiment* (Chicago: Chicago City Junior College, August, 1960).

28. Mary Howard Smith, *"Midwest Program on Airborne Television Instruction," Using Television in the Classroom* (New York: McGraw-Hill, 1961).

29. *Broadcasting Yearbook, 1975*, A–7.

30. Gene R. Stebbins, "PTV Membership Support: Its Beginnings," *Public Telecommunications Review* (July/August 1979): 52–54.

31. Carnegie Commission on Public Television, *Public Television: A Program for Action* (New York: Harper and Row, 1967).

32. The Public Broadcasting Act of 1967, Public Law 90–129, 90th Congress (November 7, 1967).

33. Robert M. Pepper, *The Formation of the Public Broadcasting Service* (New York: Arno Press, 1979).

34. Arthur Shulman and Roger Youman, *The Television Years* (New York: Popular Library, 1973), 270–303.

35. *The Public Television Station Program Cooperative* (Alexandria, Va.: Public Broadcasting Service, 1986), 1–3.

36. "CPB, PBS Strike Truce in Atlanta," *Broadcasting* (February 14, 1977): 31.

37. "Public Broadcasting's Major Players," *Broadcasting* (May 11, 1987): 70.

38. "Twenty Tumultuous Years for CPB," *Broadcasting* (May 11, 1987): 60–75.

39. *Carnegie Commission of the Future of Public Broadcasting: A Public Trust* (New York: Carnegie Commission, 1979); and "Carnegie II's Revised Blueprint for Public Broadcasting," *Broadcasting* (February 5, 1979): 56–65.

40. Willard D. Rowland, Jr., "Continuing Crisis in Public Broadcasting: A History of Disenfranchisement," *Journal of Broadcasting and Electronic Media* (Summer 1986): 251–74.

41. *Made Possible by a Grant from the Corporation for Public Broadcasting* (Washington, D.C.: Corporation for Public Broadcasting, n.d.).

42. Figures come from *Status Report on Public Broadcasting, 1973* (Washington, D.C.: Corporation for Public Broadcasting, 1974).

43. *Facts About PBS* (Alexandria, Virginia: Public Broadcasting Service, 1990): 6.

44. "New Budget Bill Mixed Blessing for Pubcasters," *Variety* (August 3, 1981): 1.

45. "Annenberg Grant Comes Through," *Current* (March 16, 1981): 1.

46. "Pubcasters Get OK to Air Logos of Contributors," *Variety* (April 24, 1981): 1.

47. "More Subscribers to Public Stations Called Answer to Federal Cuts," *Broadcasting* (June 29, 1981): 57.

48. "'Pledgeless' Pledge Drives: Noncommercial Phenomenon," *Broadcasting* (December 8, 1986): 110.

49. *Facts About PBS*, 2.

50. "CPB on the Prowl for New Ways to Fund the Future," *Broadcasting* (November 9, 1981): 29.

51. "10 Pub TV Stations Granted Go-Ahead to Broadcast Ads," *Variety* (February 2, 1982): 1.

52. "PBS Revises Rules for Underwriting in Wake of FCC Decision, *Broadcasting* (April 9, 1984): 92.

53. "The Commercialization of Public TV," *Electronic Media* (June 23, 1986): 1.

54. "PBS Board Walks Many Tightropes," *Broadcasting* (March 28, 1988): 51.

55. "Reagan Okays $1 Billion CPB Fund for '91–93," *Variety* (November 9, 1988): 1.

56. *Made Possible by a Grant From the Corporation for Public Broadcasting*, 3–4.

57. "Furor Over a Death," *Time* (April 21, 1980): 32.

58. "PBS & Controversy: Mostly OK," *Variety* (April 16, 1987): 1.

59. "Editorializing Ban for Public Stations Debated Before Supreme Court," *Broadcasting* (January 23, 1984): 95.

60. "'Civil' War on Webs' Fare Proves Big Victory for PBS," *Variety* (October 1, 1990): 3.

61. *Made Possible by a Grant From the Corporation for Public Broadcasting,* passim.

62. *20 Reasons for Public Television* (Alexandria, Virginia: Public Broadcasting Service, 1986), 2.

63. *Facts About PBS,* 18–19.

64. Ibid., 16.

65. "PBS Panel Suspends Program Cooperative," *Variety* (July 25, 1990): 1.

66. *Who Watches Public Television?* (Alexandria, Virginia: Public Broadcasting Service, 1986): 1.

6

1. David L. Jaffe, "CATV: History and Law," *Educational Broadcasting* (July/August 1974): 15–16; and Thomas F. Baldwin and D. Stevens McVoy, *Cable Communication* (Englewood Cliffs, New Jersey: Prentice-Hall, 1983), 8–10.

2. *Broadcasting/Cable Yearbook, 1981* (Washington, D.C.: Broadcasting Publications, Inc., 1981), G–1.

3. Jaffe, 34.

4. For other chronicles of early cable TV, see Mary Alice Mayer Philips, *CATV: A History of Community Antenna Television* (Evanston, Ill.: Northwestern University Press, 1972); and Albert Warren, "What's 26 Years Old and Still Has Growing Pains?"*TV Guide* (November 27, 1976): 4–8.

5. The relationship between cable and government can be found in *Cable Television and the FCC: A Crisis in Media Control* (Philadelphia: Temple University Press, 1973); Steven R. Rivkin, *Cable Television: A Guide to Federal Regulations* (Santa Monica, Calif.: Rand Corporation, 1973); Martin H. Seiden, *Cable Television USA: An Analysis of Government Policy* (New York: Praeger, 1972); and Stuart N. Brotman, *Communications Policymaking at the Federal Communications Commission* (Washington, D.C.: The Annenberg Washington Program, 1987), 33–46.

6. A great deal about early franchising can be learned from Leland L. Johnson and Michael Botein, *Cable Television: The Process of Franchising* (Santa Monica, Calif.: Rand Corporation, 1973).

7. Jaffe, 17.

8. "Cable: The First Forty Years," *Broadcasting* (November 21, 1988): 40.

9. *Broadcasting/Cable Yearbook, 1981,* G–1.

10. Jaffe, 35.

11. *Broadcasting/Cable Yearbook, 1981,* G-1–G-3.

12. Two sources that deal with early local origination are Ron Merrell, "Origination Compounds Interest with Quality Control," *Video Systems* (November/December 1975): 15–18; and Sloan Commission on Cable Communications, *On the Cable: The Television of Abundance* (New York: McGraw-Hill, 1972).

13. Two sources dealing with early public access are Richard C. Kletter, *Cable Television: Making Public Access Effective* (Santa Monica, Calif.: Rand Corporation, 1973); and Charles Tate, *Cable Television in the Cities: Community Control, Public Access, and Minority Ownership* (Washington, D.C.: The Urban Institute, 1972).

14. "Cable Survey Shows Growth," *Video Systems* (January/February 1976): 6.

15. "Righting Copyright," *Time* (November 1, 1976): 92.

16. Margaret B. Carlson, "Where MGM, the NCAA, and Jerry Falwell Fight for Cash," *Fortune* (January 23, 1984): 171.

17. "Compulsory License Hit Again," *Variety* (April 23, 1981): 1.

18. *HBO Landmarks* (New York: Home Box Office, n.d.), 1.

19. Sheila Mahony, Nick Demartino, and Robert Stengel, *Keeping Pace with the New Television* (New York: VNU Books International, 1980), 61.

20. "After Six Trips to the Firing Line, Satellites Finally Put Pay-Television on the Map," *Variety* (December 9, 1980): 1.

21. Don Kowet, "High Hopes for Pay Cable," *TV Guide* (June 10, 1978): 14–18.

22. "Hanging the Crepe for STV," *Broadcasting* (October 15, 1984): 46.

23. Sheila Mahony et al., *Keeping Pace,* 131.

24. Ibid., 95.

25. "The Dimensions of Cable: 1981," *Channels* (April/May 1981): 88.

26. "CATV Stats Leapin': Pretax Net 45 percent; Feevee Revenue 85 percent," *Variety* (December 31, 1980): 16.

27. "Cable Revenues Gain Faster than Profits, Survey Finds," *Broadcasting* (November 30, 1981): 52.

28. "The Dimensions of Cable: 1981," 88.

29. Ibid., 88.

30. "The Top Line: Almost $2 Billion; The Bottom Line: Almost $200 Million," *Broadcasting* (January 5, 1981): 75.

31. "Cities Issue Guidelines for Cable Franchising," *Broadcasting* (March 9, 1981): 148; and Shelia Mahony et al., *Keeping Pace,* 93.

32. Pat Carson, "Dirty Tricks," *Panorama* (May 1981): 57–59+; "Fight for TV Franchise in New England Town Elicits Big Bribe Offer," *The Wall Street Journal* (December 22, 1981): 1; and Laurence Bergeen, "Cable Fever," *TV Guide* (June 6, 1981): 23–26.

33. For more on franchising, see "City's Ability to Regulate Cable Television Industry Questioned," *Los Angeles Times* (August 17, 1981): II, 1; "Warner Amex Loses Bid to Cablevision for Subscriber TV Franchise in Boston," *The Wall Street Journal* (August 13, 1981): 6; "The Gold Rush of 1980," *Broadcasting* (March 31, 1980): 35–56; "Simmering Cable Franchise Issue Comes to a Boil," *Variety* (July 31, 1981): 1; "New York Today Picks Its Cable-TV Winners for Four Boroughs," *The Wall Street Journal* (November 18, 1981): 1; and many other newspaper articles of the late 1970s and early 1980s.

34. "The Top 50 MSO's: As They Are and Will Be," *Broadcasting* (November 30, 1981): 37; "Entertainment Analysts Find Cable Mom-Pop Days Are Gone; Big Bucks Rule the Day," *Broadcasting* (June 15, 1981):

35. "Cable Television Is Attracting More Ads; Sharply Focused Programs Are One Lure," *The Wall Street Journal* (March 31, 1981): 46; "CBS Cable Signs Kraft as Its First Advertiser," *Broadcasting* (August 3, 1981): 59; "ARTS Snags GM as Underwriter," *Variety* (September 2, 1981): 1; and "Bristol-Myers to Furnish Series to USA Network," *Broadcasting* (February 23, 1981): 61.

36. "Scribes Back at Typewriters," *Variety* (July 16, 1981): 1; and "Pay-TV Settlements Will Hasten Debut of Feevee 1st-Run Prod'n," *Variety* (July 15, 1981): 8.

37. Keith Larson, "All You Ever Wanted to Know About Buying an Earth Station," *TVC* (May 15, 1979): 16–19; "Going Once, Twice, Gone on Satcom IV," *Broadcasting* (November 16, 1981): 27; and "Greenlight for Satcom IV Plan," *Broadcasting* (March 29, 1982): 40.

38. "Viacom Becomes Second Satellite Pay Cable Network," *Broadcasting* (October 31, 1977): 64.

39. "GalaVision Reaches First Anniversary, Free Preview Scheduled for October," *TVC* (October 15, 1980): 33–36; "RCA Pay-Cable Plunge Via 50 Percent Slice of RCTA," *Variety* (May 11, 1981): 1; "Times Mirror Sat'lite Programs Join Feevee Networks on May 1," *Variety* (March 26, 1981): 3; and "Disney Previews Pay-TV Channel," *Variety* (April 13, 1983): 1.

40. Some of the articles summarizing basic cable are Michael O'Daniel, "Basic-Cable Programming: New Land of Opportunity," *Emmy* (Summer 1980): 26–30; "The Cable Numbers According to Broadcasting," *Broadcasting* (November 30, 1981); and Peter W. Bernstein, "The Race to Feed Cable TV's Maw," *Fortune* (May 4, 1981): 308–18.

46; and "FCC Okays Sale of TPT to Westinghouse, the Largest Communications Merger Ever," *Variety* (July 13, 1981): 1.

41. For five different views of access programming see Don Kowet, "They'll Play Bach Backwards, Run for Queen of Holland," *TV Guide* (May 31, 1980): 15–18; Susan Beonarcyzk, "Mid-West Cablers Accent on Access," *Variety* (March 17, 1981): 28; "Cable Interconnects: Making Big Ones Out of Little Ones," *Broadcasting* (March 1, 1982): 59; Ann M. Morrison, "Part-Time Stars of Cable TV," *Fortune* (November 30, 1981): 181–84; and the entire July 1981 issue of *Community Television Review* published by the National Federation of Local Cable Programmers.

42. "Week One for Warner's Qube," *Broadcasting* (December 12, 1981): 62; and "Second Qube Facility Launched by Warner-Amex Near Cincinnati," *TVC* (October 15, 1980): 58–59.

43. "The Two-Way Tube," *Newsweek* (July 3, 1978): 64.

44. "Home Security Is a Cable TV, Industry Bets," *The Wall Street Journal* (September 15, 1981): 25.

45. "Pay Cable TV Is Losing Some of Its Sizzle as Viewer Resistance, Disconnects Rise," *The Wall Street Journal* (November 19, 1982): 22.

46. "Cable's Lost Promise," *Newsweek* (October 15, 1984): 103–5.

47. "Warner Amex's Lowering Expectations," *Broadcasting* (March 19, 1984): 37; "Buyers Study How to Divide Group W Cable," *Electronic Media* (January 6, 1986): 3; and "TCI Makes Its Equity Play for Showtime," *Broadcasting* (October 23, 1989): 40.

48. "Viacom's Rise to Stardom," *Newsweek* (November 25, 1985): 71.

49. "HBO Takes Best Seat in the House while Rivals Still Wait in Line," *Los Angeles Times* (June 5, 1983): V, 1.

50. "Basic Cable Networks," *Channels* (Field Guide, 1987): 73–74; "Pay-Cable Networks,"

Channels (Field Guide, 1987): 81; "RCA Venture for Cable-TV Programs to End," *The Wall Street Journal* (February 23, 1983): 8; and "Spotlight Pay-TV Venture Is Seen Ending with Subscribers Being Shifted to Rivals," *The Wall Street Journal* (September 2, 1983): 9.

51. Mark Frankel, "Can Playboy Save Its Skin?" *Channels* (November 1986): 37–40.

52. "Programming Setback for HBO," *Variety* (December 19, 1983): 1; and "HBO Signs $500 Million Film Deal with Paramount," *Broadcasting* (June 30, 1987): 21.

53. "CBS to Drop Its Cultural Cable-TV Service After Failing to Draw Needed Ad Support," *The Wall Street Journal* (September 14, 1983): 7.

54. "Turner the Victor in Cable News Battle," *Broadcasting* (October 17, 1983): 27.

55. "Viacom Agrees to Buy MTV Networks," *Variety* (August 27, 1985): 1; and "Untimely End for Odyssey," *Broadcasting* (January 6, 1986): 160.

56. "Buyer Sought to Acquire Time's USA Network Interest," *Variety* (April 24, 1987): 1; "ABC Video Ent. Acquires ESPN," *Variety* (May 1, 1984): 1; "Daytime + CHN: Joining Together to Stay Afloat," *Broadcasting* (June 20, 1983): 37; "The Reach of MSOs," *Channels* (May 1987): 49; and "Sorting through the Fallout of Cable Programming," *Broadcasting* (October 17, 1983): 29.

57. "Discovery Channel Sets Sail," *Broadcasting* (June 24, 1985): 53.

58. "NBC Introduces CNBC at CTAM," *Broadcasting* (August 8, 1988): 21.

59. "TNT Takes Its First Step," *Broadcasting* (October 3, 1988): 48.

60. Janet Stilson, "Inside Cable's Laugh Track," *Channels* (April 23, 1990): 16–21.

61. "Home Is Where The Mart Is," *Channels* (Field Guide, 1987): 77–78; and "Cable Value Network and COMB Restructuring," *Broadcasting* (March 30, 1987): 163.

62. "City Council Votes to Maintain Status Quo on Cable TV Access," *Variety* (April 29, 1987): 1.

63. "Warner Amex Cable Cuts Interactive Programming Feed in Six Major Cities," *Variety* (January 19, 1984): 1.

64. Patricia E. Bauer, "Young and Impulsive," *Channels* (May 1987): 50–51; "Hollywood Tries for a New Hit," *Newsweek* (April 8, 1985): 56–58; "Playboy Ent. Joins Growing Ranks of Pay-per-view Hopefuls," *Variety* (June 5, 1985): 1; "PPV Service Changes Name: Ready to Go," *Electronic Media* (November 4, 1985): 30; and "Pay-Per-View Seems a Sure thing Despite Marketing, Technical Obstacles," *The Wall Street Journal* (February 6, 1989): A9A.

65. "Two Major Restraints on Cable Television Are Lifted by the FCC," *The Wall Street Journal* (July 23, 1980): 1.

66. "Appeals Court Upholds FCC Repeal of Distant-Signal, Exclusivity Rules," *Broadcasting* (June 22, 1981): 32.

67. "Syndex Redux: FCC Levels the Playing Field," *Broadcasting* (May 23, 1988): 31.

68. "Free at Last: Cable Gets Its Bill," *Broadcasting* (October 15, 1984): 38; and "Cable TV Gets a Deregulation Gift From FCC," *Variety* (April 12, 1985): 1

69. "GAO Says Basic Cable Rater Up 10% in 1989," *Broadcasting* (June 18, 1990): 23.

70. "Cable: Up Against the (Senate) Wall," *Broadcasting* (March 12, 1990): 27.

71. "Court Frees Cable TV Firms of Obligation to Carry Local Channels," *Los Angeles Times* (July 20, 1985): IV-1; "Few Surprises in FCC's Must-Carry Order," *Broadcasting* (December 1, 1986): 43; "Must Carry Back on Track," *Broadcasting* (March 30, 1987): 55; "Tracking 10 Years of Must-Carry Machinations," *Broadcasting* (November 13, 1989: 36–38; and Frank Lovece, "Muddling Through the Must-Carry Mess," *Channels* (September 1988): 48–51.

72. "Black Tuesday Descends on Cable Industry," *Broadcasting* (May 21, 1983): 61.

73. "Court Tells Cable to up Copyright Ante," *Broadcasting* (January 11, 1988): 40; and "Cable Royalties System Upheld," *Variety* (July 1, 1988): 1.

74. "Judge Says LA Erred in Refusing Second Cable Franchise," *Broadcasting* (January 15, 1990): 58; and Thomas W. Hazlett, "The Policy of Exclusive Franchising in Cable Television," *Journal of Broadcasting and Electronic Media* (Winter 1987): 1–20.

75. "Supreme Court Upholds Pole-Rate Regulations," *Variety* (February 26, 1987): 1.

76. "Backyard Satellite Dishes Spread But Stir Fight With Pay-TV Firms," *The Wall Street Journal* (April 2, 1982): 25.

77. "Lofty Bid for First DBS System," *Broadcasting* (December 22, 1980): 23.

78. *Enter DBS: The Story of Direct Broadcast Development in 1984* (Washington, D.C.: Television Digest, Inc., 1985); "Slow Liftoff for Satellite-To-Home TV," *Fortune* (March 5, 1984): 100; and "Another Nail in the DBS Coffin: Comsat Bows Out," *Broadcasting* (December 3, 1984): 36.

79. "Next: A Dish for Every Home?" *Newsweek* (May 21, 1990): 67–68.

80. "More Time for DBS Applicants," *Broadcasting* (December 5, 1990): 65.

81. "Dishing Out TV in U.S. Outback," *Broadcasting* (July 11, 1988): 66.

82. "SMATV: The Medium That's Making Cable Nervous," *Broadcasting* (June 21, 1982): 33–43.

83. "SMATV Free From All But Federal Regs," *Variety* (November 10, 1983): 1.

84. "SMATV," *Channels* (Field Guide, 1987): 70.

85. "MDS/Wireless Cable—Boon or Bane?" *TVC* (December 15, 1980): 195–201.

86. "FCC Reassigns Eight ITFS Channels," *E-ITV* (July 1983): 6.

87. "PacWest Sues Turner Over Dropping of TNT to MMDS," *Broadcasting* (April 30, 1990): 54.

88. "Explosive Growth of MMDS Predicted for Next Year," *Variety* (November 30, 1989): 6.

89. "FCC Wants to Loosen Cable-Telco Prohibitions," *Broadcasting* (July 25, 1988): 31.

90. "GTE Cerritos Launches 28 Channels of PPV," *Broadcasting* (July 9, 1990): 54–55.

7

1. John Barwich and Stewart Kranz, *Profiles in Video* (White Plains, New York: Knowledge Industry Publications, 1975).

2. *International Television Association: The Organization for Professional Video Communicators* (Irving, Texas: ITVA, n.d.).

3. Judith M. Brush and Douglas P. Brush, *Private Television Communications: Into the Eighties* (Berkeley Heights, New Jersey: International Television Association, 1981): 51–67.

4. Charles Callaci, *Learning Through Television* (Chino, CA: Ramo II Publishers, 1975): 2–14.

5. Nancy Nelson, "Around the Nation with ITFS," *E-ITV* (June 1985): 28–29.

6. Richard C. Morgan, "Soy Group Gets World TV Report," *The Atlanta Constitution* (August 14, 1979): 21.

7. "Holiday Inn, Comsat Venture Would Create Communications Network," *Broadcasting* (November 5, 1984): 75.

8. "Videoconference Use Expands to Meet Rising Business Needs," *Aviation Week and Space Technology* (July 22, 1985): 157–64; "Teleconferencing: Nobody's Baby Now," *E-ITV* (November 1987): 26–29; and "Broadcast News, Inc." *Newsweek* (January 4, 1988): 34–35.

9. "FCC Reassigns Eight ITFS Channels," *E-ITV* (July, 1983): 6.

10. "Firms Are Cool to Meetings by Television," *The Wall Street Journal* (July 26, 1983): 2.

11. Leonard Shyles, "The Video Tape Recorder: Crown Prince of Home Video Devices," *Feedback* (Winter 1981): 1–5.

12. Bruce Cook, "High Tech: The New Videocassettes," *Emmy* (Summer 1980): 40–44.

13. David Lachenbruch, "Video-cassette Recorders—Here Comes the Second Generation," *TV Guide* (October 28, 1978): 2–6.

14. Ivan Berger, "Videocassette Recorders: Rising Stars of Home Entertainment," *Popular Electronics* (June 1980): 54.

15. "RCA Predicting $1 Bil Retail Year for the VCR Industry," *Variety* (May 13, 1981): 1.

16. Cook, 40–44.

17. "Betamax Decision Overturned," *Variety* (October 20, 1981): 1.

18. "Hollywood Loses to Betamax," *Variety* (January 18, 1984): 1.

19. "3d-Quarter VCR U.S. Population Pegged at 13 Mil," *Variety* (October 25, 1984): 1.

20. "Sony Adds VHS to VCR Format," *Variety* (January 12, 1988): 1.

21. "Camcorders Zoom into Video Products Picture," *Los Angeles Times* (May 18, 1987): IV–1.

22. "Many Field Acquisition Formats Compete for Attention in 1989," *Television Broadcast* (March 1989): 16–17.

23. "Broadcasting's Rush to the West," *Broadcasting* (October 23, 1989): 37.

24. "Success of 'Home Videos' Spurs Clones," *Broadcasting* (April 16, 1990): 38.

25. "Home Editing: Video's Final Frontier," *Video* (July 1990): 37–39.

26. Robert Nicksin, "Future Trends in Corporate Video," *Video Manager* (February 1989): 29–31.

27. "U.S. VCR Homes Pass 70% Mark," *Variety* (May 7, 1990): 9.

28. "Video Disks," *Broadcasting* (February 2, 1981): 35.

29. Neil Hickey, "Take the Videodisc, Please," *TV Guide* (December 26, 1981): 12–14.

30. "RCA Reports Big Success for Its Vidisk Launch," *Variety* (May 1, 1981): 1.

31. "RCA Gives Up on Videodisc System," *Los Angeles Times* (April 5, 1984): IV, 1.

32. "Laserdisk Players Expand Capabilities," *Variety* (December 27, 1983): 7.

33. "Videodisks Make a Comeback as Instructors and Sales Tools," *The Wall Street Journal* (February 15, 1985): 25.

34. "Not Really Combat but It Sure Seems Like It," *Los Angeles Times* (July 7, 1987): IV, 1.

35. James A. Lippke, "Interactive Video Discs: Entering the Mainstream of Business," *E-ITV* (August 1987): 12–17.

36. "The Interactive Traveller," *Corporate Video Decisions* (September 1988): 27.

37. CD-V Format Given Boost From Music, Homevid Firms," *Variety* (April 24, 1987): 8.

38. "A New Spin on Videodiscs," *Newsweek* (June 5, 1989): 68–69.

39. Some of the books that give history and major developments of the telephone are: J. Brooks, *Telephone: The First Hundred Years* (New York: Harper and Row, 1975); *Events in Telecommunication History* (New York: AT&T, 1979); John

R. Pierce, *Signals: The Telephone and Beyond* (San Francisco: Freeman, 1981); and Gerald W. Brock, *The Telecommunications Industry: The Dynamics of Market Structure* (Cambridge, MA: Harvard University Press, 1981).

40. "Ma Bell's Big Breakup," *Newsweek* (January 18, 1982): 58–59.

41. Robert Britt Horwitz, "For Whom the Bell Tolls: Causes and Consequences of the AT&T Divestiture," *Critical Studies in Mass Communication* (June 1986): 119–53.

42. "High Rollers with High Hopes," *Channels* (Field Guide, 1987): 86–87; and A. Michael Noll, "Videotex: Anatomy of a Failure," *Information and Management* (1985): 99–109.

43. "Push-Button Age," *Newsweek* (July 9, 1990): 56–57; and "Getting the Facts on Buying a Fax," *Newsweek* (April 24, 1989): 80.

44. Speech by Barry Cole, Professor of the Annenberg School of Communications, University of Pennsylvania, at the Summer Faculty Workshop, Washington, D.C., June 15, 1987.

45. "Is ISDN Worth Waiting For?" *Business Communications Review* (March/April 1986): 13–17; and *FTS2000: The Most Advanced Telecommunications System in the World* (Washington, D.C.: U.S. General Services Administration, n.d.).

46. "RCC," *Broadcasting* (October 4, 1982): 39–47.

47. "Motorola Proposed 77 Lightsats for Global Mobile Phone Service," *Aviation Week and Space Technology* (July 2, 1990): 29.

48. "Car Phones: From Toy to Valued Tool," *Los Angeles Times* (July 7, 1987): I–1.

49. "CBS News Buys Electronic Newsroom System," *Broadcasting* (January 21, 1985): 92.

50. Some sources on computers are: Gary B. Shelly and Thomas J. Cashman, *Introduction to Computers and Data Processing* (Brea, Ca.: Anaheim Publishing Company, 1980); James W. Morrison, *Principles of Data Processing* (New York: Arco Publishing Company, 1979), and Fred D'Ignazio, *Messner's Introduction of the Computer* (New York: Julian Messner, 1983).

51. Daniel Seligman, "Life Will Be Different When We're All On-Line," *Fortune* (February 4, 1985): 68–71.

52. Interview with Ann Ward, Engineering Librarian, General Dynamics, Pomona, Calif., July 21, 1987.

53. Daniel Seligman, "Life Will Be Different," 68.

54. *Hayes Smartcom II* (Norcross, Georgia: Hayes Microcomputer Products, 1984): 9.2–9.18.

55. "The 'Write' Stuff," *Online Today* (April 1987): 22–23.

56. "Technology in the Workplace," *The Wall Street Journal* (June 12, 1987): 9D.

57. *1986 DIALOG Database Catalog* (Palo Alto, Calif.: DIALOG Information Services, Inc., 1986).

58. "Zapping the Mailman," *Newsweek* (August 20, 1984): 64.

59. Peter W. Bernstein, "Atari and the Video Game Explosion," *Fortune* (July 27, 1981): 40–46; and "Nintendo and Beyond," *Newsweek* (June 18, 1990): 62.

60. "Ganging Up on Big Blue," *Newsweek* (September 26, 1988): 50; and "Apple Computer's Bruising Times," *Newsweek* (May 27, 1985): 46–47.

61. "Audiovisual Device—The Next Wave in Home Electronics?" *Los Angeles Times* (June 5, 1990): D–1.

8

1. Some materials that deal with general management include *The Small Market Television Manager's Guide* (Washington, D.C.: National Association of Broadcasters, 1987); John M. Lavine and Daniel B. Wackman, *Managing Media Organizations* (New York: Longman, 1988); Richard F. Vancil, *Passing the Baton: Managing the Process of CEO Succession* (Cambridge, MA: Harvard Business School Press, 1987); and *Great Expectations: A Television Manager's Guide to the Future* (Washington, D.C.: National Association of Broadcasters, 1986).

2. Materials dealing with finance include *Operational Guidelines and Accounting Manual for Broadcasters* (Des Plaines, IL: Broadcast Financial Management Association, 1985); Neal Koch, "Power in the Shadows," *Channels* (June 1988): 66–71; Judith and Douglas Brush, "Building a Budget," *Corporate Video Decisions* (September 1988): 33–36; Larry Nicholson, "We Lost How Much?" *Video Manager* (July 1988): 27–29; "Banking in the Fifth Estate," *Broadcasting* (November 13, 1989): 47–52; and "Efficient Financial Reporting," *Broadcast Financial Journal* (May/June 1989): 39–42.

3. Materials that deal with human resources include Dan Bellus, "4-Letter Word (EEOC) Mystifies Managers," *Video Manager* (June 1988): 42–46; Bollier, David, "Unions Learn to Go With the Flow," *Channels* (May 1989): 44–45; Vernon A Stone and John M. Quarderer, "Staff Benefits Vary Widely," *RTNDA Communicator* (May 1988): 30–33; *A Guide to Developing Your Equal Employment Opportunity Program* (Washington, D.C.: National Cable Television Association, n.d.); and *Television Employee*

Compensation and Fringe Benefit Report (Washington, D.C.: National Association of Broadcasters, 1988).

4. Materials that deal with programming include Susan Tyler Eastman, Sydney W. Head, and Lewis Klein, *Broadcast/Cable Programming* (Belmont, CA: Wadsworth Publishing Company, 1989); Claude and Barbara Hall, *This Business of Radio Programming* (New York: Billboard Publications, 1977); Michael C. Keith, *Radio Programming: Consultancy and Formatics* (Boston: Focal Press, 1987); "TV Program Director: Endangered Species?" *Broadcasting* (August 27, 1990): 32; and "Jumping From PD to GM," *R&R* (April 8, 1988): 71.

5. Material on news departments includes John M. Quarderer and Vernon A. Stone, "NDs and GMs Define News 'Profitability,' " *Communicator* (April 1989): 10–12; Frederick Shook, *The Process of Electronic News Gathering* (Englewood, CO: Morton Publishing Company, 1982); Fielder Smith, "Who Owns the News?" *Emmy* (September/October 1987): 20–25; and Joseph Vitale, "The Newsroom's Revolving Door," *Channels* (September 1987): 46–48.

6. Material on engineering includes *Hiring Guidebook for Broadcast Technical Personnel* (Washington, D.C.: National Association of Broadcasters, 1982); Stephen Lowe, "Training for a Changing Industry," *International Broadcasting* (September 1988): 70–73; "Study Says TV General Managers Make Most Equipment Decisions," *Television Broadcast* (May 1988): 38; and Lorne Parket, "Keeping Up Is Hard to Do," *E-ITV* (April 1987): 8–10.

7. Material on sales and marketing includes Jeri Baker, "The Maturing of Marketing," *Channels* (October 1987): 18; I. Martin Jacknis, "Multiple Choice Selling," *RAB Sound Management* (April 1988): 29;

Gary W. McIntyre, "Turnover in Radio Sales Staffs: A Census of Kansas Commercial Stations," *Feedback* (Spring 1988): 19–24; "Legal Checklist for the Sales Department," *Info-Pak* (August 1988): 1–5; "Cable MSO's: Back to the Future with Tiering," *Broadcasting* (May 21, 1990): 35–36; Karl Sjodahl, "The Station's Image," *BPME Image* (March 1988): 12–18; and Charles Warner, *Broadcast and Cable Selling* (Belmont, CA: Wadsworth Publishing Company, 1986).

8. Material on public relations includes Rolf Gompertz, *Promotion and Publicity Handbook for Broadcasters* (Blue Ridge Summit, PA: TAB Books, 1977); Susan Tyler Eastman and Robert A. Klein, *Strategies in Broadcast and Cable Promotion* (Belmont, CA: Wadsworth Publishing Company, 1982); Fred L Bergendorff, Charles H. Smith, and Lance Webster, *Broadcast Advertising and Promotion* (New York, Hastings House, 1983); Cecilia Capuzzi, "In Search of a Killer Spot," *Channels* (June 1988): 23; "Radio's Promotional Efforts on the Rise," *Broadcasting* (October 3, 1988): 53; and "Networks: Blowing Billion-Dollar Horn," *Broadcasting* (October 24, 1988): 30.

9. Listings of these companies can be found in *Broadcasting Yearbook 1990* (Washington, D.C.: Broadcasting Publications, Inc., 1990): sections F, G, H, and I.

10. Bob Shanks, *The Cool Fire* (New York: W. W. Norton, 1976): 113–14.

11. For a more complete list of unions, see *Broadcasting Yearbook 1990*, I-76–I-77.

12. *Sixteenth Report: Broadcast Programs in American Colleges and Universities, 1986–1987* (Washington, D.C.: Broadcast Education Association, 1986): iii; and U.S. Department of Labor, *Supplement to Employment and Earnings* (Washington, D.C.: Bureau of Labor Statistics, 1986): 119.

13. For more information see Lynne S. Gross, *The Internship Experience* (Prospect Heights, IL: Waveland Press, 1987): and Ronald H. Claxton and B. A. Powell, *The Guide to Mass Media Internships* (Boulder, CO: University of Colorado Press, 1980).

14. *Television Employee Compensation and Fringe Benefits Report* (Washington, D.C.: National Association of Broadcasters, 1987); and *Radio Employee Compensation and Fringe Benefits Report* (Washington, D.C.: National Association of Broadcasters, 1987).

15. Many books exist to give advice on how to get a job in the electronic media industries. Some of them are: Paul Allman, *Exploring Careers in Video* (New York: Rosen Publishing Group, 1985); David W. Berlyn, *Exploring Careers in Cable/TV* (New York: Rosen Publishing Group, 1985); Dan Blum, *Making It in Radio: Your Future in the Modern Medium* (Hartford, CT: Continental Media Co., 1983); Jan Bone, *Opportunities in Telecommunications* (Lincolnwood, IL: National Textbook Company, 1984); Jon S. Denny, *Careers in Cable TV* (New York: Harper and Row, 1984); Tom Logan, *How to Act and Eat at the Same Time: The Business of Landing a Professional Acting Job* (Washington, D.C.: Communications Press, 1984); Donn Pearlman, *Breaking in Broadcasting: Getting a Good Job in Radio or TV Out Front or Behind the Scenes* (Chicago: Bonus Books, 1986); Maxine Reed, *Career Opportunities in Television, Cable, and Video* (New York: Facts on File, 1986); and Rod Vahl, *Exploring Careers in Broadcast Journalism* (New York: Richards Rosen Press, 1983).

9

1. Excellent information about advertising practices can be found in Kim B. Rotzoll and James E. Haefner, *Advertising in Contemporary Society: Perspectives Toward Understanding* (Cincinnati: South-Western Publishing, 1986); David Samuel Barr, *Advertising in Cable: A Practical Guide for Advertisers* (Englewood Cliffs, New Jersey: Prentice-Hall, 1985); Charles Warner, *Broadcast and Cable Selling* (Belmont, Calif.: Wadsworth Publishing Company, 1986); and Elizabeth J. Heighton and Don R. Cunningham, *Advertising in the Broadcast and Cable Media* (Belmont, Calif.: Wadsworth Publishing Company, 1984).

2. These are actual rate cards, but the stations did not want to be identified because their rates change so often.

3. "Future of Radio: Local Advertising," *Broadcasting* (August 29, 1988): 50.

4. "Media Watch: Market Facts and Stats," *Info-Pak* (December 1988): 1–2.

5. "Following the :15's," *Broadcasting* (May 22, 1989): 65.

6. "Television Networks Fatten Commercial Calf," *Broadcasting* (June 18, 1990): 21.

7. Andy Kazen, "Working Smart with Your Rep," *Sound Management* (April 1988): 18.

8. Thomas Moor, "Culture Shock Rattles the TV Networks," *Fortune* (April 14, 1986): 22–27.

9. *Broadcasting Yearbook 1990* (Washington, D.C.: Broadcasting Publications, Inc., 1990), G-1–G-7.

10. For more information on commercial production, see Norman D. Cary, *The Television Commercial: Creativity and Craftsmanship* (New York: Decker, 1971); Lincoln Diamant, *The Anatomy of a Television Commercial* (New York:

Hastings House, 1971); and "Playing to Your Fears—and Your Funny Bone," *TV Guide* (March 10, 1984): 18–22.

11. Lincoln Diamant, *Television's Classic Commercials* (New York: Hastings House, 1970); Jim Hall, *Mighty Minutes: An Illustrated History of Television's Best Commercials* (New York: Harmony Books, 1984); and "Remembrance of Ads Past," *Newsweek* (July 30, 1990): 42.

12. "Kids' Advertisers Play Hide-and-Seek, Concealing Commercials in Every Cranny," *The Wall Street Journal* (April 30, 1990): B1; "NAB Asks FCC to Clarify Its TV Dereg Order," *Broadcasting* (November 12, 1984): 58; "Senate Unanimously Okays Blurb-Curbing Kidvid Bill," *Variety* (July 20, 1990): 1; "Double Standard for Kids' TV Ads," *The Wall Street Journal* (June 10, 1988): 25; and "Interactive Toys Hit Animation Programing," *Broadcasting* (February 9, 1987): 110.

13. "Nielsen Offers Tempered Approach to Alcohol Ads," *Broadcasting* (February 4, 1985): 31.

14. "Harris Survey Shows Viewers Think Contraceptive Ads OK," *Broadcasting* (March 30, 1987): 178.

15. "Infomercials: Can Viewers Tell the Difference?" *Broadcasting* (April 16, 1990): 61.

10

1. Several accounts of early ratings can be found in Frederick Lumley, *Measurement in Radio* (Columbus, Ohio: Ohio State University Press, 1934); Paul F. Lazarfield and Frank N. Stanton, *Radio Research* (New York: Duell, Sloan, and Pearce, 1942); Mark James Banks, "History of Broadcast Audience Research in the United States, 1920–1980" (Ph.D. dissertation, Knoxville: University of Tennessee, 1981); William S. Rubens, "A Personal History of TV Ratings, 1929 to

1989 and Beyond," *Feedback* (Fall 1989): 3–15; and Hugh Malcolm Beville, Jr., *Audience Ratings: Radio, Television and Cable* (Hillsdale, NJ: Lawrence Erlbaum Associates, 1988).

2. *Nielsen: The System for Success* (New York: Nielsen Media Research, 1988); and "Nielsen on Nielsen," *Broadcasting* (July 6, 1981): 57.

3. *What TV Ratings Really Mean* (Northbrook, IL: Nielsen Media Research, 1989).

4. *Nielsen: The System for Success,* 12–13.

5. *Nielsen Television Index* (Northbrook, IL: Nielsen Media Research, 1987).

6. *Nielsen Station Index* (New York: Nielsen Media Research, 1990).

7. *Nielsen Television Index,* special releases.

8. *Arbitron Ratings Today* (New York: Arbitron Ratings Company, n.d.): 6.

9. *Broadcasting Yearbook, 1990* (Washington, D.C.: Broadcasting Publications, Inc., 1990): C-129–C-212.

10. *How to Read Your Arbitron Television Market Report* (New York: Arbitron Ratings Company, 1987).

11. "More Stations Opting for Just One Ratings Service," *Broadcasting* (September 3, 1990): 52.

12. "Arbitron to Go With Peoplemeter," *Broadcasting* (June 27, 1988): 62.

13. *Arbitron Ratings Radio* (New York: Arbitron Ratings Company, 1987).

14. "A Ratings Infant Makes Marked Progress," *Broadcasting* (September 20, 1982): 63.

15. "Ratings for PBS," *Broadcasting* (July 21, 1986): 51; "Tracking Hispanic Viewing Is a Three-Way Race," *Broadcasting* (December 10, 1990): 78; and *Broadcasting Yearbook, 1990,* I-39–I-44.

16. "Nielsen Increases Ratings Point Value," *Variety* (August 14, 1989): 1.

17. *Turning the Numbers into Sales Strategies* (New York: Arbitron Ratings Company, 1987): 3–5.

18. "Vanishing Viewers Cause a Stir Over TV Ratings System," *Los Angeles Times* (June 27, 1990): D–1.

19. "Net Ratings Plunge to a Record Low," *Variety* (July 11, 1990): 1.

20. "Three Networks' Rating Drop to Worst-Ever Share," *Variety* (July 7, 1988): 6.

21. Susan Tyler Eastman and Robert A. Klein, *Strategies in Broadcast and Cable Promotion* (Belmont, CA: Wadsworth Publishing Company, 1982): 47–50.

22. Steve Behrens, "What is Good for You?" *Channels* (May 1988): 28.

23. Eastman and Klein, 48.

24. Michael Couzens, "TVs Missing in Action," *Channels* (January 1990): 22.

25. "Cable Contrast," *Broadcasting* (September 13, 1982): 84.

26. Eric Taub, "Who's Watching Now?" *Emmy* (February 1990): 32–35.

27. Pros and cons of ratings (mostly cons) can be found in many cleverly written articles or sections of books, including Dick Adler, "The Nielsen Ratings—and How I Penetrated Their Secret Network," *New York Times* (October 1, 1974): 74B; "Roundtable: Research and Ratings," *Emmy* (Spring 1981): 13–16; Bob Shanks, *The Cool Fire* (New York: W. W. Norton and Company, 1976): 244–58; Michael Wheeler, "Life and Death in the Little Black Box," *Television Quarterly* (May/July 1976): 5–14; Robert M. Ogles and Herbert H. Howard, "Keeping Up with Changes in Broadcast Audience Measurement: Diaries and People Meters," *Feedback* (Winter 1990): 8–11; and Dennis Kneale, "How to Shun Friends and Influence Shows: Notes From a Week as a Nielsen Household," *The Wall Street Journal* (March 3, 1989): B–1.

11

1. *A Short History of Radio Music Licenses* (Washington, D.C.: All-Industry Radio Music Licensing Committee, 1987).

2. "Supreme Court OK's Blanket Music Licenses," *Broadcasting* (February 25, 1985): 38–39.

3. "ASCAP, Broadcasters Face Court Battle," *NAB News* (December 3, 1990): 15.

4. "First Low-Power Station Licenses Arriving at BMI," *Variety* (July 25, 1986): 7.

5. "Cable Whistling a New Tune," *Variety* (November 15, 1989): 1.

6. "Pornographic Rock Lyrics Issue Gets Airing at Radio Convention as PMRC Calls for Record Warning Labels," *Broadcasting* (September 6, 1985): 38.

7. "Dick Clark Spins Another Record," *Los Angeles Times* (August 26, 1987): pt. V, B.

8. "Supreme Court Ruling on Programming Formats Big Victory for Broadcasters," *Variety* (March 25, 1981): 1.

9. "The Beat Goes on TV: Music Videos, Sign-On to Sign-Off," *Broadcasting* (June 25, 1984): 50–52.

10. "MTV Changing as Video Novelty Wears Off," *Electronic Media* (June 30, 1986): 26.

11. Specific references to drama include: Linda S. Lichter and S. Robert Lichter, *Prime Time Crime: Criminals and Law Enforcers in TV Entertainment* (Washington, D.C.: Media Institute, 1983); Richard Levinson and William Link, *Stay Tuned . . . An Inside Look at the Making of Prime-Time Television* (New York: St. Martin's Press, 1981); Dave Kaufman, "10 Best Programs in Television History," *Variety* (43rd Anniversary Issue): 148–85; Saul N. Scher, "Anthology Drama: TV's Inconsistent Art Form," *Television Quarterly* (Winter 1976–77): 29–34; Tim Brooks and Earle Marsh, *The Complete Directory to Prime Time Network TV Shows: 1946–Present* (New York: Ballantine Books, 1985); Louis Solomon, *The TV Doctors* (New York: Scholastic Book Services, 1974); and "A Solution to the Murder of Laura," *Newsweek* (April 30, 1990): 6.

12. "Next in Line for Moral Scrutiny: Cable Programming," *Broadcasting* (August 31, 1981): 40.

13. Eli A. Rubinstein, "The TV Violence Report: What Next?" *Journal of Communication* (Winter 1974): 34–40.

14. "TV on Trial," *Broadcasting* (November 5, 1984): 35.

15. "The New Right's TV Hit List," *Newsweek* (June 15, 1981): 101.

16. "Simon Introduces TV Violence Bill," *Broadcasting* (March 30, 1987): 148.

17. "More Violence than Ever Says Gerbner's Latest," *Broadcasting* (February 28, 1977): 20; and "Schneider Attacks Gerbner's Report on TV Violence," *Broadcasting* (May 2, 1977): 57.

18. Max Gunther, "All That TV Violence: Why Do We Love/Hate It?" *TV Guide* (November 6, 1976): 6–10.

19. Studies on violence include: Seymour Feshbach and Robert D. Singer, *Television and Aggression: An Experimental Field Study* (San Francisco: Jossey-Bass, 1970); Melvin S. Heller and Samuel Polsky, *Studies in Violence and Television* (New York: American Broadcasting Company, 1976); Murray Feingold and G. Timothy Johnson, "Television Violence Reactions from Physicians, Advertisers, and the Network," *The New England Journal of Medicine* (February 24, 1977): special article; and George Gerbner, "Science or Ritual Dance? A Revisionist View of Television Violence Effects Research," *Journal of Communication* (Summer 1984): 164–73.

20. Other material dealing with violence includes "A Blizzard of Paper on Violence," *Broadcasting* (May 16, 1977): 22–24; Harriet Steinberg, "Must Night Fall So Hard?" *Television Quarterly* (Winter 1976–77): 61–64; and "Top Prime-Time Advertisers Reject AMA's TV Violence Rx: Baretta Takes Gloves Off," *Variety* (March 4, 1977): 1.

21. For additional information on situation comedies, see Vince Waldron, *Classic Sitcoms* (New York: Macmillan, 1987); David Handler, "Which Old Favorites Are Still Choice and Which Aren't," *TV Guide* (April 4, 1987): 18–23; Larry Wilde, "The Genesis of Comedy," *Television Quarterly* (Winter 1976–77): 70; Alan Alda, "My Favorite Episodes," *TV Guide* (February 12, 1983): 6–21; and Richard Adler, ed. *All in the Family: A Critical Appraisal* (New York: Praeger, 1979).

22. For additional information on variety and reality shows, see Arthur Schulman and Roger Youman *How Sweet It Was: Television, a Pictorial Commentary* (New York: Bonanza Books, 1966); Ted Sennett, *Your Show of Shows* (New York: Collier Books, 1977); and "ABC, Parton Try to Add Spice of Life to Defunct Variety Format," *Variety* (November 19, 1987): 16; and "Success of 'Home Videos' Spurs Clones," *Broadcasting* (April 16, 1990): 38.

23. For additional information on movies, see Richard Zacks, "Picture Window," *Channels* (May 1986): 40–41; Neil M. Malamuth and Victoria Billings, "Why Pornography? Models of Functions and Effects," *Journal of Communication* (Summer 1984): 117–29; "Coloring the Black and White Past," *Broadcasting* (May 24, 1986): 92; Tom Allen, "TV: Father of the Film," *America* (October 30, 1976): 286–88; Saul Kaufman, "Films: TV Versions of Moving Pictures," *New Republic* (August 30, 1975): 20; "TV Movies:

Exploiting the Drama of Real Life," *The Wall Street Journal* (July 30, 1985): 22; and Jon Krampner, "In the Beginning: The Genesis of the Telefilm," *Emmy* (December, 1989): 30–35.

24. For additional information on talk shows, see Murray B. Levin, *Talk Radio and the American Dream* (Lexington, Ma: Lexington Books, 1987); Steve Benes, "Tabloid TV: It's Brash, It's Crass, Is It More Than Trash?" *Emmy* (January, 1989): 20–29; Michael J. Arlen, "Hosts and Guests: Hospitality on Talk Shows," *New Yorker* (January 3, 1977): 62–65; and Glenn Esterly, "Talk Show Hosts: Who Are the Best?" *TV Guide* (December 8, 1979): 26–32.

25. For additional information on game shows, see Arleen Francis, "I Was There from First to Last: What's My Line," *Saturday Evening Post* (September 1976): 43; Dick Russell, "The Return of the TV Outcast," *TV Guide* (December 1, 1979): 21–24; "What a Deal," *Newsweek* (February 9, 1987): 62–68; and David Schwartz, Steve Ryan, and Fred Wostbrock, *The Encyclopedia of TV Game Shows* (New York: Zoetrope, 1987).

26. For additional information on soap operas, see Ien Ang, *Watching Dallas: Soap Opera and Melodramatic Imagination* (New York: Methuen, 1985); Charles Derry, "Television Soap Opera: Incest, Bigamy, and Fatal Disease," *Journal of the University Film and Video Association* (Winter 1983): 4–16; Ellen Seiter, "Men, Sex, and Money in Recent Family Melodramas," *Journal of the University Film and Video Association* (Winter 1983): 17–27; Bradley S. Greenberg and Dare D'Allessio, "Quantity and Quality of Sex in the Soaps," *Journal of Broadcasting and Electronic Media* (Summer 1985): 309–21; Rhoda Estep and Patrick T. Macdonald, "Crime in the Afternoon: Murder and Robbery in Soap Operas," *Journal of Broadcasting and*

Electronic Media (Summer 1985): 323–31; and Michael James Intintoli, *Taking Soaps Seriously: The World of Guiding Light* (New York: Praeger, 1984).

27. Elizabeth M. Perse, "Soap Opera Viewing Patterns of College Students and Cultivation," *Journal of Broadcasting and Electronic Media* (Spring 1986): 175–93; and Lenore Silvian, "Spinoffs from Soapland," *Television Quarterly* (Winter 1976–77): 37–41.

28. George W. Woolery, *Children's Television: The First Thirty-Five Years, 1946–1981. Part I: Animated Cartoon Series* (Metuchen, New Jersey: Scarecrow Press, 1983); George C. Woolery, *Children's Television: The First Thirty-Five Years, 1946–1981. Part II: Live, Film and Tape Series* (Metuchen, New Jersey: Scarecrow Press, 1985); and "TV Crowds Children Out of Daily Schedule," *USA Today* (July 16, 1984): 11A.

29. George Comstock et al., *Television and Social Behavior: A Technical Report to the Surgeon General's Scientific Advisory Committee on Television and Social Behavior* (Washington, D.C.: Government Printing Office, 1972).

30. Some of the studies concerning children and TV are: Barbara J. Wilson, "The Effects of Two Control Strategies on Children's Emotional Reactions to a Frightening Movie Scene," *Journal of Broadcasting and Electronic Media* (Fall 1989): 397–418; Gerald S. Lesser, *Children and Television: Lessons from Sesame Street* (New York: Random House, 1974); *Learning While They Laugh: Studies of Five Children's Programs on the CBS Television Network* (New York: Columbia Broadcasting System, 1977); Thomas Skill and Sam Wallace, "Family Interactions on Primetime Television: A Descriptive Analysis of Assertive Power Interactions," *Journal of Broadcasting and Electronic Media* (Summer

1990): 243–62; and Lynne Schafer Gross and R. Patricia Walsh, "Factors Affecting Parental Control Over Children's Television Viewing: A Pilot Study," *Journal of Broadcasting* (Summer 1980): 411–19.

31. "Kidvid Program Standards Get Brush-Off from Mark Fowler," *Variety* (December 28, 1983): 2.

32. Jan Cherubin, "Toys Are Programs, Too," *Channels* (May/June 1984): 31–33.

33. "Kidvid: A National Disgrace," *Newsweek* (October 17, 1983): 81–83.

34. "A Legislative Showdown for Children's TV," *Los Angeles Times* (June 21, 1990): F1.

35. Other material dealing with children's television includes: John P. Murray and Gabriel Salomon, eds., *The Future of Children's Television: Results of the Markel Foundation/Boys Town Conference* (Boys Town, Nebraska: Boys Town Center, 1984); Aimee Door, *Television and Children: A Special Medium for a Special Audience* (Beverly Hills, Ca.: Sage, 1986); Hilde Himmelweit, *Television and the Child* (New York: Television Information Office, 1961); Robert M. Liebert, *The Early Window* (New York: Pergamon Press, 1973); William Melody, *Children's Television: The Economics of Exploitation* (New Haven, Connecticut: Yale University Press, 1975); Fred Rogers, *Mr. Rogers Talks About . . .* (Bronx, N.Y.: Platt and Munk, 1974); Marie Winn, *The Plug-In Drug* (New York: Viking, 1975); Walter Karp, "Where the Do-Gooders Went Wrong," *Channels* (March/April 1984): 41–51; and Bruno Bettelheim, "A Child's Garden of Fantasy," *Channels* (September/October 1985): 54–56.

12

1. For more on news sources see, Thomas L. Jacobson, "Use of Commercial Data Bases for Television News Reporting," *Journal of Broadcasting and Electronic Media* (Spring 1989): 203–8; Max Robbins, "Camcorder Assaults," *Channels* (January 1989): 30–31; Emerson Stone, "Too Much and Too Little," *RTNDA Communicator* (November 1987): 20–22; and "Ted's Global Village," *Newsweek* (June 11, 1990): 48–52.

2. "Newsroom Computer Systems Gain Strength," *Television Broadcast* (August 1988): 67.

3. "TV News: The Rapid Rise of Home Rule," *Newsweek* (October 17, 1988): 94; Kim Standish, "Satellite News Gathering: The Next Round," *RTNDA Communicator* (September 1986): 37; Phillip Keirstead, "A Sea of New Goodies for Television News Production," *Television Broadcast* (June 1986): 30; and "Satellite Trucks Gain Real News Acceptance," *Television Broadcast* (June 1988): 79.

4. Chuck Reece, "The Home Town Report," *Channels* (September 1989): 57–62; "Local Stations: Survival of the Fittest," *Broadcasting* (August 31, 1987): 41–48; Eric Levin, "How the Networks Decide What's News," *TV Guide* (July 2, 1977): 6–11; Joseph Vitale, "The Newsroom's Revolving Door," *Channels* (September 1987): 46–48; and Dennis McDougal, "Broadcast News, L.A.," *Los Angeles Times Magazine* (January 24, 1988): 7–17.

5. Barbara Matusow, *The Evening Stars: The Making of the Network News Anchor* (New York: Ballantine Books, 1984).

6. For more information about controversies and news, see: "Invasion of the Public's Sense of Privacy?" *RTNDA Communicator* (February 1988): 34; Dan Nimmo and James E. Combs, *Nightly Horrors: Crisis Coverage by Television Network News* (Knoxville, Tennessee: University of Tennessee Press, 1985); "Rethinking TV News in the Age of Limits," *Newsweek* (March 16, 1987): 79–80; and "TV Journalists Debate Their 'Art'," *Broadcasting* (June 23, 1986): 53; Fielder Smith, "Who Owns the News?" *Emmy* (September/October 1987): 20–25; and Whitney D. Charles, Marilyn Fritzler et al., "Source and Geographic Bias in Network Television News," *Journal of Broadcasting and Electronic Media* (Spring 1989): 159–74.

7. For more information about documentary programs, see Daniel Einstein, *Special Edition: A Guide to Network Television Documentary Series and Special News Reports 1955–1979* (Metuchen, NJ: The Scarecrow Press, 1987); "Westmoreland Lawyer Calls CBS Vietnam Documentary 'Powerful Work of Fiction'," *Variety* (October 12, 1984): 1; Mike Wolverton, *Reality on Reels: How to Make Documentaries for Video/Radio/Film* (Houston, Texas: Gulf Publishing Company, 1983); Axel Madsen, *Sixty Minutes: The Power and the Politics of America's Most Popular TV News Show* (New York: Dodd, Mead, and Company, 1984); Jon Krampner, "Around the World in Twenty-Five Years," *Emmy* (April 1990): 36–41; and Charles Montgomery Hammond, Jr., *The Image Decade: Television Documentary 1965–1975* (New York: Hastings House, 1981).

8. For more information on editorials, see L. S. Feuer, "Why Not a Commentary on Sevareid?" *New Republic* (August 15, 1975): 874–76; and Karl E. Meyer, "Uncommon Commentary of ABC's Howard K. Smith," *Saturday Review* (December 11, 1976): 12.

9. "Regional Cable Sports Are on a Winning Streak," *Broadcasting* (April 16, 1990): 40.

10. "No End Seen for Rising TV
Sports Fees," *Variety* (January
26, 1990): 45.

11. " CBS Whiffs With Baseball,"
Variety (October 17, 1990): 1;
"Emerging from Slump, TV
Sports Regains Favor Among
Advertisers," *The Wall Street
Journal* (August 26, 1987): 23;
and "Broadcasting/Cable Plot
Thickens," *Broadcasting*
(December 26, 1988): 27.

12. Stanley Frank, "What TV Has
Done to Sports," *TV Guide*
(February 4, 1967): 4–8; Don
Kowet, "For Better or for
Worse," *TV Guide* (July 1, 1978):
2–6; Melissa Ludtke, "Big
Scorers in the Ad Game," *Sports
Illustrated* (November 7, 1977):
50; and Jim Baker, "Why TV
Needs Sports," *TV Guide*
(August 11, 1990): 3–14.

13. For more information on
magazine shows, see
"Westinghouse Puts Magazine
Show on All Its Stations in
Access Periods," *Broadcasting*
(January 24, 1977): 34; "Winds
of Change Blow Through Group
W's 'PM Magazine' Strip,"
Variety (December 27, 1988): 1;
Tom Shales, "Networks Love TV
Newsmagazines," *Los Angeles
Times* (May 31, 1985): VI, 1;
"TV's Local Magazines,"
Newsweek (December 31, 1979):
50–51; Nancy R. Casplar, "Local
Television: the Limits of Prime-
Time Access," *Journal of
Communication* (Spring 1983):
124–31; and Michelle R. Serra
and Richard A. Kallan, "A
Sexual Egalitarianism in TV: An
Analysis of 'P.M. Magazine,'"
Journalism Quarterly (Autumn
1983): 56–63.

14. For additional information on
educational programming, see
"Cable Ready to Go Back to
School," *Broadcasting*
(September 3, 1990): 38–40;
Voices and Values (New York:
Television Information Office,
1984): 37–44; "The Second
Blackboard," *Newsweek* (April 2,
1990): 46–47; Gregory Jordahl,
"The Television Classroom," *E-
ITV* (October 1987): 38–41; and
"Schools Giving TV Warmer

Reception," *Wall Street Journal*
(December 26, 1989): B1.

15. For more information on religious
programming, see George H.
Hill, *Airwaves to the Soul: The
Influence and Growth of
Religious Broadcasting in
America* (Saratoga, CA: R&E
Publishers, 1983); Peter G.
Horsfield, *Religious Television:
The American Experience* (New
York: Longman, 1984); "Getting
Time on the Tube," *Christianity
Today* (May 7, 1976): 27–28; "If
You Can't Beat 'Em . . . .,"
Newsweek (February 9, 1981):
101; Roderick Townley, "It's
Pitch and Pray and Wish the
Scandal Away," *TV Guide*
(August 15, 1987): 5–9; and
Frank Razelle, *Televangelism:
The Marketing of Popular
Religion* (Carbondale, Illinois:
Southern Illinois University Press,
1987).

16. For more information on public
affairs, see Roger Simon, "Those
Interview Shows—They're
Tougher Now, but Are They
Better?" *TV Guide* (March 14,
1987): 4–7; "Public Affairs: A
Chance to Serve," *Broadcasting*
(November 16, 1987): 117; David
Bollier, "Raise the Halo High,"
Channels (April 1989): 32–38;
and Michael Bruce MacKuen and
Steven Lane Coombs, *More than
News: Media Power in Public
Affairs* (Beverly Hills, CA: Sage,
1981).

17. For more information on political
broadcasting, see Patricia Sellers,
"The Selling of the President in
'88," *Fortune* (December 21,
1987): 131–36; Gerald R. Ford,
"10 Ways to Improve TV's
Campaign Coverage," *TV Guide*
(February 8, 1984): 6–10; "Webs
To Rethink Vote Projection After
Election, *Variety* (November 2,
1988): 1; Eric Engberg, "The
Way We Choose the President,"
RTNDA Communicator
(November 1987): 36–38;
Kathleen Hall Jamieson,
*Packaging the Presidency: A
History and Criticism of
Presidential Campaign
Advertising* (New York: Oxford
University Press, 1984); Austin

Kanney, *Channels of Power: The
Impact of Television on
American Politics* (New York:
Basic Books, 1983); John Tebbel
and Sarah Miles Watts, *The
Press and the Presidency: From
George Washington to Ronald
Reagan* (New York: Oxford
University Press, 1985); "Election
Night Coverage Sees Net
Manpower Halved, *Variety*
(November 2, 1988): 15; and
Gladys E. Lang and Kurt Lang,
*Politics and Television Re-
Viewed* (Beverly Hills, CA: Sage,
1984).

18. For more information on special
interest programming, see "Home
Is Where the Mart Is," *Channels*
(Field Guide, 1987): 77–78;
Joanmarie Kalter, "Now You Can
See How to Call a Goose, Pan for
Gold, or Fix Your Horse's Teeth,"
TV Guide (November 21, 1987):
38–40; "Hispanic Programming is
Cable's Latest Trend,"
Broadcasting (August 6, 1990):
50; "TV in the Senate: One Year
Later," *Broadcasting* (June 1,
1987): 36; and Al Jaffe, "Show
and Sell," *Channels* (September
24, 1990): 10.

13

1. "The Expanding Universe of
Radio Networks," *Broadcasting*
(February 19, 1990): 37–45.

2. "Program Syndication: Pushing
Radio's Hot Buttons,"
Broadcasting (July 23, 1990):
35–55.

3. "Radio's New Golden Age,"
Newsweek (August 3, 1987): 41.

4. These are some of the formats
listed in *Broadcasting/
Cablecasting Yearbook, 1990*
(Washington, D.C.: Broadcasting
Publications, Inc., 1990): F–72.

5 These are some of the types of
material listed in *Syndicated
Radio Programming Directory*
(Washington, D.C.: National
Association of Broadcasters,
1986): 10–39.

6. "Public Radio's Programming
Explosion," *Broadcasting* (May
28, 1989): 57; and *American*

Public Radio: Profile 1989 (St. Paul, MN: American Public Radio, 1989).

7. "TV Networks, Producers Battle Over Fees," *Los Angeles Times* (March 17, 1986): IV, 1.

8. "Coming In Off the Network's Bench," *Los Angeles Times/Calendar* (October 21, 1990): 5; and "CBS To Slice Up Pilot Season," *Variety* (August 24, 1989): 1.

9. "TV Networks Anxious to Make Their Own Shows," *Broadcasting* (May 2, 1988): 40.

10. " 'Cosby': Off-network's Biggest Deal Ever," *Broadcasting* (September 12, 1988): 76.

11. "Upfront Sales for Syndicates Programing Off to Robust Start," *Broadcasting* (June 18, 1990): 65.

12. "The 'Other' Independents: Sharing the Success of Fox," *Broadcasting* (October 15, 1990): 74.

13. "Theatricals Find Strength in Ad Hoc Numbers," *Broadcasting* (September 30, 1985): 93.

14. "WOR-TV Bags 'Cosby,'" *Broadcasting* (November 10, 1986): 53.

15. "The World of TV Programming, 1984," *Broadcasting* (October 22, 1984): 54–82.

16. "First-run Programming Fueling Syndication Market," *Broadcasting* (January 19, 1987): 106.

17. "Barter Joins the Big Leagues," *Variety* (August 15, 1988): 1.

18. "Syndicators Look to Access," *Broadcasting* (October 12, 1987): 70.

19. Steve Behrens, "Will Temptation Undo the Tie That Binds?" *Channels* (May 1987): 41–43; "Industry Divided Over 'Reverse Compensation,' " *Broadcasting* (October 31, 1988): 45; and "NBC Cuts Compensation in Line with Audience Drop," *Broadcasting* (August 20, 1990): 50.

20. "ABC-TV to Affiliates: Use Us or Lose Us," *Broadcasting* (June 13, 1988): 27.

21. *Broadcasting/Cablecasting Yearbook, 1987,* A-21–A-23.

22. *The Public Television Station Program Cooperative* (Alexandria, VA: Public Broadcasting System, 1986): 1–3.

23. "Cable Subscribes to Original Programming," *Broadcasting* (October 12, 1987): 50; and "Cable-syndication's Two-Way Pipeline," *Broadcasting* (February 22, 1988): 160.

24. "Pegasus's Basic Cable Flight Upsets Some Broadcasters," *Broadcasting* (August 29, 1988): 48.

25. "TCI and Fox in Affiliation Talks," *Broadcasting* (July 9, 1990): 29–31.

26. Scott Carlberg, "Outside Vs. Inside Production," *Video Manager* (May 1988): 28–30.

27. Peter Utz, "For Love of Money," *AV Video* (January 1990): 67.

14

1. "CCD vs. Three-Tube TV Cameras: How to Choose?" *E-ITV* (June 1987): 20–29.

2. "Choosing the TV of the Future," *Technology Review* (April 1989): 30–40.

3. Nora Lee, "HDTV: The Artists Speak," *American Cinematographer* (September 1987): 85–90.

4. "Robotic Cameras: Cutting Out the Middleman," *Broadcasting* (December 7, 1987): 96.

5. "High Wire Video," *Broadcasting* (December 31, 1984): 108.

6. "Many Field Formats Compete for Attention in 1989," *Television Broadcast* (March 1989): 16–17.

7. "Audio and the New VTRs," *AV Video* (June 1990): 51–55.

8. "Digital VTRs Fuel Audio Post," *Television Broadcast* (May 1990): 1, 35.

9. Lynne Gross and David E. Reese, *Radio Production Worktext* (Boston: Focal Press, 1990): 12–18.

10. Lynne S. Gross and Larry W. Ward, *Electronic Moviemaking* (Belmont, CA: Wadsworth Publishing Company, 1991): 101–22.

11. "Home Editing: Video's Final Frontier," *Video* (July 1990): 37–39.

12. Michael H. Stanton, "Ediflex: An Electronic Editing System," *American Cinematographer* (July 1987): 101–3.

13. Patricia Mozzillo, "The Trusty Time Base Corrector," *Television Broadcast* (May 1990): 51–52.

14. Manfred Dorn, "Understanding Monitoring Equipment," *Video Systems* (December 1979): 14–21.

15. Gross and Reese, 108–15.

16. Clay Gordon, "Pixels at an Exhibition," *AV Video* (June 1990): 42–47.

17. "DVE (Digital Video Effects) Defined," *E-ITV* (November 1985): 44–45.

18. "The LPs Wobbly Future," *Newsweek* (February 9, 1987): 52.

19. "DAT Players Arrive: Opponents Vow Suit," *Variety* (June 25, 1990): 1.

20. Mel Lambert, "Digital Audio Workstations: Wave of the Future," *Television Broadcast* (May 1989): 51.

21. Frank Beacham, "Desktop Audio Boosts Production," *TV Technology* (March 1989): 20.

15

1. Joseph Kerman, *Listen* (New York: Worth Publishers, Inc., 1972): 13.

2. David Lachenbruch, "A Field Guide to the Airwaves," *Channels* (November/December 1983): 48–49.

3. Milton Kiver, *FM Simplified* (Princeton, N.J.: Van Nostrand, 1960).

4. Stan Prentiss, "AM Stereo Wins FCC Approval," *Popular Electronics* (August 1980): 59.

5. "Lid's Off FM SCA's," *Broadcasting* (April 11, 1983): 35.

6. "DAB: The Next Generation of Radio Broadcasting," *Broadcasting* (June 4, 1990): 62.

7. *Broadcasting/Cablecasting Yearbook, 1990* (Washington, D.C.: Broadcasting Publications Inc., 1990): A–4.

8. "Extended AM Band: Problem or Solution?" *Broadcasting* (August 28, 1989): 47–48.

9. *Broadcasting/Cablecasting Yearbook, 1990,* A–5.

10. K. R. Sturley, *Radio Receiver Design* (New York: Barnes and Noble, 1965).

11. David Lachenbruch, "They're Getting Cheaper—and More Bossy," *TV Guide* (January 9, 1988): 30–33; and Gary H. Arlen, *Tomorrow's TVs* (Washington, D.C.: National Association of Broadcasters, 1987).

12. Tom Sutherland, "How to Choose a Projection System," *Video Manager* (December 1988): 40–43; and Peter Utz, "The Big Picture: A Look at TV Projectors," *AV Video* (September 1990): 36–50.

13. Herbert Shuldiner, "Flat Screen Color TV," *Popular Science* (November 1983): 100–102; "Improvements in LCDs," *E/ITV* (October 1985): 66–67; and "Flat is Where It's At," *Broadcasting* (June 12, 1989): 48–49.

14. John Hoglin, "Do-It-Yourself Three-Dimensional Video Magic!" *Videomaker* (January 1990): 100; "Brits Announce 3-D Process," *Variety* (June 5, 1990): 8; and David Lachenbruch, "Here Come Deepies—Maybe," *TV Guide* (March 14, 1981): 22–26.

15. Donald G. Monteith and Nicholas F. Hamilton-Piercy, "Building Plant to 499 MHz and Beyond," *TVC* (December 15, 1980): 132–50; and John P. Taylor, "Not Enough Channel Capacity? 'Supercable' to the Rescue," *Cable Age* (May 18, 1981): 21–32.

16. William B. Johnson, "The Coming Glut of Phone Lines," *Fortune* (January 7, 1985): 96–99; and Phillip Kurz, "DS3: An Alternative to Satellite Distribution," *Television Broadcast* (December 1987): 58–61.

17. Joel Dreyfuss, "The Coming Battle," *Fortune* (February 13, 1989): 104–7; "Tulsa Firm High In Fiber," *Broadcasting* (July 16, 1990): 89; "TCI Commits to Fiber Optic Delivery," *Broadcasting* (November 28, 1988): 56; "Bell Plans to Deliver Video, Telephone Via Fiber Optics," *Broadcasting* (July 25, 1988): 90; and Christopher Podmore and Denise Faguy, "The Challenge of Optical Fibers," *Telecommunications Policy* (December 1986): 341–51.

18. Jay C. Lowndes, "Optical Fiber Hardware Threatens Telecommunications Market Share," *Aviation Week and Space Technology* (October 20, 1986): 115.

19. Chris Bowick and Tim Kearney, *Introduction of Satellite TV* (Indianapolis, In.: Howard W. Sams and Co., 1983).

20. "Satellites!" *Broadcasting* (July 20, 1987): 33–44.

21. "Domestic Satellites: Dawning of a New Generation," *Broadcasting* (July 18, 1988): 39–56.

22. "Satellite Operators Reap Mass Allocations for 1990s," *Broadcasting* (November 21, 1988): 21.

23. "Communications Satellites: The Birds Are in Full Flight," *Broadcasting* (November 19, 1979): 36–47.

24. William D. Houser, "Satellite Interconnection," *Television Quarterly* (Fall 1976): 78–80.

25. Robert N. Wold, "Satellite Distribution Is No Pie in the Sky," *Broadcast Communications* (December 1979): 49.

26. "Demand Exceeds Supply of Westar Transponders," *Broadcasting* (February 4, 1980): 23.

27. "Keeping Up with the Satellite Universe," *Broadcasting* (August 17, 1981): 32.

28. "Satellite Delivery Set for Two More Group W Programs," *Variety* (October 30, 1981): 1; "Radio Joins Television in Move to Sat'Lite Transmission," *Variety* (April 13, 1981): 1; and "ABC Targets Switchover to Satellite Delivery," *Broadcasting* (April 27, 1981): 94.

29. "Satellites: Flying Higher Than Ever," *Broadcasting* (July 14, 1986): 41–57.

30. "Clear Advantages to High Resolution," *Broadcasting* (February 16, 1981): 30.

31. "HDTV Wows 'Em in New York," *Broadcasting* (February 15, 1982): 33.

32. "Compact Video's Contribution to HDTV State of the Art," *Broadcasting* (November 2, 1981): 25.

33. "U.S. Industry Adopts NHK Parameters for HDTV," *Broadcasting* (March 25, 1985): 68.

34. "CCIR Puts End of Hope for HDTV Standard," *Broadcasting* (May 19, 1986): 70.

35. "The Thomson View on HDTV," *International Broadcasting* (September 1988): 32–36.

36. "Pentagon Seeks to Spur U.S. Effort To Develop 'High-Definition' TV," *The Wall Street Journal* (January 5, 1988): B4.

37. "Congress Declares Itself in on HDTV," *Broadcasting* (October 12, 1987): 35.

38. "FCC Writes a First Draft for HDTV," *Broadcasting* (September 5, 1988): 32–33.

39. "PSI, MUSE-6 Pulled from ATV Tests. DigiCipher Added, MIT Holding On," *Television Broadcast* (June 1990): 4; "NBC Unveils New HDTV Standard," *Broadcasting* (October 17, 1988): 31; "Test of Faroudja ATV System Set," *Variety* (October 17, 1989): 20; "Zenith Introduces HDTV Production System," *Broadcasting* (November 20, 1989): 64; "General Instrument's

11th Hour Digital Entry: Will It Revolutionize HDTV?" *Broadcasting* (June 4, 1990): 33.

40. "Compression: Changing the World of Television?" *Broadcasting* (June 11, 1990): 68.

41. Tanrat Mereba and Dana Ulloth, "The HDTV Political Debate," *Feedback* (Spring 1990): 6–7.

42. "U.S. May Only Get to Watch TV's Leap into the Future," *Insight* (July 4, 1988): 40–41; and "HDTV: Can the U.S. Get Its Act Together?" *Electronics* (March 1989): 71–79.

16

1. Four sources from the FCC that give general information about its organization are: *The FCC and Broadcasting* (Washington, D.C.: Federal Communications Commission, 1988); *Federal Communications Commission: Spectrum Management* (Washington, D.C.: Federal Communications Commission, 1987); *Federal Communications Commission: Private Radio Services* (Washington, D.C.: Federal Communications Commission, 1986); and *The FCC In Brief* (Washington, D.C.: Federal Communications Commission, 1989).

2. "Worsening Jam in CB Radio Swamps FCC," *Broadcasting* (January 19, 1976): 30; "Another Deluge of LPTV Filings Inundates FCC," *Broadcasting* (February 23, 1981): 29; "FCC Gets First MMDS Deluge," *Broadcasting* (October 31, 1983): 55; and "Jockeying for DBS Slots," *Broadcasting* (April 10, 1989): 70.

3. *How FCC Rules Are Made* (Washington, D.C.: Federal Communications Commission, n.d.).

4. The Saga of Call Letters: From KAAA to WZZZ," *Broadcasting* (August 6, 1984): 67.

5. *Assignment and Transfer of Control of Broadcast Licenses Statutory and Regulatory*

Requirements (Washington, D.C.: Pierson, Ball & Dowd, 1987).

6. Public Law No. 416, June 19, 1934, 73rd Congress (Washington, D.C.: Government Printing Office, 1934), Section 326.

7. Ibid., Section 307(a).

8. "Lots of Lottery Language," *Broadcasting* (November 22, 1982): 42.

9. "Deregulation Bill Passes Senate," *Broadcasting* (April 5, 1982): 36.

10. "From Fighting Bob to the Fairness Doctrine," *Broadcasting* (January 5, 1976): 46.

11. Erik Barnouw, *The Golden Web* (New York: Oxford University Press, 1968): 227–36.

12. Don R. Pember, *Mass Media in America* (Chicago: Science Research Associates, 1974): 283.

13. Giraud Chester, Garnet R. Garrison, and Edgar E. Willis, *Television and Radio* (New York: Appleton-Century-Crofts, 1971), 133–36.

14. Pember, *Mass Media in America,* 283.

15. "Looking Back to WLBT(TV)," *Broadcasting* (April 16, 1984): 43.

16. "The Checkered History of License Renewal," *Broadcasting* (October 16, 1978): 30.

17. Sydney W. Head and Christopher H. Sterling, *Broadcasting in America* (Boston: Houghton Mifflin Co., 1982): 503–5.

18. John H. Pennybacker, "Comparative Renewal Hearings: Another Dialogue Between Commission and Court," *Journal of Broadcasting* (Fall 1980): 532–47.

19. "FCC Gives RKO Green Light to Sell Stations," *Broadcasting* (July 25, 1988); and Elizabeth Jensen, "The Trials of RKO," *Channels* (January 1990): 70–73.

20. "The Next Best Thing to Renewal Legislation," *Broadcasting* (January 10, 1977): 20; and "Comparative Renewal," *Broadcasting* (August 24, 1987): 27–33.

21. "Supreme Court Upholds Postcard Renewal," *Broadcasting* (June 25, 1984): 38.

22. *The FCC and Broadcasting,* 4.

23. Charles E. Clift, III, "Station License Revocations and Denials of Renewal, 1970–1978," *Journal of Broadcasting* (Fall 1980): 411–21.

24. John R. Bittner, *Broadcasting and Telecommunications* (Englewood Cliffs, N.J.: Prentice-Hall, Inc., 1985): 347–50.

25. "Station Flipping: Looking for a Smoking Gun," *Broadcasting* (June 30, 1986): 27–28.

26. Interview with Dan Gingold, Executive Producer, KNXT-TV, Los Angeles, California, January 11, 1979.

27. "FCC Eases Radio-TV Owner Limit," *Variety* (December 13, 1988): 1; and "Crossownership Rules for TV/Cable Should Be Repealed, Says Marsh Media," *Broadcasting* (December 22, 1980): 64.

28. *A Guide to the Federal Trade Commission* (Washington, D.C.: The Federal Trade Commission, 1989): 3.

29. "What's Going On at the FTC?" (Washington, D.C.: The Federal Trade Commission, 1989): 1–8; and Penny Pagano, "Another Set of Eyes," *Channels* (November 5, 1990): 14–15.

30. James Miller, "The President's Advocate: OTP and Broadcast Issues," *Journal of Broadcasting* (Summer 1982): 625–39.

31. Norman Black, "The Deregulation Revolution," *Channels* (September/October 1984): 52–59.

32. Erik Barnouw, *A Tower In Babel: A History of Broadcasting in the United States to 1933* (New York: Oxford University Press, 1966): 120.

33. *The Television Code* (New York: National Association of Broadcasters, 1976); and *The Radio Code* (New York: National Association of Broadcasters, 1976).

34. "Setback for NAB's Ad Code," *Variety* (March 5, 1982): 1.

35. A more complete list of broadcasting associations, their officers, and addresses can be found in *The Broadcasting Yearbook* (Washington, D.C.: Broadcasting Publications, 1990): I-63–I-77.

36. A more complete list of broadcasting awards can be found in *The Broadcasting Yearbook,* F-40–F-44.

37. This information was gathered from interviews with station and network personnel—primarily James Harden, general manager for radio station KNAC, Long Beach, California, 1976; Jay Strong, program director, KCBS, Los Angeles, California, 1987; and Walt Baker, vice-president of programming, KHJ-TV, Los Angeles, California, 1987.

38. Insightful books and articles dealing with the pros and cons of government regulation are Edwin G. Krasnow and Lawrence D. Longley, *The Politics of Broadcast Regulation* (New York: St. Martin's Press, 1972); Marvin R. Bensman, *Broadcast Regulation: Selected Cases and Decisions* (Lanhan, MD: University Press of America, 1985); K. R. Middleton and B. F. Chamberlin, *The Law of Public Communication* (New York: Longman, 1988); John R. Bittner, *Broadcast Law and Regulation* (Englewood Cliffs, NJ: Prentice-Hall, 1982); and Barton T. Carter et al., *The First Amendment and the Fourth Estate: Regulation of Electronic Mass Media* (Mineola, NY: Foundation Press, 1988).

17

1. Good sources for general information on laws and regulations are Marvin R. Bensman, *Broadcast Regulation: Selected Cases and Decisions* (Lanham, MD: University Press of America, 1985); *Political Broadcast Handbook* (Washington, D.C.: National Association of Broadcasters,

1988): John R. Bittner, *Broadcast Law and Regulation* (Englewood Cliffs, NJ: Prentice-Hall, 1982); Stuart N. Brotman, *Communications Policymaking at the Federal Communications Commission* (Washington, D.C.: The Annenberg Washington Program, 1987); and *NAB Legal Guide to Broadcast Law and Regulation* (Washington, D.C.: National Association of Broadcasters, 1988).

2. General sources dealing with electronic media and the First Amendment are: T. Carter Barton et al., *The First Amendment and the Fifth Estate: Regulation of Electronic Mass Media* (Mineola, NY: Foundation Press, 1986); Maurice R. Cullen, Jr., *Mass Media and the First Amendment: An Introduction to the Issues, Problems and Practices* (Dubuque, Iowa: Wm. C. Brown Publishers, 1981); and Lucas A. Powe, Jr., *American Broadcasting and the First Amendment* (Berkeley, CA: University of California Press, 1987).

3. Sydney W. Head and Christopher H. Sterling, *Broadcasting in America* (Boston: Houghton Mifflin Co., 1982): 471–72.

4. "New York Court Finds Shield Law Protection Not Absolute," *Broadcasting* (July 13, 1987): 51.

5. "Family Viewing's Inauspicious Demise," *Broadcasting* (January 7, 1985): 204.

6. "Briefs Filed in Preferred Case," *Broadcasting* (March 3, 1986): 40.

7. "Exit Polling Ban Ruled Unconstitutional," *Broadcasting* (February 8, 1988): 114.

8. Wayne Overbeck and Rick D. Pullen, *Major Principles of Media Law* (New York: Holt, Rinehart, and Winston, 1985).

9. "FCC Launches Attack on Indecency," *Broadcasting* (April 20, 1987): 35.

10. "FCC Fines KZKC-TV for Indecency," *Broadcasting* (June 27, 1988): 36.

11. "FCC Votes 5–0 Indecency Ban," *Broadcasting* (July 16, 1990): 30–31.

12. "FCC Pulls Chicago TV License," *Broadcasting* (September 24, 1990): 30.

13. "High Court Upholds Ruling Striking Down Utah Indecency Statute," *Broadcasting* (March 30, 1987): 143.

14. "Supreme Court Hears Porn Video Case," *Electronic Media* (March 10, 1986): 6.

15. For an excellent overview, see the special issue "Defamation and the First Amendment: New Perspectives," *The William and Mary Law Review* (1984); and *Staying Out of Libel Stew* (Washington, D.C.: National Association of Broadcasters, 1986).

16. "Good News for Broadcasters and Libel Suits," *Broadcasting* (October 15, 1984): 70.

17. "As TV News Reporting Gets More Aggressive, It Draws More Suits," *The Wall Street Journal* (January 21, 1983): 1; "$19.2 Mil Victory for Newton in NBC Libel Suit," *Variety* (December 18, 1986): 1; "Newton's NBC Libel Damage Award Nixed," *Variety* (August 31, 1990): 1; and "Herbert Libel Win Against CBS Overturned," *Broadcasting* (January 20, 1986): 70.

18. "The General's Retreat," *Newsweek* (March 4, 1985), 59–60; and Don Kowet, *A Matter of Honor: General William C. Westmoreland Versus CBS* (New York: Macmillan, 1984).

19. "Media Wins Big Libel Victory," *Variety* (April 22, 1986): 1; and "Supreme Court Hands Press Another Victory," *Broadcasting* (June 30, 1986): 51.

20. "Providing Coverage, Protecting Victims: Rape Trials on Trial," *Broadcasting* (April 30, 1984): 134.

21. Susanna Barber, *News Cameras in the Courtroom: A Free Press-Fair Trial Debate* (Norwood, New Jersey: Ablex, 1987).

22. "Supreme Court Says Constitution Does Not Ban Cameras in Courts," *Broadcasting* (February 2, 1981): 43.

23. "Courtroom Doors Begin to Open for TV, Radio," *Broadcasting* (September 3, 1990): 25.

24. More information regarding editorializing cases can be found in "From Fighting Bob to the Fairness Doctrine" *Broadcasting* (January 5, 1976): 46; and William L. Rivers, Theodore Peterson, and Jay W. Jensen, *The Mass Media and Modern Society* (San Francisco: Rinehart Press, 1971): 228–29.

25. "Public Broadcasters Granted Right to Editorialize," *Broadcasting* (July 9, 1984): 33.

26. *Public Law No. 416,* Section 315.

27. Ibid.

28. "Appeals Court Agrees with FCC: Broadcasters May Sponsor Debates," *Broadcasting* (May 12, 1984): 69.

29. "Supreme Court Won't Review Branch 315 Challenge," *Broadcasting* (March 28, 1988): 53.

30. "Supreme Court Rules Vid Nets Aired in Not Selling Carter Air Time," *Variety* (June 2, 1981): 1.

31. "Appeals Court Tells FCC to Justify its 315 Ruling in King Case," *Broadcasting* (November 7, 1988): 53.

32. More information on Section 315 can be found in "From Fighting Bob to the Fairness Doctrine"; "Section 315: Be Prepared," *Broadcasting* (March 29, 1976): 44; David H. Ostroff, "Equal Time: Origins of Section 18 of the Radio Act of 1927," *Journal of Broadcasting* (Summer 1980): 367–80; "Supreme Court Rules Vid Nets Erred in Not Selling Carter Campaign Air Time," *Variety* (July 2, 1981): 1; and "Attack on 315," *Broadcasting* (October 12, 1987).

33. *In the Matter of Editorializing by Broadcast Licensees,* 13 FCC 1246, June 1, 1949.

34. *Public Law No. 416,* Section 315.

35. "The President Stands Firm," *Broadcasting* (December 28, 1987): 31; "Fairness Returns to Senate," *Broadcasting* (March 20, 1989): 67; and "High Court Nixes Fairness Ruling," *Variety* (January 9, 1990): 1.

36. Material regarding the fairness doctrine can be found in David Schoenbrun, "Is Perfect Fairness Possible?" *Television Quarterly* (Special Election Issue, 1976): 77–79; "From Fighting Bob to the Fairness Doctrine"; Fred W. Friendly, *The Good Guys, The Bad Guys, and the First Amendment: Free Speech vs. Fairness in Broadcasting* (New York: Random House, 1976); Steven J. Simmons, *The Fairness Doctrine and the Media* (Berkeley: The University of California Press, 1978); Louis F. Cooper and Robert E. Emeritz, *Pike and Fischer's Desk Guide to the Fairness Doctrine* (Bethesda, Maryland: Pike and Fischer, 1985); "Fairness Doctrine: Gone But Not Forgotten," *Current* (September 22, 1987): 6; and Timothy J. Brennan, "The Fairness Doctrine as Public Policy," *Journal of Broadcasting and Electronic Media* (Fall 1989): 419–40.

37. The United States Constitution, Section 8.

38. Robert Nicksin, "Legal Issues of Video Management," *Video Manager* (June 1988): 36–38.

39. "A Scourge of Video Pirates," *Newsweek* (July 27, 1987): 40–41.

40. *Lotteries and Contests: A Broadcasters Handbook* (Washington, D.C.: National Association of Broadcasters, 1985).

41. "High Court Upholds Minority Preference Policies," *Broadcasting* (July 2, 1990): 25.

A

ACE An award for cable TV programming excellence given out by the National Cable Television Association.

actuality An event that really occurs, primarily an interview in which someone talks about what actually happened.

actual malice In libel suits, knowing that something was false or harmful and saying or printing it anyhow.

ad hoc networks Groups of independent stations that agree to air the same movies at the same time in order to capitalize on national promotion.

advanced-definition TV Another term for high definition TV.

advanced television Another term for high definition TV.

advertising agency An organization that decides on and implements an advertising strategy for a customer.

affiliates Stations or systems that receive programming from a broadcast or cable network.

all-channel receiver bill A law passed by Congress in 1962 that said that all sets manufactured from 1964 on had to be able to receive both UHF and VHF.

alternator A device for converting mechanical energy into electrical energy in the form of alternating current (AC).

amplitude modulation (AM) Changing the height of a transmitting radio wave according to the sound being broadcast; stations that broadcast in the 535 to 1,705 kilohertz range.

analog A device or circuit in which the output varies as a continuous function of the input.

antenna A wire or set of wires or rods used both to send and to receive radio waves.

anthology dramas Plays of the 1950s that probed character and emphasized life's complexities.

antitrust act An order that is used to prevent monopoly.

areas of dominant influence (ADIs) Arbitron's name for the individual parts of the country that it surveys.

ascertainment A process stations used to undertake to keep their licenses that involved interviewing community leaders to learn what they believed were the major problems in the community.

aspect ratio The relationship of the length of a TV picture to its width.

audience flow The ability to hold an audience from one program to another.

audimeter An electronic device that used to be attached to TV sets by the A. C. Nielsen Company in order to determine, for audience measurement purposes, when a set was turned on and to what station it was tuned.

audio board A mixer used to combine various sound inputs, adjust their volume, and then send them to other pieces of equipment.

audio frequency amplifier The part of the radio receiver that strengthens the sound signal and sends it to a speaker.

audion A three-electrode vacuum tube invented by Lee DeForest that was instrumental in amplifying voice so it could be sent over wireless.

GLOSSARY

Automated Measurement of Lineups (AMOL) A method that picks up special codes from programs so that Nielsen can determine what programs are on what channels in each market.

average quarter-hour ratings (AQR) A calculation based on the average number of persons listening to a particular station for at least five minutes during a fifteen-minute period.

B

back light A light placed behind an object to make it stand out from the background.

backyard satellite reception A process through which individual households receive satellite signals by putting a dish on their property.

balance sheet A financial form that lists assets and liabilities.

band compression Putting more information in a certain amount of spectrum space by including only the information that changes in each frame of video.

bandwidth The number of frequencies within given limits that are occupied by a particular transmission such as one radio station.

barndoor A black metal flap placed on a lamp to block off some of the light.

barriers In the Bradley model, anything that reduces the effectiveness of the communication process.

barter To give goods or services (such as advertising time) for other goods or services; to receive a program for free because much of the advertising time is sold by the seller.

barter syndication Various forms for selling commercial time within programs that occur between seller and buyer.

basic cable Channels, often supported by advertising, for which the cable subscriber does not pay a large extra fee.

Betacam A professional half-inch videocassette format.

Betacam SP An improvement on the Betacam videocassette format.

Betamax The first consumer half-inch videocassette recorder format.

bicycling A distribution method by which material is mailed, flown, or driven from one place to another.

bidirectional A microphone that picks up sound on two sides.

Biltmore Agreement A settlement between newspapers and radio worked out at the Biltmore Hotel that stipulated what radio stations could and could not air in the way of news.

blacklisting A phenomenon of the 1950s when many people in the entertainment business were accused of leaning toward communism and, as a result, could not find work.

blackouts Not allowing a particular program, usually sports, to be shown in a particular area because the event is not sold out.

blanket licensing Obtaining the right to use a large catalogue of musical selections by paying one set fee.

block programming Airing one type of programming, such as comedies, for an entire evening.

Blue Book The nickname given to a 1946 FCC document that gave principles regarding license renewal.

boom A long counterweighted pole used to hold a microphone.

bridging Programming a show so that it runs over the time when a new program is starting on a competing network.

broadcasting The sending of radio and television programs through the airwaves.

broadcast standards The department at a network or station that decides the general standards of acceptability of program content.

brokers People who aid in the buying and selling of radio and TV stations and cable systems.

C

call letters A series of government-assigned letters that identify a transmitting station.

camcorder A single unit that consists of both a camera and a videocassette recorder.

Canon 3A An American Bar Association policy that let individual states and judges decide whether or not cameras should be allowed in their courtrooms.

Canon 35 An American Bar Association policy that banned cameras in the courtrooms.

capacitance disc A videodisc system that used a stylus that moved over grooves in a manner similar to a record player.

carbon microphones Early radio microphones that used carbon elements to respond to sound.

cardioid A microphone that picks up in a heart-shaped pattern.

carrier wave A high-frequency wave that can be sent through the air and is modulated by a lower-frequency wave containing information.

cartridge A continuous loop of audiotape wound onto a spool.

cash Paying a set amount for a program and then being able to sell all the commercial time.

cash plus barter Paying a reduced amount for a program but letting the seller fill some of the advertising time.

cassettes Closed containers that hold tape so that it can be threaded automatically in a recorder.

cathode-ray tube A tube in which an electronic beam can be focused to a small cross section on a luminescent screen and can then be varied in position and density to produce what appears to be a moving picture.

cause and effect A method of research that tries to determine the relationship between two or more modes of behavior.

C-band A satellite frequency band between four and six gigahertz.

Ceefax The BBC's teletext system.

cellular phones Mobile phones that operate by relaying signals from one small area to another through the use of low-powered transmitters.

censorship Not allowing certain information to be disseminated.

chain regulations FCC rulings of the 1940s that related to network operations, such as giving stations the right to reject a network program.

channel reservations Reserving particular station allocations for a particular purpose, such as reserving space for public TV stations and not allowing it to be used by commercial interests.

channels In the Bradley model, the elements through which messages are transmitted; the frequencies that carry individual stations or networks.

character generator (c.g.) A piece of equipment that is used to type letters electronically onto a screen.

charged-coupled devices (CCDs) A solid-state imaging device that translates an image into an electronic signal within a TV camera.

charter A right to operate that is given to a broadcasting corporation (such as the BBC) by the government.

chief executive officer (CEO) The highest ranking person within an organization who actively makes decisions for the organization.

chief financial officer (CFO) A high-ranking person within a company who is in charge of monetary matters.

clear and present danger A threatening of the security of the nation if certain news information is made public.

Clio An award given for outstanding commercials.

clock A circle that shows all the segments that appear in an hour's worth of programming by a radio station.

closed captioning Material broadcast on the vertical interval that is intended for the hearing impaired.

closed-circuit Television signals that are transmitted via a self-contained wire system rather than broadcast through the air.

clutter Having too many advertisements on at one time.

coaxial cable A transmission line in which one conductor completely surrounds the other, making a cable that is not susceptible to external interference from other sources.

coincidental telephone technique A method of audience research that asks people what programs they are watching or listening to at the moment.

color corrector A piece of equipment used to make subtle color changes in a video picture.

colorization Using a computerized process to put color into black-and-white films.

Commentaries Personal viewpoints by individuals or groups that do not necessarily represent the viewpoint of the station on which they are aired.

Communications Act of 1934 The Congressional law that established the FCC and set the guidelines for the regulation of telecommunications.

community relations Another name for a public relations department.

compact disc players (CDs) Pieces of equipment that play back discs on which sound has been recorded digitally.

comparative license renewal A procedure by which a group could challenge the license of incumbent station owners.

composite week Seven days over a three year period that were randomly selected by the FCC so that a station's programming record could be judged for license renewal purposes.

compressors Equipment that boosts audio signals that are too soft and lowers signals that are too loud.

compulsory licenses Copyright fees that must be paid and that are usually a set fee, such as a percentage of income.

computer A device used to perform rapid, complex calculations and to store, manipulate, and retrieve data.

computer graphics Lines and forms used to create digitally-generated pictures that can then be manipulated in various ways.

consent decree An order, usually from the Justice Department, that lists things some company or companies need to do in order not to have a monopolistic situation.

consent order A type of ruling, sometimes initiated by the FTC, whereby a company stops some particular practice but does not admit it violated any law by undertaking the practice.

construction permit (CP) A document issued by the FCC that allows the recipient to begin building a radio or television station.

controller A person who oversees profit and loss and expenditures for a company or other organization; equipment that is used to mark edit points.

co-op advertising Joint participation in the content and cost of a commercial by two entities, usually a national company that manufactures the product and a local company that sells it.

co-productions Programs or series produced by two different entities, both of which contribute and gain (or lose) financially.

copyright The exclusive right to publication, production, or sale of rights to a literary, dramatic, musical, or artistic work.

copyright fees Money paid for the right to use a literary, dramatic, musical, or artistic work created by someone else.

cost per thousand (CPM) The price that an advertiser pays for each thousand households a commercial reaches.

counterprogram Programming something different from what is being programmed on another network in order to attract a different audience.

crane A mechanism on which a camera can be mounted so that it can move from close to the floor to a fairly high distance above the floor.

Crossleys An early rating system that used the recall method.

cross-ownership A situation in which one company owns various media in one market.

cume The average number of households or people that tune in to a particular station at different times.

D

data banks Information within computers that can be accessed by customers.

dbx An audio noise reduction system.

decoder boxes Devices that can descramble scrambled television signals.

deintermixture An attempt to strengthen UHF stations by making some markets all UHF and some all VHF.

demographics Information pertaining to vital statistics of a population, such as age, sex, marital status, and geographic location.

deregulation The removal of laws and rules that spell out government policies.

designated market areas (DMAs) Nielsen's term for the various individual markets it surveys.

detector A part of a radio receiver that demodulates or separates the carrier wave from the information it carries.

diary A booklet used for audience measurement in which people write down the programs they listen to or watch.

digital A device or circuit in which the output varies in discrete "on-off" steps.

digital audio broadcasting (DAB) A proposed form of transmission that involves pulse code modulation.

digital audiotape (DAT) Tape that records information in numerical data bits that yield high-quality sound.

digital audio workstation (DAW) Equipment that can store, edit, mix, process, and manipulate sounds all in digital form.

digital effects generator A device to produce unusual forms of video pictures.

direct broadcast satellite (DBS) A process of transmission and reception whereby signals sent to a satellite can be received directly by TV sets in homes that have small satellite receiving dishes mounted on them.

directional antennas Antennas that send out radio waves more effectively in some directions than in others.

disc jockey Someone who plays music and announces on a radio station.

dissolve The gradual fading in of one picture while another picture is fading out.

distant signal importation The bringing in of stations from other parts of the country by a cable system.

docudramas A fictionalized production that is based on fact.

dolby An audio noise reduction system.

dolly Wheels placed under a tripod so it can move.

drop-out compensator A piece of processing equipment that fills in gaps of video information in places where the head loses contact with the tape.

duopoly rule A rule stating that a network organization cannot operate two or more networks that cover the same territory at the same time.

E

earphones Small devices placed over the ear in order to hear audio signals or radio stations.

edit deck The videocassette recorder that holds the material that is being edited.

educational broadcasting An early name for public broadcasting.

Educational Broadcasting Facilities Act A 1962 act that authorized money for the construction of educational television facilities.

electron gun The part of the TV tube that shoots a steady stream of electrons that scan the target from top to bottom.

electronic mail The sending of letters and other information through a computer network.

electronic media A variety of dissemination forms based on electrons, such as radio, broadcast television, cable TV, and MMDS.

electronic publishing Providing material such as that traditionally supplied by newspapers through a network that usually includes phone lines and computers.

electronic scan Putting information into a computer directly from a page of print.

electronic scanning A method that analyzes the density of areas to be copied and translates this into a moving arrangement of electrons that can later reproduce the density in the form of a picture.

Emmys An award for programming excellence given out by the Academy of Television Arts and Sciences.

enhanced definition TV Another term for high definition television.

enhanced underwriting A public broadcasting practice that allows corporate logos and products of companies that contribute to programs to be mentioned on the air.

episodic serialized dramas Televised plays that have set characters and problems that are dealt with in each program.

equal opportunity Giving the same treatment to political candidates. *See* equal time.

equal time A rule stemming from Section 315 of the Communications Act stating that TV and radio stations should give the same treatment and opportunity to all political candidates for a specific office.

Eureka The European system of high definition TV.

exit polls Surveys conducted by the media asking people who have just voted how they voted.

F

fade To go from a picture to black or vice versa.

fairness doctrine A policy that evolved from FCC decisions, court cases, and congressional actions stating that radio and TV stations had to present all sides of the controversial issues they discussed.

fair trial Making sure someone accused of a crime is treated equitably by the courts.

fair use Allowing part of a copyrighted work to be used without copyright clearance or payment.

family hour A policy that stated all programs aired between 7:30 P.M. and 9:00 P.M. should be suitable for children as well as adults.

feedback In the Bradley model, anything that gives information back to the source.

fiber optics Glass strands through which information can be sent.

field research Study that is conducted in environments where people naturally are, rather than in special controlled environments.

fill light Light that reduces the shadows cast by the key light.

filters Equipment that is used to take out particular audio frequencies.

financial interest-domestic syndication (fin-syn) A policy that precludes networks from having any monetary remuneration in programs aired on the network or any rights to distribute those programs within the country after they have aired on the network.

fireside chats Radio talks given by President Franklin Roosevelt.

First Amendment The part of the United States Constitution that guarantees freedom of speech and freedom of the press.

first-run syndication Programs produced for distribution to stations rather than for the commercial networks.

fixed buy When the advertiser states a specific time that a commercial should air and the station abides by that time.

fixed lenses Lenses with a single focal length that cannot be used to zoom in or out on a subject.

flat-screen TV A form of TV reception that does not use a picture tube but can involve sets that are hung on the wall like a picture.

flying spot scanner A machine that produces a high quality, gentle transfer of film to tape.

focus groups In-depth discussion sessions to determine what people do and don't like about various aspects of the media.

footprints The section of the earth that a satellite's signal covers.

format The type of programming a radio station selects, usually described in terms of the music it plays.

frame The 525 lines of picture information that the TV camera system produces every one-thirtieth of a second.

franchises Special rights granted by a government or corporation to operate a facility such as a cable TV system.

free-lance To work on a per-project basis rather than as a full-time employee.

freeze Immobilization or cessation of an activity, such as a stop in the assigning of radio or TV station frequencies.

frequencies The number of recurrences of a periodic phenomenon, such as a carrier wave, during a set time period, such as a second.

frequency modulation (FM) Placing a sound wave upon a carrier wave in such a way that the number of recurrences is varied; stations that broadcast between 88 and 108 megahertz.

full-service agency An advertising agency that handles all the advertising needs for a particular client.

futures Committing to buying programs before they are actually available.

G

gatekeepers People who make important decisions regarding what will or will not be communicated through the media.

Gateway A videotext system operated by Times-Mirror.

grazing Using the remote control of a TV set to change channels frequently.

Great Debates The televising of presidential candidates John Kennedy and Richard Nixon opposing each other face-to-face.

grid Metal bars that are suspended from a ceiling and are used to hang lights.

ground station A receiving station for information transmitted by satellite.

group owner A company that owns a number of radio and/or TV stations.

H

hackers People who break the security of computer systems and access the information in them.

hammocking Programming a new or weak program between two successful programs.

hard documentaries Investigative programs dealing with controversial subject matter.

helical A method of videotape recording whereby video information is placed on the tape at a slant.

hertz A frequency unit of one cycle per second that is abbreviated as Hz.

Hi-Band 8 An improvement on the Sony Video-8 format.

hi-fi An audio recording method that records sound in the same area of the tape as picture.

high definition television (HDTV) Television that scans at above one thousand lines a frame and has a wider aspect ratio than present TV sets.

HMI light Hydrogen medium-arc-length iodide light that is used as a supplement to the sun.

home shopping Programs that show products enabling viewers to buy them instantly by calling a particular phone number.

homes using TV (HUT) The percentage of homes that have a TV set tuned to any channel.

Hooper An early audience measurement service that used the coincidental telephone technique.

hyping Trying to increase audience size temporarily in order to obtain high ratings.

I

iconoscope The earliest form of TV camera tube in which a beam of electrons scanned a photoemissive mosaic screen.

image enhancer A piece of equipment that corrects for smearing of color and loss of sharpness that result from recording a video picture.

improved-definition TV Another term for high definition TV.

indecency Something that is offensive in relation to the standards of a particular community.

independent networks Organizations that offer specialized programming, such as sports, usually to independent TV stations.

independents TV or radio stations that are not affiliated with one of the major networks; small film and television production companies.

infomercials Program length material with content that extols the virtues of a particular product.

in-house Performing some service within a company rather than hiring outside help.

instant analysis Remarks made by commentators right after political pronouncements that draw conclusions regarding what has just been said.

instructional television fixed service (ITFS) A form of over-the-air broadcasting that operates in the 2,500 megahertz range and is used primarily by educational institutions.

integrated circuit Multiple active electronic components on a single semiconducting chip, most often made of silicon.

Integrated Services Digital Network (ISDN) A worldwide standard that will allow voice, data, graphics, and video to be processed, stored, and transported.

interactive cable Two-way capability that allows interaction between the subscriber and sources provided by the cable TV system.

internships The chance to work and learn within a company without actually being hired.

invasion of privacy Not leaving someone alone who wishes to be left alone, or divulging facts about a person that he or she does not wish divulged.

iris The adjustable opening in a camera lens that admits light.

islands Still pictures at the end of commercials aimed at children to help them differentiate between the commercial and the program.

ITVs Independent television production companies that supply programming for the British Independent Broadcasting Authority.

K

keyboarding Typing information into a computer.

key light The main source of light in a typical lighting setup.

kinescope An early form of poor-quality TV program reproduction that basically involved making a film of what was shown on a TV screen.

kiosk A small booth that sometimes contains a TV information screen that operates in conjunction with a videodisc system and/or computer.

Ku-band A satellite frequency band between eleven and fourteen gigahertz.

L

laboratory research A form of research conducted in a controlled environment that is not a person's usual environment.

laser disc A videodisc system that uses a laser to read the information on the disc.

lavalieres A small microphone, most often attached to a person's clothing.

least objectionable fare Programming watched, not because someone really wants to watch it, but because it is the best of what is on at a particular time.

lens The part of the camera that gathers the light and focuses the image onto the pickup tube or CCD.

libel To broadcast or print something unfavorable and false about a person.

license fee The amount a network pays a producing company in order to air the program the company produces.

linear Recording material on tape in a straight horizontal line.

liquid crystal display (LCD) TV sets that employ chemical elements that show black or a particular color when power is applied.

local buying Purchasing advertising time on stations in a limited geographic area.

local origination Programs produced for the local community, particularly as it applies to programming created by cable TV systems.

long waves Radio waves at low frequencies.

lottery The involvement of chance, prize, and consideration (money) for a game or contest; a random selection method.

loudspeakers Devices that amplify sound so that a large group of people can hear it.

low-power TV (LPTV) Television stations that broadcast to a very limited area because they do not transmit with much power.

M

M-II An improvement on the M-format videocassette recorders.

made-fors Movies that are specially produced for television.

magazine concept Placing ads of various companies within a program rather than having the entire program sponsored by one company.

majors Large film and television production companies.

marketing Making a product attractive to buyers.

Mayflower Decision An FCC statement that said broadcasters should not editorialize.

mechanical scanning An early form of scanning by which a rotating device, such as a disc, broke up a scene into a rapid succession of narrow lines for conversion into electrical impulses.

medium waves Radio waves at medium frequencies.

message In the Bradley model, that which is communicated.

metro survey area (MSA) Arbitron's division for radio ratings of areas that receive radio stations clearly.

M-format A professional half-inch videocassette format.

microwave Radio waves, 1,000 megahertz and up, that can travel fairly long distances.

mini-docs Short documentaries, usually within a newscast.

miniseries Plays shown on TV over a number of different nights.

mobile telephone service High-powered transmitters that are used to relay phone messages, primarily from and to cars or other moving vehicles.

modem A device that allows computer-generated data to be sent over phone lines.

modular agencies Advertising agencies that handle only a limited number of tasks for a client.

modulate To place information from one wave on another so that the wave that is carrying represents the signal of the original wave.

monaural One channel of sound coming from one direction.

monitors TV sets that do not receive signals sent over the air but can receive signals from a camera or VCR.

multichannel multipoint distribution service (MMDS) An over-the-air service of several channels that operates at higher frequencies than broadcast TV.

multiple system ownership (MSO) The owning and operation of several cable systems in different locations by one parent company.

multi-track tape recorders Audiotape recorders that can record and play back a large number of separate sounds on separate tracks.

MUSE The Japanese system of high definition TV.

music preference research Determining what music people would be likely to stay tuned for by playing them samples over the phone or in an auditorium and soliciting their reactions.

music videos Songs accompanied by visuals that are heard and seen on TV.

must-carry A ruling that has stated that cable TV systems have to put certain broadcast stations on their channels.

N

narrowcasting Airing program material that appeals to a small segment of the population.

national buying Buying advertising time on individual stations throughout the whole country.

needle-drop A one-time fee paid to use music for a particular program.

network-affiliate contract The agreement between network and station regarding how much the network will pay the station and how programs are to be aired.

network buying Buying ads that are seen everywhere that the network program is seen.

Nielsen Station Index A ratings report that covers local areas.

Nielsen Television Index (NTI) A ratings report on programs that are shown nationally.

noncommercial broadcasting An early name for public broadcasting.

NTSC (National Television Systems Committee) The electronic scanning and color system used in the United States and some other parts of the world.

O

O and Os Stations that are owned and operated by a network.

obscenity Something that depicts sexual acts in an offensive manner, appeals to prurient interests of the average person, and lacks serious artistic, literary, political, or scientific value.

off-line Video editing that uses workprints rather than the original recorded footage.

off-net Syndicated shows that at one time played on a network.

omnidirectional A microphone that picks up from all directions.

on-line Video editing that uses the original material and executes all the edits into a finished master tape.

open reel Recorders with two reels that necessitate the threading of the tape from one reel to the other.

opposing viewpoints Opposite sides presented in regard to a station editorial.

Oracle The teletext system of Britain's Independent Broadcasting Authority.

outboard equipment Equipment such as turntables and cartridge players that feeds into an audio board.

out-of-house Having material produced by a company that is independent from the company that needs to have the material produced.

overbuild The process by which a company installs a second cable system in an area where another company already has a cable system.

overnight reports Ratings reports that are delivered to customers in less than twenty-four hours.

P

packages A group of films, usually offered to independent TV stations.

PAL (Phase Alternative Line) The electronic scanning and color system used in western Europe and some other parts of the world.

payola The practice of paying a disc jockey under the table so that he or she will air music that might not otherwise be aired.

pay-per-view Charging the customer for a particular program watched.

pay-TV A method by which people pay in order to receive television programming free of commercials.

Peabody An award for distinguished broadcasting programming.

pedestal A mechanism that holds a camera and allows it to be raised or lowered, usually by hydraulic or pneumatic means.

peoplemeter A machine used for audience measurement that includes a keypad to be pushed by each person to indicate when they are watching TV.

people using radio (PUR) The percentage of people who have a radio tuned to any station.

people using television (PUT) The percentage of people who have a TV set tuned to any channel.

per-program fee A right to use music that is obtained by paying a fee for each program that utilizes the music.

petition to deny A process by which citizen groups expressed the desire for a license to be taken away from a station.

phosphor screen A layer of material on the inner face of the TV tube that fluoresces when bombarded by electrons.

photosensitive surface An area of a camera tube that is capable of emitting electrons when it is hit by light.

pilot A film or tape of a single program of a proposed series that is prepared in order to obtain acceptance and commercial support.

piracy Obtaining electronic signals or programs by illegal means and gaining financially from them.

pirate ships Boats that broadcast rock music off the coast of Great Britain during the 1960s.

pixel A light-sensitive picture element usually made of silicon.

play lists Information that tells what records were played on a radio station and when they were played.

postproduction The time during which a TV program is edited.

potted palm music Soft music played on early radio that derived its name because this type of music was usually played at teatime by orchestras in restaurants that had many potted palm trees.

preproduction The time before a TV show is filmed or taped that is used to plan and budget for production.

press-radio war A dispute of the 1930s between radio and newspapers over news that could be broadcast on radio.

pretesting Determining ahead of time what programs or commercials people are likely to respond to in a positive manner.

prime-time access rule An FCC ruling declaring that stations should program their own material rather than network fare during one hour of prime time.

production The time during which all the shots for a TV program are filmed or taped.

profanity Irreverent use of the name of God.

profit and loss statement A financial form that shows what a company received and spent over a period of time.

program buying When one company pays the costs of an entire program and has its ads inserted within that program.

projection TV A television screen that is very large with the signal coming from the rear or front.

promise vs performance A procedure whereby broadcasters promised what they would do during a license period and then were judged on the fulfillment of their promises when their license came up for renewal.

promotion Methods of enhancing an image that a station or an organization must pay for.

promotional spots (promos) Advertisements for a station's own programs that are broadcast on its own air.

public access Programming conceived and produced by members of the public for cable TV channels.

Public Broadcasting Act of 1967 A Congressional bill that implemented most of the recommendations of the Carnegie Commission regarding public broadcasting.

public convenience, interest, or necessity A phrase in the Communications Act of 1934 that the FCC has used as a basis for much of its regulatory power.

public domain Works that are not copyrighted, such as very old material or publications from the government.

public figures Well-known people who have to prove actual malice in order to win libel suits.

public inspection file Station documents that are placed in a location easily accessible to members of the community.

publicity Free articles or other forms of enhancement that will garner public attention.

public service announcements (PSAs) Advertisements for nonprofit organizations.

pulse code modulation (PCM) An audio recording pattern that records sound helically and digitally on its own section of a videotape.

Q

quiz scandals The discovery during the 1950s that some of the contestants on quiz shows were given the answers to questions ahead of time.

quotas Set limits on something, such as the amount of foreign programming a country will allow on its broadcasting systems.

R

Radio Act of 1912 The Congressional law that required everyone who transmitted on radio to obtain a license.

Radio Act of 1927 The Congressional law that established the Federal Radio Commission.

radio frequency amplifier The part of a radio receiver that selects the frequency and amplifies the carrier wave and its information.

radio waves The waves of the radio frequency band of the electromagnetic spectrum.

random sampling A method of selection whereby each unit has the same chance of being selected as any other unit.

rate protection   Guaranteeing a customer a certain advertising fee, even if the rate card increases.

rating   The percentage of households that are watching or listening to a particular program.

recall method   An audience research method that asks people what they have seen or heard on radio or TV at some time in the past.

receiver   In the Bradley model, the person or group that hears or sees a message; the part of a radio or TV set that reproduces the sound and/or displays the picture.

recordimeter   An audience measurement device that has been used in conjunction with diaries to validate what was recorded in the diaries.

Red Lion   A legal case that upheld the constitutionality of the concept of fairness.

reflector   A light-colored surface used to bounce light back into a scene.

regional buying   Buying advertising time on stations in a particular part of the country.

remotes   Productions shot out in the field and not in the studio.

renewal expectancy   An FCC statement that said that if a station had provided favorable service, its license was not likely to be in jeopardy.

reruns   Programs that are shown on TV again after their initial airing.

residuals   Payments made to those involved in a production when the program is rerun.

retainer   An amount of money paid to a company or consultant to handle whatever might occur that is within their expertise.

ribbon microphones   Microphones that create representations of sound through the use of a metallic ribbon, a magnet, and a coil.

rights fees   Money paid in order to broadcast something, such as a sports contest.

robotic cameras   Cameras that do not require an operator because they are controlled remotely.

roster-recall method   An audience research methodology that helps people remember what they have seen or listened to by supplying them with a list of call letters, station slogans, and the like.

run-of-schedule (ROS)   When a station, rather than the advertiser, decides on the time that a particular commercial will air.

S

satellite   A human-made object that orbits the earth and can pick up and transmit radio signals.

satellite master antenna TV (SMATV)   A system that incorporates signals received by satellite dish with broadcast signals received by a master antenna and distributes them, usually to apartments within large apartment complexes.

satellite news gathering (SNG)   News gathered with trucks equipped with satellite uplinks that can then send that news almost anywhere.

satellite news vehicle (SNV)   Trucks that have satellite uplinks that can transmit news stories great distances.

ScanAmerica   A peoplemeter service Arbitron is developing that will tie ratings to the groceries that people buy.

scanners   Radio devices that can be used to listen in on police and fire communications.

scarcity theory   Reasoning that broadcasters should be regulated because there are not enough station frequencies for everyone to have one.

scrambling   Changing transmitted electronic signals so they cannot be received properly without some sort of decoding system.

script   The word-by-word text for a program or movie.

season   The period between when new network programs used to begin (usually September) and when they finished their run (usually June).

SECAM (Sequence Couleur a Memoire)   The electronic scanning and color system used in Eastern Europe and some other parts of the world.

Section 315   The portion of the Communications Act that states that political candidates running for the same office must be given equal treatment.

share   The percentage of households or people watching a particular program in relation to all programs available at that time.

shield laws   State laws that have the effect of allowing reporters to keep sources of information secret.

shortwaves   Radio waves at high frequencies.

shotgun   A microphone with a narrow pickup pattern that is used to gather sound from a great distance.

single-dial tuning   Radio receiver operation that requires only one dial in order to tune in a station.

siphoning   A process whereby pay-TV systems might drain programming from networks by paying a higher price for it initially.

slander   To say something harmful to a person's character or reputation.

soft documentaries   Programs that give in-depth material but deal with subject matter that is not overly controversial.

source In the Bradley model, the total number of people needed in order to communicate a message.

source deck The videocassette recorder that holds the material to be edited.

spin-offs Series that are created out of characters that have appeared on other series.

split-30 Two fifteen-second commercials in one thirty-second time.

spot buying When a company purchases advertising time within or between programs in such a way that the company is not specifically identified with the program.

standupper A story delivered by a reporter live from a remote location.

station representative A company that sells advertising time for a number of stations.

Steadicam A camera mounting device that attaches to the body to allow for very smooth camera operation.

stereo Sound reproduction using two channels through two speakers to give a feeling of reality.

stereotypes A fixed impression of a person or group of people that does not allow for individuality.

Storage Instantaneous Audimeter (SIA) Audience measure equipment used by Nielsen that sent data over phone lines from homes to Nielsen's central headquarters.

storyboard A chart that contains step-by-step pictorialization of a commercial or program.

stringers People who are paid for the stories or footage they gather if that material is used by some news-gathering organization.

strip To air programs, usually old network shows, at the same time every day of the week.

stunt broadcasts Radio broadcasts of the 1930s that involved unusual locations such as gliders and balloons.

subcarrier authorization (SCA) The ability to carry a secondary signal in addition to the main signal of a radio or TV station.

subscription TV Scrambled programs that were broadcast over the air and could be descrambled when a subscriber paid a fee for the service.

superstation A broadcast station that is put on satellite and shown by cable systems.

Super-VHS A half-inch videocassette format that is an improvement on VHS.

sweeps Ratings reports for local areas.

switcher A piece of equipment used to select the video input that will be taped or shown live.

synchronous satellite A satellite that appears to hang motionless in space.

syndicated exclusivity A cable TV rule that stated that if a local station was carrying a particular program, the cable system could not show that same program on a service imported from a distant location.

syndicators Companies that produce or acquire programming and sell it to stations, cable networks, and other electronic media.

T

table stands Short devices that sit on a table to hold a microphone.

tabloid TV Programs, mainly talk shows, that discuss exploitative subjects that are often sexually oriented.

take A quick cut from one shot to another.

target The portion of a TV camera tube that is scanned by the electron beam.

telcos Telephone companies.

telecommunications An umbrella term that covers broadcasting, electronic media, telephone, and computer technologies.

teleconferencing Transmitting material over satellite from one or various points to other points, usually instead of having a meeting or conference.

telephone An electrical device that transmits voice or data.

teletext Words, numbers, and graphics placed in the vertical blanking interval of a broadcast television signal.

televangelism Using television to promote religion.

tentpoling Programming a successful program between two weak or new programs.

ten-watters Public radio stations that once broadcast with low power and were used mainly to train students, although most have now increased power to at least 100 watts.

testimonials Positive statements about a commercial product usually given by a well-known person.

three-dial tuning An early radio receiver system whereby three different dials had to be turned in order to tune in a radio station.

three-dimensional television (3-DTV) Material that is produced in such a way that it approximates the way the eyes see two images and can then be projected back so that it appears to have multi-dimensions.

360 meters The frequency on which all early radio stations broadcast.

tiering Charging cable subscribers different rates for different services.

time base corrector   A device that places a videocassette recorder frame or part of it in temporary digital storage so that its scanning timing will be the same as that from other VCRs.

time code   Numbers recorded on tape that give the address of each particular frame.

toll station   A name for the type of radio programming WEAF initiated in 1922 that allowed anyone to broadcast a public message by paying a fee, similar to the way that one pays a toll to communicate a private message by telephone.

topless radio   A term used to describe sexually-oriented talk shows that were broadcast by radio stations.

total survey area (TSA)   Arbitron's division for radio ratings of geographically large areas that may not receive all stations equally clearly.

toy-based programming   Children's programming that was designed around having toys as the main characters.

trade-out   To give goods or services (such as advertising time) in exchange for other goods or services.

traffic   Keeping track of programs and commercials that are on the air.

transactional services   Interactive methods that allow someone with a computer to take some action such as buying a ticket.

transistors   A solid-state electronic device that controls current flow without the use of a vacuum.

translators   A low-powered TV transmitter usually used to send a signal into an area with poor reception.

transmitter   A piece of equipment that generates and amplifies a carrier wave and modulates it with information that can be radiated into space.

transponder   The part of a satellite that carries a particular signal.

transverse quadraplex   A type of videotape recorder that uses two-inch tape and places the signal on the tape vertically.

treasurer   A person who handles cash for a company or other organization.

treatment   Written paragraphs that tell the general idea for a series or a particular movie or program.

tripod   A three-legged structure on which a camera can be mounted.

tube   The part of a TV camera that creates the electronic impulses that represent the picture.

tungsten-halogen lamps   Lights that have tungsten filaments, are filled with halogen, and have quartz bulbs.

turntable   A device for spinning a record at a precise speed.

turret   A turning device mounted onto the front of a camera to hold fixed lenses so that they can be changed during production.

TVQ   A measurement that indicates the degree to which people are aware of and like particular programs or personalities.

Type C   A one-inch videotape format.

U

ultrahigh frequency (UHF)   The area in the spectrum between 300 and 3,000 megahertz; TV stations above channel 13.

U-matic   The standardized three-fourth-inch videocassette format.

umbrella   Shiny material in an umbrella shape that bounces light back into a scene.

unbundled   A program selling process of National Public Radio wherein individual programs can be bought and aired by affiliated stations rather than their having to pay a set fee for all the programming.

unidirectional   A microphone that picks up on only one side.

uses and gratification   A method of research that tries to determine the perceived value people feel by becoming involved with certain actions.

V

vacuum tube   An electron tube evacuated of air to the extent that its electrical characteristics are unaffected by the remaining air.

vast wasteland   A term coined by FCC Chairman Newton Minow to refer to the lack of quality TV programming.

vectorscope   A piece of equipment that displays color information about a video picture.

vertical blanking interval   The retrace of the electron beam from the bottom to the top of the TV screen.

vertical integration   A process by which one company produces, distributes, and exhibits its products without decision making from any other sources.

very high frequency (VHF)   The area in the spectrum between 30 and 300 megahertz; TV stations that broadcast on channels 2 to 13.

VHS   The consumer grade half-inch videocassette format developed by Matsushita.

videocassette recorder (VCR)   Tape machines that use magnetic tape enclosed in a container that is automatically threaded when the machine is engaged.

videoconferencing   Transmitting material over satellite from one or various points to other points, usually instead of having a meeting or conference.

videodiscs Devices shaped somewhat like phonograph records that contain video and audio information and can display this information on a TV screen connected to a videodisc player.

Video-8 A consumer videocassette format introduced by Sony that uses tape that is 8mm wide.

video games A form of play that involves interaction between the person or people playing and a TV screen.

videotext A system to deliver information to a consumer's home through the use of computers and phone lines.

viewfinder A monitor on a TV camera that allows the camera operator to see the picture that the camera is taking.

Viewtron A videotext system operated by Knight-Ridder.

VLSI Very large scale integration; a multiplication of the number of elements that could be placed on a chip.

voice-activated computers Computers that can have material inputted into them by someone reading the material out loud.

W

waveform monitor A piece of equipment that displays luminance information about a video picture.

Wheeler-Lea Amendment The 1938 Congressional action that enabled the FTC to look into deceptive advertising.

window The period between the time a movie is shown in a theater and the time it is released for showing on other media such as pay-TV or network TV.

wireless mike A microphone that operates on FM frequencies, so it does not need audio cable.

wire recording An early form of magnetic recording that used wire rather than tape.

wire services Organizations that supply news to various media.

word processing Using a computer for inputting, storing, manipulating, retrieving, and reproducing written material.

workprint A copy of the original tape footage that is used for editing.

Z

zoom lenses Complex lenses with variable focal lengths adjustable in a range from wide to close.

INDEX

A

X

Xerox, 143

Y

Young, Owen, 54–55
Young, Robert, 293
"Young and the Restless, The," 302
"Young People's Concerts," 288
"Young People's News," 306
"Your Hit Parade," 68, 288
"Your Show of Shows," 96

Z

Zenith, 408
Zworykin, Vladimir, 91–92